Interactive Media with Next-Gen Technologies and Their Usability Evaluation

Interactive media are a human-machine interface that allows people to connect with each other by making them active participants in the media they consume through text, graphics, audio, and video. This book presents the challenges and opportunities presented by emerging media technologies to explore usability evaluation.

It covers the current trends in interactive media technologies such as Social Media, Dark Patterns, Internet of Things (IoT), Android Development, Assistive Technologies, and Augmented Reality (AR)/Virtual Reality (VR). It explores various application areas such as Education, Film and Television, Agriculture, Cyber Security, Bird Conservation, Smart Vehicles, Fashion Technology, and e-Learning.

Key features of this edited book are as follows:

- Evaluates related Interactive Media Technologies and Applications for assessment and enhancement of their usability
- Illustrates current discussions on interactive media technologies such as Social Media, Dark Patterns, Internet of Things (IoT), Android Development, Assistive Technologies, and Augmented Reality (AR)/Virtual Reality (VR)
- Includes various case studies from application areas such as Education, Film and Television, Agriculture, Cyber Security, Bird Conservation, Smart Vehicles, Fashion Technology, and e-Learning, which are helpful for researchers
- Presents concept illustrations with appropriate figures, tables, and suitable descriptions in a reader-friendly way

This book is ideal for both beginners and experts working in the fields of HCI, Multimedia Techniques, and Next-Gen Technologies.

Interactive Media with Next-Gen Technologies and Their Usability Evaluation

Edited by
Chhaya Santosh Gosavi, Ganesh Bhutkar,
Abhijit Banubakode, and Emmanuel Eilu

CRC Press
Taylor & Francis Group
Boca Raton London New York

CRC Press is an imprint of the
Taylor & Francis Group, an **informa** business

A CHAPMAN & HALL BOOK

Designed cover image: Chhaya Santosh Gosavi, Ganesh Bhutkar, Abhijit Banubakode, and Emmanuel Eilu

First edition published 2025
by CRC Press
2385 NW Executive Center Drive, Suite 320, Boca Raton FL 33431

and by CRC Press
4 Park Square, Milton Park, Abingdon, Oxon, OX14 4RN

CRC Press is an imprint of Taylor & Francis Group, LLC

© 2025 selection and editorial matter, Chhaya Santosh Gosavi, Ganesh Bhutkar, Abhijit Banubakode, Emmanuel Eilu; individual chapters, the contributors

ISBN: 9781032664798 (hbk)
ISBN: 9781032664804 (pbk)
ISBN: 9781032664828 (ebk)

DOI: 10.1201/9781032664828

Typeset in Times
by codeMantra

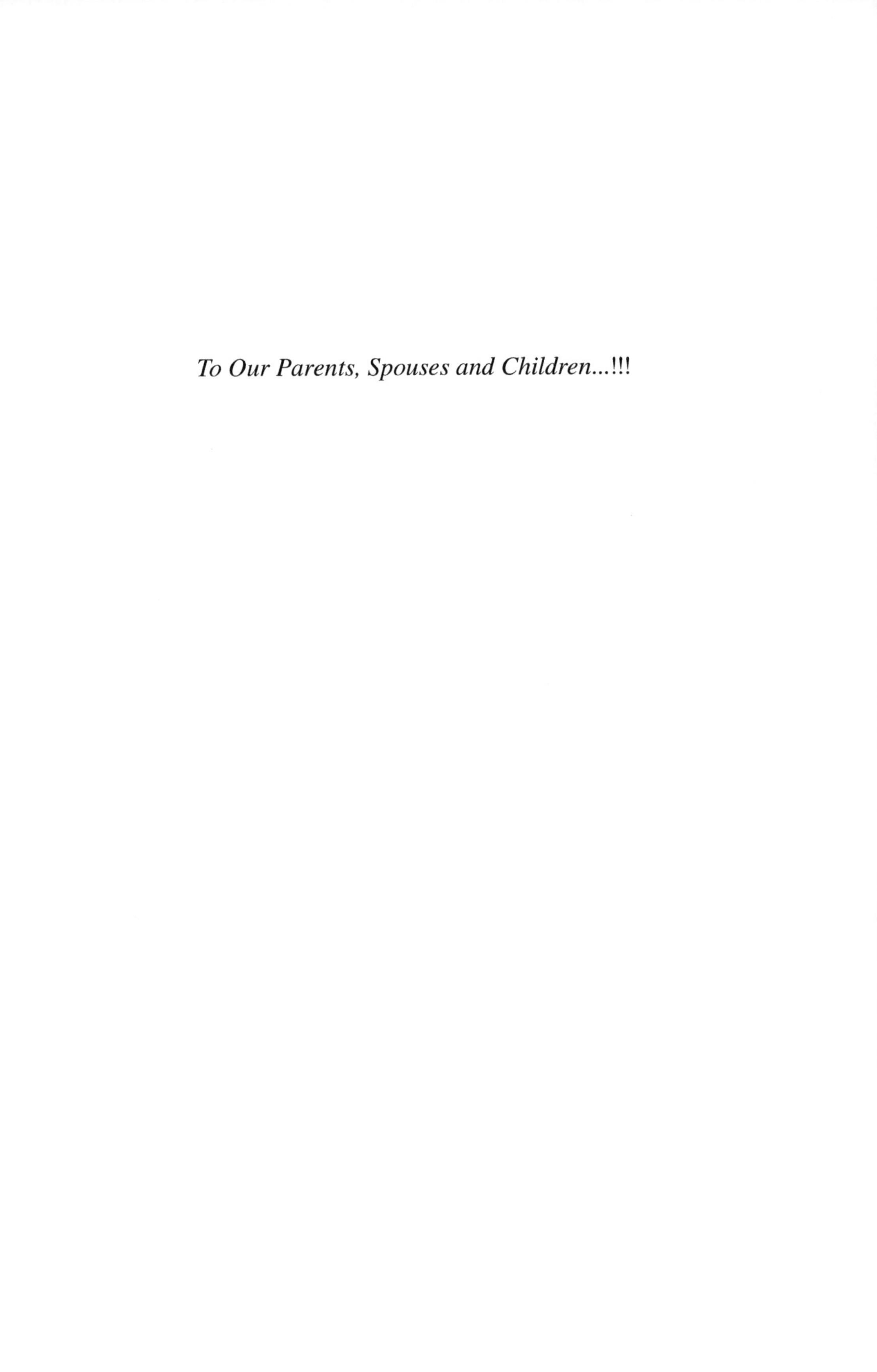

To Our Parents, Spouses and Children...!!!

Contents

SECTION A   Interactive Media Technologies

SECTION B   Next-Gen Technologies

SECTION C   *Educational Technologies*

Contents ix

Foreword

It is indeed my pleasure to be writing the foreword for this book *Interactive Media with Next-Gen Technologies and Their Usability Evaluation*, which is offering valuable glimpses into the local contexts and perspectives of the hand-picked authors from various developing countries like India, Indonesia, Morocco, Uganda, and Vietnam. The diverse perspectives represented through the chapters are helpful in understanding the multifaceted cultural and socio-economic parameters of usability in various application domains such as agriculture, fashion technology, education, social media, accessibility, film, and television.

So far, the field of usability is predominantly shaped by research, theories, and case studies originating from Western countries. This book makes a very relevant contribution to enriching the discourse on usability comprising more inclusive and culture-specific research contributions coming from developing countries from various parts of the globe.

This book includes exhaustive case studies and usability evaluations in the form of sixteen chapters spanning a vast array of cutting-edge technological applications such as interactive media, over-the-top (OTT) platforms, computer-aided learning, addictive interfaces, augmented and virtual reality technology, Computing, personalized and adaptive computer-aided learning, IoT-based water system, Smart Birdnest for bird conservation, AI-powered smart vehicles, sustainable fashion technology, cyber forensic accounting, and assistive technologies. The chapters of this book dig deeper into the reasons, motivations, emotions, and psychological aspects of various types of users including the complex learning needs of visually challenged students.

This book shall be useful to technologists, researchers, usability professionals, and engineering and design students as the research articles are replete with well-documented case studies, lab experiments, theoretical frameworks, and literature review.

I extend my heartfelt congratulations to the esteemed editors of the book, Dr. Chhaya Gosavi, Dr. Ganesh Bhutkar, Dr. Abhijit Banubakode, and Dr. Emmanuel Eilu, for their concerted efforts in collecting, curating, and editing the manuscripts and bringing together a wealth of insights from diverse developing countries.

Dr. Dinesh Katre
March 2024

Dr. Dinesh Katre is the recipient of the Emmett Leahy Award in 2020 and the National Digital Stewardship Alliance's Individual Innovation Award in 2019, both prestigious honors established in the USA. He has served as Vice Chair of TC WG 13.6 on Human-Work Interaction Design (HWID) of IFIP. He is former Senior Director & Head and founder of the Human-Centred Design & Computing Group at the Centre for Development of Advanced Computing (C-DAC), India.

Preface

This edited book is intended to be used by university students, academic researchers, industry professionals, and technology users from a range of backgrounds pursuing academic or research work in the fields of Human-Computer Interaction (HCI), Multimedia Techniques, Interactive Media Technology, and Augmented Reality and Virtual Reality (AR/VR).

Initially, a 'Call for Chapters' was announced for submission of abstracts, and interested authors submitted abstracts relevant to the topic of this book. Received abstracts reviewed by the editorial team and authors informed about selected abstracts for full chapter submissions. These submitted full chapters were reviewed by two expert reviewers. This chapter review round was followed by an additional round of editorial review. Papers with at least two positive reviews were accepted for publication in the edited book.

HCI has expanded rapidly and steadily for three decades, attracting professionals from many other disciplines, and incorporating diverse concepts and approaches. The study of HCI can be applied in real-world application areas such as Education, Film and Television, Agriculture, Cyber Security, Bird Conservation, Smart Vehicles, Fashion Technology, and e-Learning. HCI promotes user-centered application design, incorporating user experiences.

Interactive Media Technologies and Multimedia Techniques have been playing a vital role in the development of software and hardware applications. Interactive media is a human-machine interface that allows people to connect with each other, by making them active participants in the media they consume through text, graphics, audio, and video. Interactive Media Technologies include the latest technologies such as social media, Dark Patterns, Internet of Things (IoT), Android Development, Assistive Technologies, and Augmented Reality (AR)/Virtual Reality (VR). Interactive media facilitates interaction with a broader spectrum of users including specialized, marginalized, and differently abled users.

Related Interactive Media Technologies and Applications need to be evaluated for assessment and enhancement of their usability. This usability evaluation can be formative or summative, expert-based or user-based evaluation. This book is an attempt to cover most of the topics related to Interactive Media Technologies, Next-Gen Technologies, and their Usability Evaluations.

This edited book will be ideal for both beginners and experts related to HCI, Multimedia Techniques, and Next-Gen Technologies. This book has received a good response from five countries including Morocco, Vietnam, Uganda, Indonesia, and India. It covers a vast range of relevant topics in an easy-to-understand language.

Acknowledgments

When a dream is fulfilled, it is always a glorious feeling. That's our sentiment while presenting this edited book. This is the only page, where we have the opportunity to express our gratitude to those who were part of our wonderful journey.

Behind every success, there is an invisible power inspiring the creators. Timely acceptance of the book proposal and constant support by CRC Press, Taylor & Francis Group have encouraged us to complete the book on time. The editors appreciate the effort and the cooperation extended by the publishing team.

We also thank our real contributors – all the authors – for their patience in submitting multiple revisions of their book chapters from time to time. Their effort has resulted in successful completion of this quality work.

We would like to thank all the users and experts, who proactively participated in the usability evaluations and surveys involved in the research work presented in thought-provoking book chapters. We extend our thanks to the sponsors and funding agencies, supporting some of the selected research projects included in this edited book.

We express our gratitude toward Cummins College of Engineering for Women – Pune, Vishwakarma Institute of Technology – Pune, MET Institute of Computer Science – Mumbai, India, and Uganda Christian University – Mukono, Uganda for motivating us to aim for such a thought-provoking book project.

The standard quality of the chapters is the result of timely and meticulous reviews provided by members of the reviewer board. We express deep sense of gratitude toward these members for their immense contribution.

The list of members of the reviewer board is as follows:

Nasim Mohammad	Emmanual Elu	Fatima Zahra Quariach
Yohannes Kurniawan	Shrikant Salve	Santosh Borde
Uday Sagale	Anil Surve	Priyadarshan Dhabe
Sandhya Arora	Rajeshwari Kanan	Nuzzhat Shaikh
Baisa Gunjal	Sneha Tombre	Geeta Navale
Girija Chiddarwar	Ashwini Deshpande	Kirti Wanjale
Madhuri Tasgaonkar	Rushali Deshmukh	Anagha Kulkarni
Laxmi Bewoor	Makarand Velankar	Anjali Kulkarni
Anjali Naik	Hitendra Khairnar	Manoj Bangare
Seema Purohit	Neha Patil	Govind Suryawanshi
Smita Chaudhari	Jyoti Chauhan	Flavia Gonsalves
Girish Navale	Meenal Nerkar	Deepmala Salunke
Gitanjali Mate	Deepali Patil	Shital Barekar
Sushila Shelke	Radhika Bhagwat	Niraja Jain

We thank the Board of Studies (BOS): Computer Engineering and BOS: Information Technology, Savitribai Phule Pune University (SPPU) – Pune, India and BOS: Master of Computer Applications, Mumbai University – Mumbai, India.

We also thank members of WhatsApp groups Doctoral Super Group – Computer Engineering of SPPU and PICT ME 2001 – 2003 group for helping us to reach out to every corner of the academic and research community.

We express love and gratitude to our beloved parents, spouses, and children. Without their tremendous support and patience, this book journey was impossible.

Finally, we are thankful to the Great God for the blessings during the journey of this memorable book.

Yours,

Chhaya Gosavi,
CCEW Pune, India
chhaya.gosavi@cumminscollege.in

Ganesh Bhutkar,
VIT Pune, India
ganesh.bhutkar@vit.edu

Abhijit Banubakode,
MET Mumbai, India
abhijitsiu@gmail.com

Emmanuel Eilu,
UCU, Uganda
emmanueleilu0@gmail.com

Editors

Dr. Chhaya S. Gosavi has completed her graduation, Post-graduation, and PhD in Computer Engineering at Savitribai Phule Pune University. She is an Associate Professor in the Computer Engineering Department of MKSSS's Cummins College of Engineering for Women, Karvenagar, Pune for the past 24 years. She is a life member of the Computer Society of India. Her areas of interest include Multimedia, Pedagogy, Machine Learning, Image Processing, Operating Systems, and Compilers. She received the Cambridge International Certificate for Teachers and Trainers from Wipro. She has more than 30 international journal and conference publications. She has a copyright on her name, two granted patents by the Government of India, and a recent publication *Designing User Interfaces with Data Science Approach*.

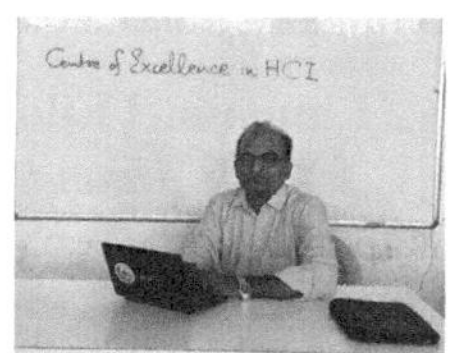

Dr Ganesh Bhutkar is a Coordinator at the Centre of Excellence in Human-Computer Interaction (HCI) and also a Professor and Assistant Head (Research) at the Department of Computer Engineering at Vishwakarma Institute of Technology (VIT), Pune, India. His research work is mainly focused on HCI, Artificial Intelligence, Medical Usability, and Assistive Technology. He has a PhD in HCI from the Indian Institute of Technology (IIT), Bombay, India. He is an active member of ACM as well as SIGCHI and also a Vice Chair of a workgroup—IFIP 13.6—Human-Work Interaction Design (HWID). He has more than 600 Google Scholar citations with about more than 100 research papers. He has delivered invited talks at research events across Europe, Africa, and Asia. He has co-authored six edited books in HCI. He is also a Founder-Secretary of Shirwal Gymkhana, a local sports academy training students.

Dr. Abhijit Banubakode is currently working in the capacity of Principal at MET Institute of Computer Science, Mumbai, and Head of the Institute of Information Technology Centre for Development of Advanced Computing (C-DAC), Mumbai, India. A has 23+ years of Academic, Training, and Industrial experience. He has received a PhD degree in Computer Studies from Symbiosis International University (SIU), Pune. He is a member of the Board of Studies and, the Academic Council of various universities. He is also a member of national and international professional bodies like the International Association of Computer Science and Information Technology (IACSIT), IEEE, ISTE, and CSI. He is a PhD guide to distinguished Indian

Universities, and he has presented more than 100 research papers at international journals and conferences. He has authored textbooks with international publishers along with one patent and one copyright to his credit.

 **Emmanuel Eilu** holds a PhD in Information Technology and a Master of Science in Information Technology. His areas of research are in Human-Computer Interaction (Usability, User Experience, and Anticipated UX), Electronic Government, and ICT for Development Research (Adoption of E-Government Systems; E-governance; E-voting; Biometric Voter Registration Systems; Electronic Civil Registers; E-Health; and M-Agriculture). Dr. Eilu has been teaching both postgraduate and undergraduate programs at the School of Computing and IT—Makerere University. He also lectured at the International University of East Africa-Uganda. Currently, he teaches at Uganda Christian University-Mbale University College and coordinates a Master's program at the Department of Computing. His major teaching areas include Research Methods, Human-Computer Interaction, Computer Networks, Computer Security, Information Systems Security, Strategic Information Systems Management, and E-governance. Dr. Eilu has multiple publications to his name. He has also worked as a Data Analyst Consultant for international NGOs.

Contributors

Mayuresh Aher
Department of Computer Engineering
Vishwakarma Institute of Technology
Pune, India

Shreya Ambekar
Vishwakarma Institute of Technology
Pune, India

Lamya Anoir
Research team in Computer Science and
 University Pedagogical Engineering
Higher Normal School
Abdelmalek Essaadi University
Tetouan, Morocco

Saniya Atalatti
Department of Computer Engineering
SCTR's Pune Institute of Computer
 Technology
Pune, India

Tanishk Babar
MIT Institute of Design
Pune, India

Abhijit Banubakode
Department of Computer Science
MET Institute of Computer Science
Mumbai, India

Ganesh Bhutkar
Department of Computer Engineering
Vishwakarma Institute of Technology
Pune, India

Kushal Birla
K. K. Wagh Institute of Engineering
 Education and Research
Nashik, India

Racheal Ddungu Mugabi
College of Humanities and Social
 Sciences
Makerere University
Kampala, Uganda

Mohamed Erradi
Research team in Computer Science and
 University Pedagogical Engineering
Higher Normal School
Abdelmalek Essaadi University
Tetouan, Morocco

Chhaya Gosavi
Department of Computer Engineering
Cummins College of Engineering for
 Women
Pune, India

Himani Kadam
Department of Mechanical Engineering
Cummins College of Engineering for
 Women
Pune, India

Snehal Kamalapur
K. K. Wagh Institute of Engineering
 Education and Research
Nashik, India

Rahul Kanade
Department of Computer Engineering
Vishwakarma Institute of Technology
Pune, India

Shreyas Kapse
Department of Computer Engineering
Vishwakarma Institute of Technology
Pune, India

Shubham Kasar
Department of Computer Engineering
Vishwakarma Institute of Technology
Pune, India

Shridhar Kedar
Department of Mechanical Engineering
Cummins College of Engineering for
    Women
Pune, India

Mohamed Khaldi
Research team in Computer Science and
    University Pedagogical Engineering
Higher Normal School
Abdelmalek Essaadi University
Tetouan, Morocco

Vu Kien Phuc
University of Economics Ho Chi Minh
    City (UEH)
Ho Chi Minh, Vietnam

Ayush Kurapati
Department of Computer Engineering
Vishwakarma Institute of Technology
Pune, India

Yohannes Kurniawan
Department of Information Systems
Bina Nusantara University
Jakarta, Indonesia

Adarsh Londhe
Department of Computer Engineering
Vishwakarma Institute of Technology
Pune, India

Kshitij Magare
Department of Computer Engineering
Vishwakarma Institute of Technology
Pune, India

Chittaranjan Mahajan
Dolphin Labs
Pune, India

Akshay Manchekar
Department of Computer Engineering
Vishwakarma Institute of Technology
Pune, India

Rishab Manocha
Department of Fashion Designing
Pearl Academy
Jaipur, India

Neelam Naik
Department of Information Technology
SVKM's Usha Pravin Gandhi College
    of Arts, Science and Commerce
VileParle-W, Mumbai, India

Rosemary Nakijoba
Department of Social Work and Social
    Admnistration
Muteesa Royal University
Kampala, Uganda

Hasifah Namatovu
Department of Information Systems
Makerere University
Kampala, Uganda

Rashmi Naveen
Department of Information Science &
    Engineering
NMAM Institute of Technology
Nitte, India

Anikka Pandey
Department of Computer Engineering
Vishwakarma Institute of Technology
Pune, India

Harshali Patil
Department of Information Technology
    and Computer Science
S. K. Somaiya College, Somaiya
    Vidyavihar University
Mumbai, India

Urmila Pawar
Department of Computer Engineering
SCTR's Pune Institute of Computer
  Technology
Pune, India

Pham Quang Huy
University of Economics Ho Chi Minh
  City (UEH)
Ho Chi Minh, Vietnam

Shirish Sane
K. K. Wagh Institute of Engineering
  Education and Research
Nashik, India

Anirudha Sonawane
Department of Computer Engineering
Vishwakarma Institute of Technology
Pune, India

Fiona Ssozi
Department of Information Systems
Makerere University
Kampala, Uganda

Madhuri Tasgaonkar
Department of Computer Engineering
Cummins College of Engineering for
  Women
Pune, India

Sachin Vanjire
Department of Computer Engineering
Automotive Vehicle Network
  Communication
Pune, India

Seema Vanjire
Vishwakarma University
Pune, India

Section A

Interactive Media Technologies

User Survey on OTT Platforms

Anikka Pandey, Ganesh Bhutkar, Shreya Ambekar, Anirudha Sonawane, and Mayuresh Aher

1.1 INTRODUCTION

The headway in the internet has profoundly changed how users consume content, prompting creators to innovate in program content and distribution platforms. A notable innovation in this regard is Over-the-Top (OTT) platforms, which are revolutionizing the content delivery chain. The leading Indian multinational Information Technology (IT) company—Tata Consultancy Services (TCS) defines OTT platforms as 'platforms that deliver film and television content, transcending traditional cable and satellite TV distribution, by directly connecting producers to consumers through online channels'. Popular OTT platforms in India and worldwide include Disney+Hotstar, Netflix, Amazon Prime Video, JioCinema, and others [1–4]. The top three OTT platforms are depicted in Table 1.1.

Mirroring global trends, India has seen a surge in internet usage, driven by increased adoption of tablets, smartphones, and laptops. The worldwide OTT market surpassed $121 billion (approx. INR 10,000 billion) in 2021, with projections indicating further growth [5]. Approximately 1.1 billion users globally accessed OTT video services in 2022, driven by high-speed internet and smartphone prevalence [5–7].

TABLE 1.1

Top Three OTT Apps in India

Disney+Hotstar	Netflix	Amazon Prime

DOI: 10.1201/9781032664828-2

In the fiscal year 2020, OTT platform revenue in India surged to around $229 million (approx. INR 19 billion), projected to reach $555 million (approx. INR 46 billion) by 2022 [8]. India's OTT video market is on a strong growth trajectory, anticipating revenues of $3,666 million (approx. INR 304 billion) in 2023. With a Compound Annual Growth Rate (CAGR 2023–2028) of 8.78%, the market is forecasted to reach a volume of $5,584 million (approx. INR 463 billion) by 2028 [6,7].

The COVID-19 pandemic significantly boosted the demand for OTT content as people sought entertainment during lockdowns. In the dynamic realm of media and entertainment, OTT platforms have emerged as a transformative force, reshaping content consumption and revolutionizing the industry [9,10]. These digital streaming services deliver diverse video and audio content via the internet, bypassing traditional cable, or satellite distribution models.

In today's digital age, intense competition among OTT platforms involves globally recognized services striving to secure and retain subscribers, including Disney+Hotstar, Netflix, Amazon Prime Video, JioCinema, and others. Traditional media companies have also entered the streaming arena, launching their platforms and escalating competition.

As OTT platforms shape the industry's trajectory, it is crucial to understand user preferences, content diversification, challenges, user experience, and desired features, while also considering accessibility for users. The user survey conducted during this research work is designed to explore these aspects comprehensively and provide invaluable insights into the ever-evolving OTT ecosystem.

1.2   RELATED WORK

The following are a few research papers dealing with user surveys and OTT apps. The first paper offers a literature review on the transformative impact of online streaming platforms in digital media, concentrating on aspects like user preferences, content quality, personalized recommendations, and platform convenience, while addressing challenges like excessive screen time and privacy concerns [11]. This research extends these findings by focusing on young users' interactions with streaming services, adding empirical insights into user behavior in the OTT landscape. This complements and broadens the understanding of digital entertainment, drawing parallels between past research and our findings in user engagement and preferences. The dynamic evolution of OTT platforms in India, a key focus of contemporary media research, is profoundly articulated in the second paper, which examines the factors influencing millennials' adoption of these services. This seminal work underscores the importance of content variety, affordability, and convenience, employing a mixed-methods approach that provides comprehensive insights into millennials' preferences and behaviors [12]. This study extends this narrative by conducting a targeted survey, aiming to add empirical depth to the understanding of millennial engagement with OTT platforms in the Indian context, thereby enriching both academic and industry perspectives in this rapidly evolving domain. The research in the next paper, dissects the growth of OTT platforms in India, examining their influence on entertainment consumption patterns. It highlights evolving media habits, and increased internet usage while addressing content variety, personalized recommendations, and adoption factors [13]. This study is particularly pertinent to research, offering insights into user engagement and preferences, while also enriching the understanding of the factors driving OTT usage, including pivotal challenges like data privacy and regulatory considerations. The study in another paper delves into Amazon Prime Video's strategic approaches within India's OTT market, focusing on enhancing customer engagement through personalized recommendations and interactive features. It investigates how these strategies boost user experience and loyalty while addressing key factors like customer satisfaction and content censorship challenges specific to India [10,14]. This paper's insights significantly augment our research by shedding light on Amazon Prime Video's unique methodologies and hurdles in the Indian context. It resonates with our exploration of user preferences and decision-making, especially regarding personalized experiences. The analysis presented in the next paper centers on Netflix's impact on consumer behavior, specifically highlighting the role of personalized recommendations and the elements driving engagement on OTT platforms. It thoroughly investigates aspects such as content diversity, social influences on user choices, and the psychological implications of binge-watching [15]. This study complements our research by providing a detailed exploration of the above-mentioned aspects. Such insights are instrumental in enriching our understanding of the intricate patterns of user behavior within the OTT landscape. The next study investigates the accelerated adoption of OTT platforms during the COVID-19 pandemic, with a focus on major players like Netflix and Amazon Prime Video. It explores their role in reshaping on-demand content delivery and the consequent shifts in global media consumption. Key aspects such as the economic ramifications of the pandemic, content production challenges, and potential enduring changes in the industry are discussed [9,16]. This research is vital in understanding the transformation of the OTT industry during and after the pandemic. It sheds light on the increased reliance on OTT services and evolving content consumption patterns, underscoring the importance of continued research into the persistent growth and adaptation of OTT platforms within the ever-changing entertainment landscape. Another research paper has an investigation, which delves into the repercussions of COVID-19 on OTT media consumption in India, employing a mixed-methods methodology. Combining a survey involving 80 participants with an extensive review of news articles, industry reports, and academic journals, the study incorporates well-established theoretical frameworks like TAM, TPB, DOI, and TRA [17]. These frameworks not only aided the structuring of the survey questionnaire but also facilitated the exploration of diverse patterns of OTT consumption during the pandemic. In this research paper, a quantitative survey involving 95 respondents examines the influence of OTT platforms like Netflix, Amazon Prime, and Hulu on viewing experiences. It specifically focuses on demographics, pricing, and consumption behaviors. By combining survey data with a literature review, this study yields valuable insights into the patterns of digital entertainment consumption. It not only illuminates gaps in existing OTT platform research but also serves as a compass for future inquiries in the digital entertainment domain [18]. These findings contribute to the formulation of new research questions and methodologies while enhancing our understanding of OTT impacts. They inspire ongoing exploration of viewer behavior and preferences within the OTT context. The next research paper investigates college-going students' perceptions of OTT platforms in Mumbai, utilizing a survey approach. Primary data is collected from 131 students through a Google Form, supplemented by secondary data from various sources. Survey questions are strategically designed to explore the merits, drawbacks, and challenges associated with OTT platforms. The Likert scale is employed for response analysis, and findings are presented using tables and graphs. The paper also includes a literature review examining the impact of OTT platforms on consumer behavior during the COVID-19 pandemic [19]. In our research, we drew inspiration from the survey methodology outlined in this paper, incorporating certain elements that suited our study, such as utilizing Google Forms for data collection. The next research paper investigates the factors driving customer preference for OTT platforms over traditional media through a quantitative approach, employing Snowball Sampling with 120 respondents. The study combines structured questionnaires with insights from secondary sources and utilizes statistical tools such as percentage analysis and ANOVA for thorough data analysis [20]. Integrating the findings from this study into the research paper has enhanced the understanding of OTT user preferences and strengthens the methodology, establishing a valuable framework for presenting research results effectively. The last paper explores strategies for emerging OTT players in India, with a focus on forming partnerships with telecom operators, regular content updates, AI integration, and innovative subscription models [10]. Drawing on industry insights from PwC and Ovum for market comprehension and customer retention strategies, it also addresses challenges within the OTT sector, such as market entry and customer retention, while recommending distinctive content offerings and flexible business models [1,2]. This paper underscores the pivotal role played by content creators in expanding the reach and attractiveness of OTT platforms, providing actionable guidance for navigating India's fiercely competitive digital media environment.

1.3   RESEARCH METHODOLOGY

The research methodology consists of several activities as discussed in this section. (A) Questionnaire Design and User Sampling: During the user survey, a structured questionnaire comprising 18 questions was crafted. The first segment in the questionnaire, focused on gathering demographic information, encompassing four questions. Following this, participants moved to the second section, consisting of five questions, which explored the usage patterns of the OTT platform, and the final section delved into user experiences and preferences, containing nine questions. This comprehensive user survey accumulated 162 responses from users who either used one or multiple OTT platforms or abstained from using any. The selection of OTT platforms for the survey was based on their popularity in the Indian market [1,3,4]. The platforms included in the survey were Disney+Hotstar, Netflix, Amazon Prime Video, JioCinema, Zee5, and Apple TV+. (B) Data Collection and Analysis: Considering the evolution of technology and the growing reliance on digital platforms, it was decided to survey an online method using Google Forms. This approach offered numerous advantages in terms of accessibility, ease of participation, and broader outreach beyond the confines of physical space [21,22]. The data collected through the forms was compiled and sorted according to the needs of the researchers using Microsoft Excel tools as depicted in Tables 1.2–1.7. In these tables, the first row depicts the number of user responses acquired for a given option. The second row conveys it in terms of percentage.

TABLE 1.2

Gender of the Respondents of the User Survey

Type	Male	Female	Non-Binary	Prefer Not to Say	LGBTQ+	Others	Total Users
User responses	77	80	2	2	0	1	162
Response percentage	47.5%	49.4%	1.2%	1.2%	0%	0.7%	100%

TABLE 1.3

Occupation of the Respondents of the User Survey

Type	Student	Employee	House-Maker	Retired	Other	Total Users
User responses	67	43	14	7	31	162
Response percentage	41.4%	26.5%	8.6%	4.3%	19.2%	100%

TABLE 1.4

Respondents' Engagement with the OTT Platforms of the User Survey

Type	Less than 6 Months	6 Months to 1 Year	One Year to 2 Years	More than 2 Years	Total Users
User responses	16	44	43	59	162
Response percentage	9.9%	27.2%	26.5%	36.4%	100%

TABLE 1.5

Usage of OTT Platforms by the Respondents of the User Survey

Type	Daily	Several Times a Week	Once a Week	Less than Once a Week	Total Users
User responses	32	71	37	22	162
Response percentage	19.8%	43.8%	22.8%	13.6%	100%

TABLE 1.6

Devices Used by the Respondents to Access the OTT Platforms of the User Survey

Type	Smart TV	Laptop/ Computer	Smartphone	Tablet	Other
User responses	58	93	111	36	2
Response percentage	35.8%	57.4%	68.5%	22.2%	1.2%

TABLE 1.7

Most Preferred OTT Platforms by the Respondents of the User Survey

Type	Netflix	Amazon Prime Video	Disney+	JioCinema	Apple TV+	Zee 5	Total Users
User responses	64	36	27	13	7	15	162
Response percentage	39.5%	22.2%	16.7%	8%	4.3%	9.3%	100%

How long have you been using OTT platforms?
162 responses

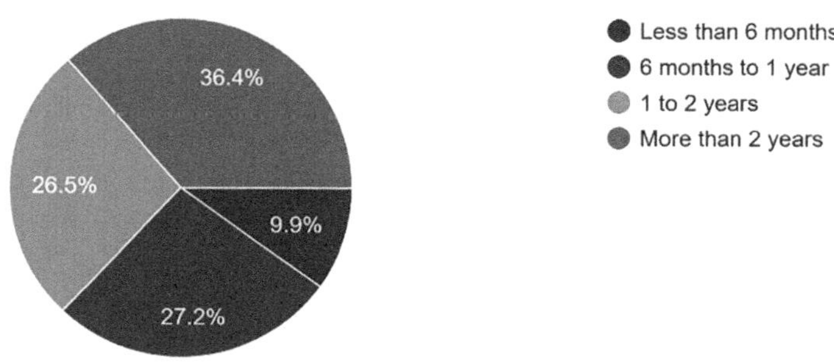

FIGURE 1.1   Graph depicting OTT usage.

The user survey data provide a comprehensive insight into the demographics and usage patterns of OTT platforms among users [21,22]. These findings underscore the diverse user base of OTT platforms across genders, occupations, duration of usage, frequency of engagement, devices used, and platform preferences. The prevalence of specific platforms over others can offer insights into user preferences, potentially guiding content creation or service enhancement strategies for these platforms [16].

1.4   RESULTS AND DISCUSSION

This user survey, focused on OTT platforms, attracted diverse users, providing a comprehensive understanding of preferences and trends. The various participant user profiles ensure a holistic exploration of how different demographic segments engage with and appreciate OTT platforms, allowing for nuanced analyses and targeted decision-making in the evolving landscape of digital streaming services.

The pie chart in Figure 1.1 illustrates the duration of OTT platform usage among 162 participant users. A plurality, 36.4%, have been using OTT services for more than two years, indicating a well-established user base. The second-largest segment, at 27.2%, reported usage between one and two years. Those who have used OTT platforms for six months to a year account for 26.5% of the respondents. The smallest group, at 9.9%, includes users who have adopted OTT platforms within the last six months, suggesting a slower rate of new user acquisition recently.

How often do you use OTT platforms?
162 responses

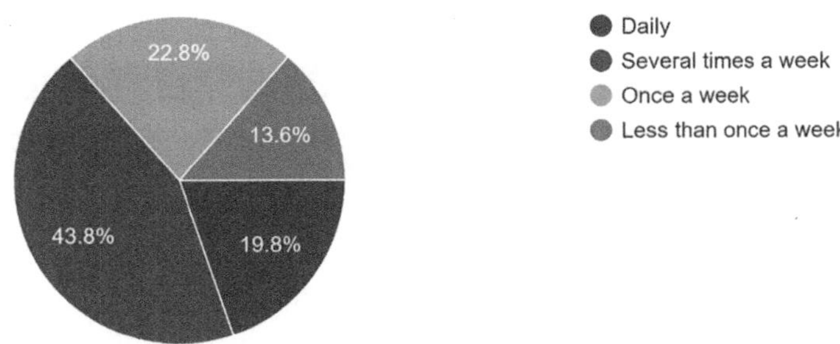

FIGURE 1.2   Graph depicting user usage frequency.

The chart's indication that over one-third of users have been using OTT platforms for over two years suggests a stable and engaged user base, likely comfortable with the platform's content and features [23]. This longevity in usage could reflect well on the platforms' ability to retain subscribers, possibly due to quality content or a robust user experience [4]. On the flip side, the smallest segment of users being those who have adopted OTT platforms within the last six months points to a slowdown in new user acquisition, which could be due to market saturation, competition, or a shift in consumer entertainment preferences [24]. The OTT services may need to explore new content strategies or innovative features to reignite growth and appeal to a fresh audience segment.

The next pie chart in Figure 1.2 depicts the usage frequency of OTT platforms, showing that the largest portion of users, 43.8%, engage with these services on a daily basis. A substantial segment, 22.8%, uses them several times a week, while those accessing the platforms once a week represent 19.8%. The smallest group, at 13.6%, engages with OTT content less than once a week [2].

The prevalence of daily usage highlights the embedded role of OTT platforms in users' daily lives, potentially rivaling traditional television [3]. The regular engagement of users several times a week underscores the importance of OTT services in the current entertainment landscape. However, the presence of users engaging less frequently suggests a diverse user base with varying content consumption habits [24]. OTT services could leverage this insight by tailoring their content release schedules and recommendations to cater to both the habitual daily viewers and the less frequent users, aiming to increase overall platform engagement [25].

The bar graph in Figure 1.3 illustrates the devices used to access OTT platforms, with smartphones leading at 68.5%. Laptops and computers are also commonly used, with 57.4% of users accessing OTT services through them. Smart TVs are used by 35.8% of the respondents, while tablets are used by 22.2%. Dedicated streaming devices like Roku or Fire Stick are the least used, with only 2.1% of users choosing them as their primary means to access OTT content.

The high usage of smartphones for accessing OTT platforms indicates a preference for mobile and on-the-go content consumption [1]. The significant use of laptops

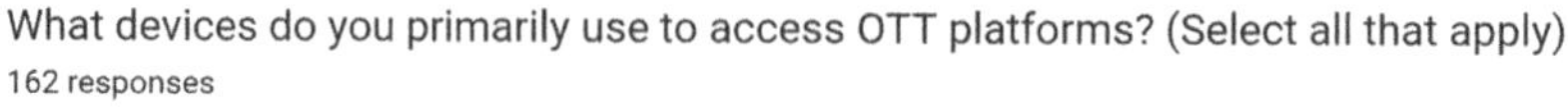

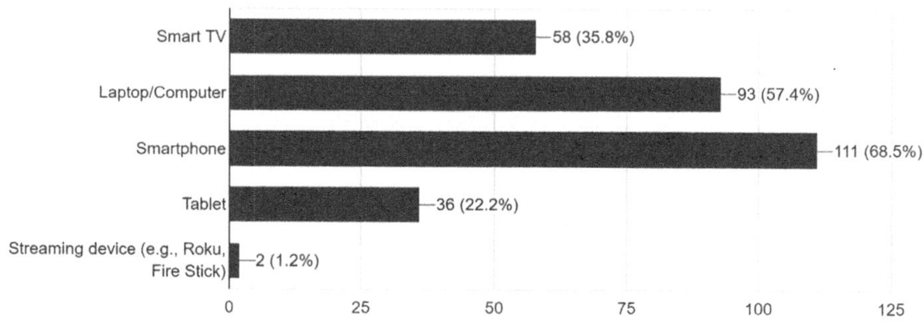

FIGURE 1.3   Graph representing various devices used to access the OTT platform.

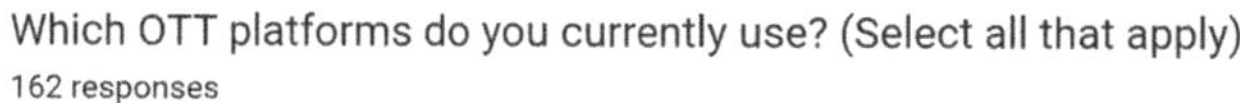

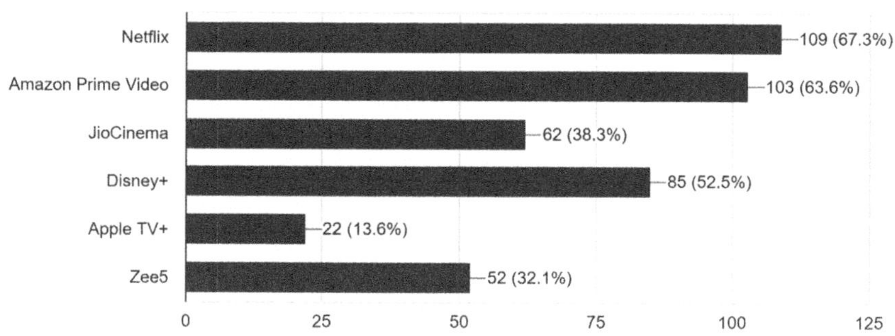

FIGURE 1.4   Graph illustrating usages of various OTT platforms.

and computers may reflect a user base that values larger screens and possibly multi-tasks while consuming content. The moderate use of smart TVs suggests that while traditional television viewing is being supplemented by OTT platforms, there's still room for growth in this area [3]. The limited use of dedicated streaming devices could point to user's preference for multipurpose devices over single-function ones. OTT services could leverage these insights to optimize their apps and interfaces for the most commonly used devices to enhance user experience.

The bar graph in Figure 1.4 presents the OTT platforms users are currently subscribed to. Netflix leads with 67.3% of users, closely followed by Amazon Prime Video at 63.6%. Disney+ has a substantial share with 52.5% of users, while JioCinema is used by 38.3% of users. Zee5 is chosen by 32.1% of the participants. Apple TV+ has the fewest users at 13.6%, suggesting it is less popular among the surveyed group [6].

Netflix and Amazon Prime Video's dominance reflects their strong market presence and possibly a more extensive content library or better user experience [26].

Which platform do u prefer/use the most?
162 responses

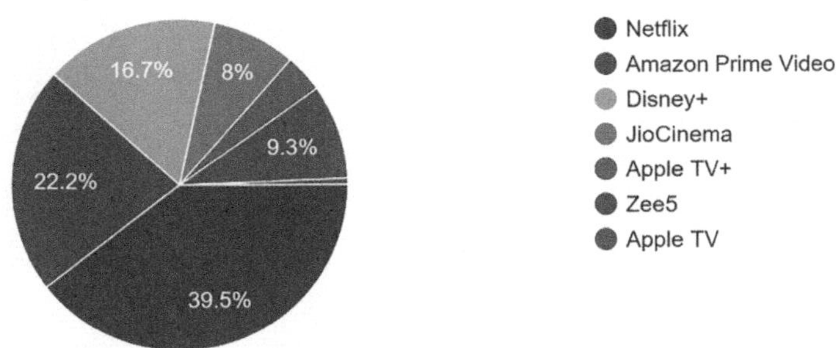

FIGURE 1.5   Graph representing user's most preferred OTT platform.

Disney+'s significant user base may be attributed to its family-friendly content and strong brand. The moderate usage of JioCinema and Zee5 could indicate a regional preference or niche content offerings [7]. The lower adoption of Apple TV+ might suggest a need for a broader content selection or more competitive pricing strategies. These trends offer valuable insights for OTT platforms looking to expand their user base or retain current subscribers by understanding consumer preferences and competitive positioning.

The pie chart in Figure 1.5 indicates the user's most preferred OTT platform, with Netflix being the most favored at 39.5%. Amazon Prime Video is the second choice at 22.2%, followed by Disney+ at 16.7%. JioCinema is preferred by 9.3% of users, while Zee5 and Apple TV+ are the least favored, with 8% and 3.3%, respectively.

Netflix's position as the most preferred OTT platform could be attributed to its diverse content library and strong brand loyalty [11,26]. Amazon Prime Video's significant preference share suggests that its combination of services, including fast shipping and a video library, is highly valued [14]. Disney+'s strong performance might be due to its exclusive content and family-oriented offerings [7]. The less preference for JioCinema, Zee5, and Apple TV+ could point to a more niche audience or the need for these platforms to increase their content variety or improve user experience to compete with the leaders. Understanding these preferences is crucial for OTT platforms as they strategize to attract and retain subscribers in a highly competitive OTT market [3].

The bar graph in Figure 1.6 portrays users' ratings for the quality of content on their most preferred OTT platform. These user ratings are ranged from 1 (Very Ineffective) to 5 (Very Effective). A majority of users rate the quality highly, with 49.4% giving it a four out of five, and 32.7% rating it as a perfect 5. A smaller group, 11.1%, feels neutral with a rating of 3. Only a minimal number of users express dissatisfaction, with 3.1% rating the quality as 2% and 3.7% giving the lowest rating of 1.

The high ratings indicate that content quality is a significant factor in user preference and satisfaction with OTT platforms. The substantial number of users who rate content quality at 4 or above reflects a general satisfaction with the content provided

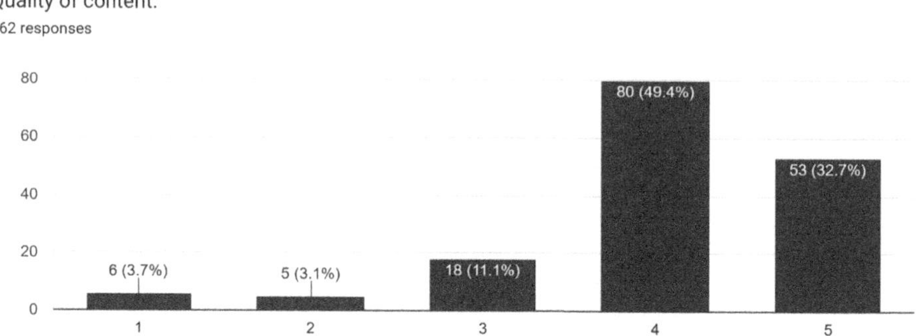

FIGURE 1.6 Graph illustrating user rating with respect to 'quality of content'.

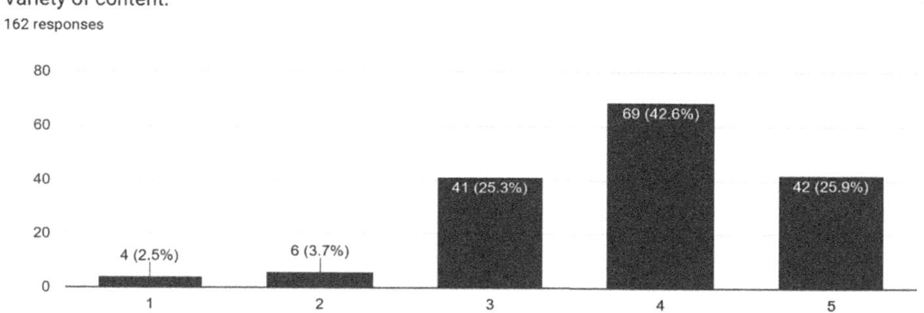

FIGURE 1.7 Graph illustrating user rating with respect to 'variety of content'.

by their preferred service. This satisfaction could be due to the originality, variety, or production values of the content. The low ratings may represent an opportunity for OTT services to identify gaps in content offerings or quality expectations. Content providers need to maintain high standards and possibly enhance the content quality to satisfy those who rated them neutrally or poorly.

The bar graph in Figure 1.7 reflects users' perceptions of the variety of content on their most preferred OTT platform. The majority of users, 42.6%, rate the variety as good with a score of four out of five. A significant number, 25.9%, consider the variety to be excellent, giving it the highest score of 5. A neutral stance is taken by 25.3% with a score of 3, suggesting room for improvement. Very few users rate the variety poorly, with 2.5% giving a score of 1 and 3.7% a score of 2.

The predominance of high ratings underscores the importance of a diverse content library in user satisfaction with OTT platforms. Users' strong positive ratings suggest that their preferred platforms offer a mix of genres, themes, and formats that cater to varied tastes. A quarter of users rated content variety as neutral, and it may signal a desire for more niche or specialized content, indicating an opportunity for OTT platforms to broaden their content spectrum. The minimal negative ratings could point to specific gaps in content variety that, if addressed, could improve overall user experience.

The bar graph in Figure 1.8 shows user ratings of the value for money of their most preferred OTT platform. A plurality of users, 39.5%, rate the value for money as good with a score of 4. The highest rating, a perfect 5, is given by 14.8%, while the neutral rating of 3 is chosen by 27.2%. Few users consider the value for money to be low, with 11.7% rating it as 2, and 6.8% providing the lowest rating of 1.

The concentration of users rating the value for money of their OTT service as good indicates a general satisfaction with the cost-to-content ratio. The significant number of users scoring the value as neutral or below suggests some concerns over pricing or perceived value, which could be a reflection of the cost sensitivity among certain user segments. For OTT platforms, understanding the importance of pricing about content quality and variety is crucial. Striking the right balance between pricing and perceived value could help to convert those with neutral sentiments into more positive advocates, potentially increasing overall satisfaction and loyalty [27].

The bar graph in Figure 1.9 reflects the satisfaction levels of users with the user experience on the OTT platform of their choice. The majority of users, 42.6%, rate the user experience as good with a score of 4. A significant portion, 19.1%, rated it as excellent, giving the highest score of 5. Neutral satisfaction is reported by 27.2% with

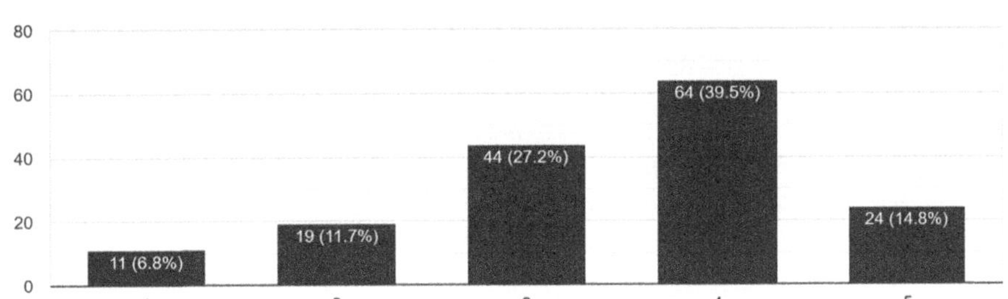

FIGURE 1.8   Graph illustrating user rating with respect to 'subscription cost'.

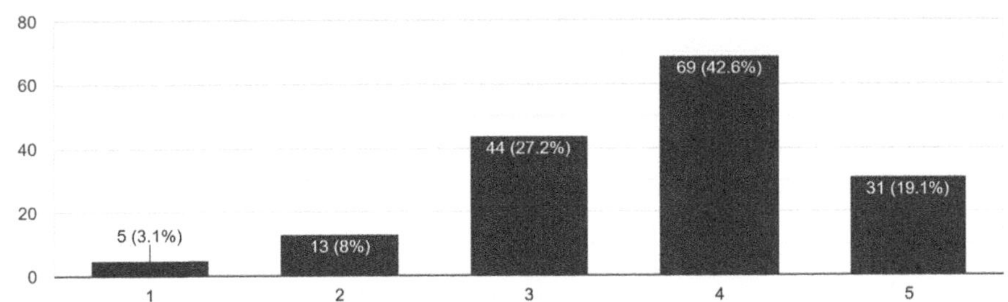

FIGURE 1.9   Graph illustrating user rating with respect to 'user experience'.

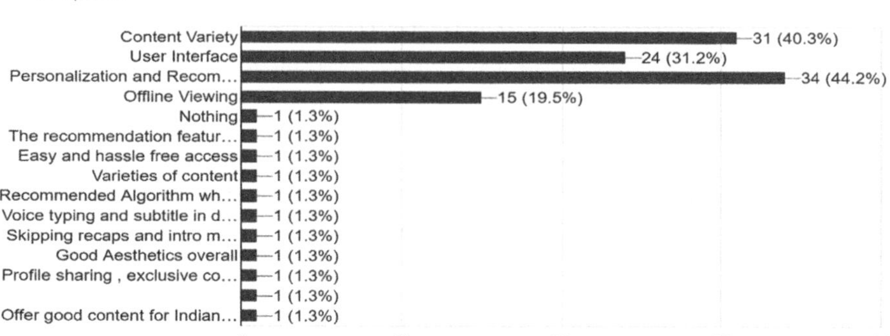

FIGURE 1.10   Graph illustrating user's appeal factors for the OTT platform.

a score of 3. A minority of users are dissatisfied, with 8% rating it 2 and only 3.1% giving the lowest rating of 1.

The user experience is a critical aspect of OTT platforms, and the high ratings reflect that platforms are generally meeting user expectations. The predominance of scores on the higher end suggests that most users find the platforms easy to navigate and content discovery to be satisfactory [28]. However, the presence of a neutral group indicates that there is a potential for improvement, possibly in areas such as personalization, streaming quality, or app responsiveness [24]. Addressing the concerns of the users who rated the user experience lower could lead to enhanced overall user satisfaction, reducing churn and fostering brand loyalty.

The bar graph in Figure 1.10 illustrates user preferences for the most appealing or user-friendly features on OTT platforms. Personalization and recommendations are highly valued, with 44.2% of users finding these features most appealing. Content variety comes second, with 40.3% of users appreciating this aspect. The user interface's ease and functionality are important for 31.2% of users, while offline viewing capability is essential for 19.5%. All other listed features, such as hassle-free access, subtitle, and dubbing options, and aesthetics, are each favored by 1.3% of the respondents.

The data indicates a strong preference for personalized content and content variety, suggesting that users value a tailored and seamless viewing experience. The importance placed on user-friendly interfaces and offline viewing options highlights the demand for accessible and diverse content consumption options [28]. Although less frequently mentioned, the presence of specific features like subtitle and dubbing options suggests that even niche preferences can significantly impact user satisfaction. OTT platforms can leverage this feedback to prioritize feature development and improve user engagement by focusing on customization and ease of use.

The bar graph in Figure 1.11 outlines the challenges and frustrations users experience with OTT platforms. The most significant issue is ads and interruptions, identified by 35.8% of users. Close behind are content availability concerns and, user interface and navigation difficulties, each cited by 33.3% and 32.1% of users, respectively. Buffering and streaming issues are a challenge for 27.2%, while device

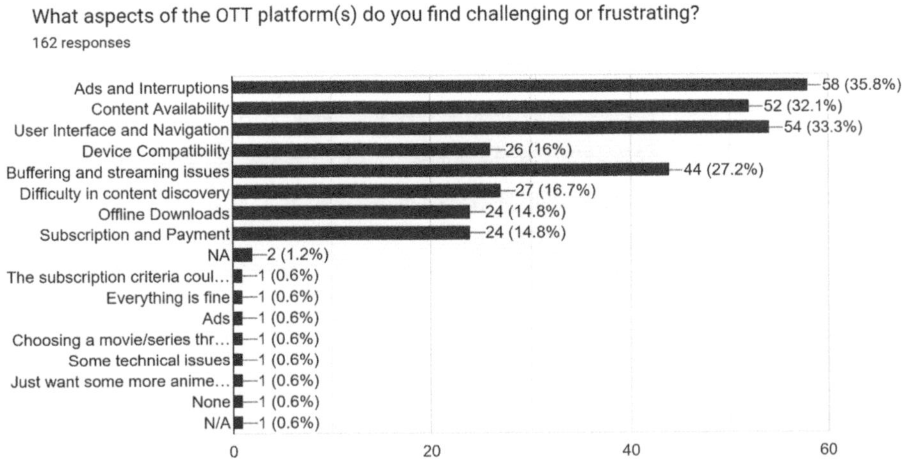

What aspects of the OTT platform(s) do you find challenging or frustrating?

162 responses

FIGURE 1.11 Graph depicting user's challenges with the OTT platform.

compatibility issues concern 16% of the users. Difficulty in content discovery and offline downloads are frustrating for 16.7% and 14.8% of users, respectively, and similar percentages find subscription and payment issues troublesome.

The prominence of ads and interruptions as a primary challenge indicates a user preference for uninterrupted viewing experiences, which may affect satisfaction and retention. The significant frustration with content availability could reflect users' desire for a more comprehensive selection of titles or dissatisfaction with regional licensing restrictions [7]. User interface complexity and navigation issues highlight the importance of a seamless and intuitive design in retaining user engagement [28]. To improve user satisfaction, OTT platforms might consider reducing ad frequency, enhancing content libraries, streamlining their interfaces, and addressing streaming and compatibility issues.

The bar graph in Figure 1.12 shows the methods users typically use to discover new content on OTT platforms. The most common method is through trending or popular content, with 66.7% of users selecting this option. Friend's recommendations are also a significant source, with 51.9% of users relying on them. Recommendations based on viewing history are used by 47.5% of the respondents. Social media plays a role for 25.3% of users, while the search function within the platforms is used by 9.3%. Self-search, which implies independent discovery without the use of platform features or external suggestions, is the least utilized at 0.6%.

The data highlight the importance of social influence and algorithm-driven recommendations in content discovery on OTT platforms. The popularity of trending content suggests that users are inclined to watch what is current and widely discussed, which could help OTT services in curating their featured lists [13,14]. The reliance on friends' recommendations and viewing history indicates that personal and tailored suggestions are valued. However, the lower utilization of the search function could suggest either satisfaction with the presented options or a potential area for improvement in search functionality. OTT platforms might benefit from

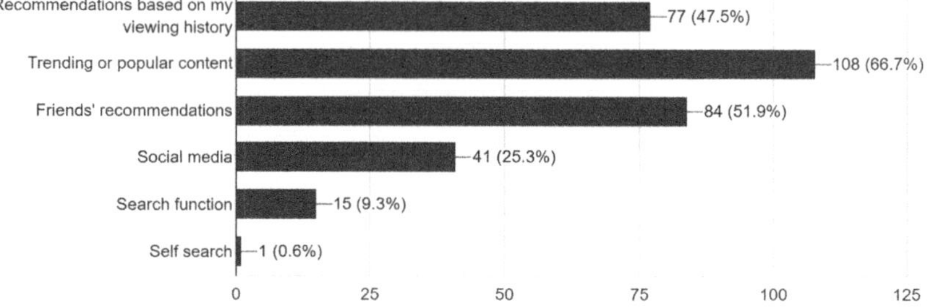

FIGURE 1.12 Graph illustrating user's content discovery pattern.

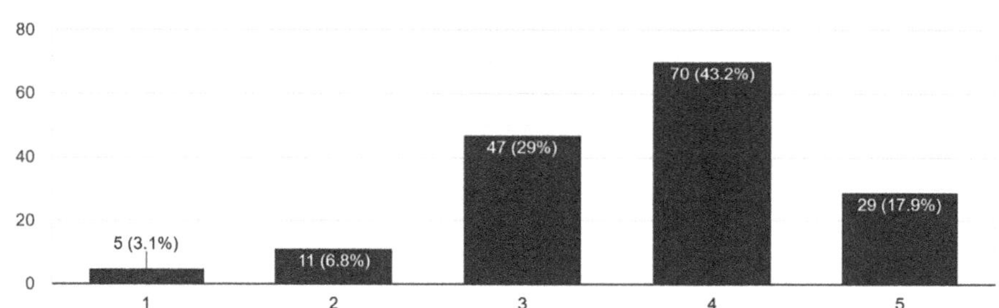

FIGURE 1.13 Graph depicting user rating with respect to recommendation algorithms.

enhancing search tools and algorithms to better support users, who wish to discover content independently.

The bar graph in Figure 1.13 depicts user ratings regarding the effectiveness of content recommendation algorithms on OTT platforms. A majority of users find the algorithms to be good, with 43.2% giving a rating of 4. A substantial 17.9% of users rate the effectiveness as excellent, with a score of 5. The neutral rating of 3 is given by 29% of users, suggesting that while many are satisfied, there is a notable portion that sees room for improvement. Few users find the algorithms to be ineffective, with 6.8% rating them as 2 and only 3.1% giving the lowest rating of 1.

The general tilt towards higher ratings demonstrates that most users are fairly satisfied with the content recommendation algorithms, finding them to be useful tools in discovering content that aligns with their preferences. The presence of nearly a third of users rating the algorithms as neutral, however, indicates an opportunity for OTT platforms to refine their recommendation systems, perhaps by incorporating more nuanced user feedback or behavioral data. Improving these algorithms could lead to increased viewer satisfaction and engagement, as well as provide a more personalized user experience.

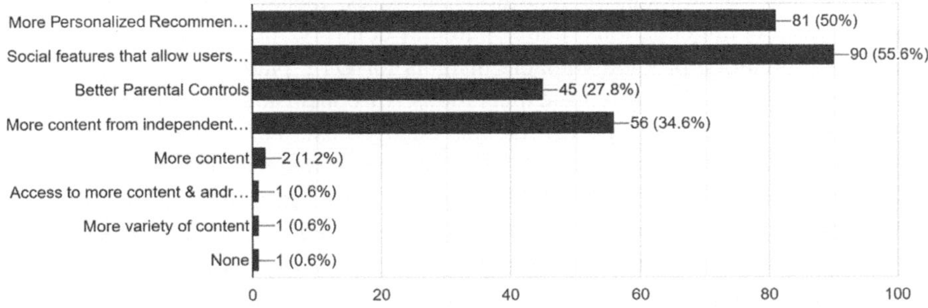

FIGURE 1.14 Graph illustrating user's desired new features for OTT platforms.

The bar graph in Figure 1.14 reflects user preferences for new features and content on OTT platforms. The most requested feature is social features that allow users to interact and about 55.6% of users are interested in this vital feature. More personalized recommendations are desired by 50% of users. A significant 34.6% of users want more content from independent creators. Better parental controls are also a priority for 27.8% of users. The interest in more content and variety is marginal, and very few users feel that no new features or content is needed.

The data show a clear demand for more social interaction and personalized content, suggesting that users are looking for a more customized and connected viewing experience. The desire for content from independent creators indicates an appetite for diverse and potentially niche content offerings. The need for improved parental controls points to the importance of user safety and content appropriateness on these platforms [29]. OTT services could take this feedback into account to enhance user engagement and satisfaction, potentially increasing their competitive edge by fostering community aspects and enhancing content discoverability and safety features.

1.5 CONCLUSION

The user survey conducted on OTT platforms provides a comprehensive understanding of the evolving landscape of digital streaming services, particularly in the Indian context. The observations gleaned from the survey shed light on user preferences, engagement patterns, and the factors influencing their choices in a market characterized by intense competition and rapid growth. The prolonged engagement of a significant portion of users, with 36.4% using OTT platforms for more than two years, underscores establishing a robust, long-term user base. The survey reveals a diverse frequency of platform usage, with a substantial number of respondent users integrating OTT services into their daily and weekly routines. The prevalence of smartphones and laptops/computers as primary access devices, along with the popularity of platforms like Netflix and Amazon Prime Video, indicates the dynamic ways in which users integrate OTT platforms into their lifestyles. Content variety and personalized recommendations drive user satisfaction, reflecting a growing demand for diverse and tailored entertainment options. While the majority expresses satisfaction with content quality, concerns such as ads and interruptions, content availability, and user interface/navigation issues highlight areas for improvement in the overall user experience. User aspirations for additional features, including social interactions, personalized recommendations, better parental controls, and diverse content from independent creators, provide valuable insights for OTT platforms to enhance user experience and cater to evolving expectations. In conclusion, the OTT ecosystem in India is thriving, driven by a diverse and engaged user base. The findings of this survey contribute to a deeper understanding of user preferences and challenges, providing a roadmap for OTT platforms to navigate the evolving landscape successfully. As the industry continues to innovate and adapt, user feedback remains an invaluable resource for shaping the future of digital streaming services, ensuring they stay at the forefront of entertainment consumption in an ever-evolving digital era.

1.6   FUTURE SCOPE

The user survey on OTT platforms, conducted in this research work, provides valuable insights into the dynamics of OTT platform usage among a sample of 162 respondent users. However, the scope for further exploration and expansion is substantial, enabling a deeper understanding of India's evolving landscape of OTT services. The following avenues can be considered for future research:

a. **Increased Sample Size**: Expanding the user survey to a larger and more diverse sample size could yield more comprehensive and representative insights. With a broader participant base encompassing varied demographics, regions, and socio-economic backgrounds, a more nuanced understanding of user preferences, consumption patterns, and platform choices can be attained. This expansion might involve reaching a wider audience through targeted outreach programs or collaborations with diverse institutions.

b. **Varied Survey Methodologies**: While the current survey was conducted using Google Forms, employing multiple methodologies for data collection could offer a richer perspective. Incorporating interviews, focus group discussions, or ethnographic studies could provide qualitative depth to complement the quantitative data obtained. Such methodologies could reveal nuanced motivations, behaviors, and preferences of users that might not be captured through conventional survey forms.

1.6.1   CHALLENGES AND AREAS OF IMPROVEMENT IN OTT PLATFORMS

The following are the major challenges for OTT platforms in the near future.

1. Sustaining Long-Term User Engagement:
   - **Challenge:** While many users have been using OTT platforms for over two years, maintaining their engagement remains a concern.
   - **Course of Action:** Platforms should consistently innovate on content offerings, enhance user interfaces, and provide personalized recommendations to sustain long-term user interest. Regularly updating content libraries and introducing exclusive content can increase user engagement.

2. Enhancing Usage Frequency:
   - **Challenge:** A notable portion of users engages with OTT platforms several times weekly, indicating high engagement. However, there is room for improvement in catering to users who use the platforms less frequently.
   - **Course of Action:** Platforms should explore strategies to enhance occasional or selective usage. This may involve implementing targeted promotions, offering personalized content suggestions, or organizing special events to attract users with varying viewing habits.
3. Tailoring to User Preferences:
   - **Challenge:** User preferences vary, and identifying factors influencing their preferences is essential for platform differentiation.
   - **Course of Action:** OTT Platforms should leverage insights from user preferences to tailor their offerings. This involves investing in high-quality and diverse content, improving subscription costs, enhancing user experience, and integrating appealing features to cater to varied user tastes.
4. Addressing User Ratings Concerns:
   - **Challenge:** Dissatisfaction with pricing and user experience highlights potential areas for improvement.
   - **Course of Action:** OTT Platforms should reassess subscription costs, considering competitive pricing and perceived value. Addressing user experience concerns, including content discovery, buffering issues, and overall platform usability, is crucial for enhancing satisfaction.
5. Improving Content Discovery:
   - **Challenge:** Users primarily rely on popular content and recommendations for content discovery, indicating a potential need for improved search functionalities.
   - **Course of Action:** OTT platforms should enhance search functionalities to ensure ease of content discovery. Integrating trending and popular content with personalized recommendations can provide a balanced and user-friendly approach.
6. Incorporating Desired Features:
   - **Challenge:** Users desire additional features such as social features, personalized recommendations, better parental controls, and more content from independent creators.
   - **Course of Action:** OTT platforms should prioritize developing social features to foster user interaction, enhance personalized recommendation algorithms, improve parental controls, and collaborate with independent creators to diversify content offerings.

REFERENCES

1. Strategies (2020) https://www.financialexpress.com/business/brandwagon/how-new-ott-players-can-win-indian-customers/2020753/. Accessed on Jan 10, 2024.
2. OTT Market Overview (2022) https://www.marknteladvisors.com/research-library/ott-platform-market.html. Accessed on Jan 15, 2024.

3. OTT Usage (2019) https://brandequity.economictimes.indiatimes.com/news/research/ 44audio-entertainment-users-pay-for-content-audio-series-have-high-traction survey/ 101512281. Accessed on Jan 15, 2024.

4. OTT Users in India (2019) https://www.livemint.com/news/india/young-indians-in-met rosdriving-ott-market-finds-survey-1560935637856.html. Accessed on Jan 15, 2024.

5. OTT Usage in India (2022) https://www.statista.com/outlook/amo/media/tv-video/ott-video/india#revenue. Accessed on Jan 22, 2024.

6. Market Survey and Price Comparison (2019) https://www.lsdigital.com/blog/survey-ott-the-new-age-entertainment-destination-for-indians/. Accessed on Jan 22, 2024.

7. OTT Market (2022) https://www.marketfeed.com/read/en/an-overview-of-indias-boomingott-market. Accessed on Jan 22, 2024.

8. Devadas, Menon. (2022). Purchase and continuation intentions of Over -The -Top (OTT) video streaming platform subscription: A Uses and Gratification theory perspective. *Telematics and Informatics Reports,* 5:100006–100006. doi: 10.1016/j. teler.2022.100006.

9. Patnaik, R., Shah, R. and More, U. (2021) 'Rise of OTT Platforms: Effect of the C-19 Pandemic', *Palarch's Journal of Archaeology of Egypt/Egyptology*, 18(7), 4–10.

10. Bihade, V. and Kumar, M. (2023) 'Customer Preference towards OTT Platform in India', *Global Media Journal*, 21, 3–5.

11. Mehta, K., Kothiya, P. and Hepaliya, H. (2020) 'A Study on the Usage and Awareness of Netflix among the Youth', *Journal of Emerging Technology Innovation Research*, 7(5), 236–255.

12. Singh, S. (2020) 'A Study on Factors Leading to Adoption of OTT Services among Millennial Consumers in India', *International Journal of Multidisciplinary Research and Technology*, 1(2), 30–47.

13. Yeole, S. and Bhaisare, C. (2023) 'A Study on User Perspective on OTT Platform in India', *Journal of Positive School Psychology*, 6(3), 6–13.

14. Upadhyay, P. (2023) 'Amazon Prime Video in India: Is it Customer Engagement or Media Content Strategy?' *BIMTECH Business Perspectives*, 10–20.

15. Kanniappan, S., Prasanth, N. and Jagadale, P. (2022) 'A Study on Netflix and Its Consumer Behaviour', *International Journal of Creative Research Thoughts*, 10(5), 4–8.

16. Habib, S., Hamadneh, N. and Hassan, A. (2022). 'The Relationship between Digital Marketing, Customer Engagement, and Purchase Intention via OTT Platforms', *Journal of Mathematics*, 2022, 4–9.

17. Sharma, G. and Dahiya, S. (2020) 'Role of COVID as a Catalyst in Increasing Adoption of OTTs in India: A Study of Evolving Consumer Consumption Patterns and Future Business Scope', *Journal of Content, Community & Communication Amity School of Communication*, 12(6), 8–12.

18. Raj, A. and Nair, A. (2021) 'Impact of OTT Platforms on Viewing Experience', *Journal of Research in Business and Management Volume*, 9(8), 4–11.

19. Satyanarayana, C., Kale, C. and Thakoor, M. (2023) 'OTT Platform: Pros, Cons, Challenges in India', *Journal of Survey in Fisheries Sciences*, 10(4), 5–10.

20. Vidhya, K. and Govind, A. (2022) 'A Study on Factors Influencing Customer's Adoption of Over-The-Top (OTT) Platform over Other Conventional Platforms', *Research Square*, 1, 6–9.

21. Chavan, S., Gore, P. and Bhutkar, G. (2022) User Survey of UPI-Enabled Payment Apps. In: Chakrabarti, D., Karmakar, S., and Salve, U. R. (eds) *Ergonomics for Design and Innovation*. HWWE 2021. Lecture Notes in Networks and Systems, Vol. 391. Springer, Cham, 1457–1469.

22. Patil, S., Bhutkar, G. and Vaidya, P. (2022) Psychological Survey of Color Perceptions for Indian Users. In: Rana, N. K., Shah, A. A., Iqbal, R., and Khanzode, V. (eds) *Technology Enabled Ergonomic Design*. HWWE 2020. Design Science and Innovation, Springer, Singapore, 433–445.
23. Rising OTT Market Analysis (2023) https://tele.net.in/growing-play-ott-segment-attrac tsviewer-and-investor-interest/. Accessed on Jan 23, 2024.
24. OTT Viewership Frequency (2021) https://www.medianresearch.in/short-survey-on-ottviewership/. Accessed on Jan 20, 2024.
25. OTT Statistics (2019) https://brandequity.economictimes.indiatimes.com/news/media/55-ofindians-prefer-ott-platforms-vs-41-that-still-prefer-dth-momagic-survey/70858815. Accessed on Jan 10, 2024.
26. Netflix Case Study and Report (2019) https://www.newamerica.org/oti/reports/why-am-i-seeing-this/case-study-netflix/. Accessed on Jan 23, 2024.
27. OTT Pricings (2020) https://www.businessworld.in/article/OTT-subscription-purchase-andviewing-witness-spike-during-Covid-19-FLYX-survey/12–09–2020–319580/. Accessed on Jan 20, 2024.
28. OTT Interface (2022) https://bootcamp.uxdesign.cc/ux-case-study-solving-what-to-watch on-ott-platforms-6f7df05d2083. Accessed on Jan 20, 2024.
29. Content Discovery (2018) https://www.linkedin.com/pulse/importance-content-discovery ott-ecosystem-mayank-kumar-khanna/. Accessed on Jan 24, 2024.

2 Analysis of Interactive Media Usage for Improving the Outcome of the Agricultural Sector in India

Harshali Patil and Neelam Naik

2.1 INTRODUCTION

Indian agriculture has several significant waves and trends. The trends highlight factors like increased production from existing farmlands, technology changes, government policies, and sustainable agriculture. The production within India was insufficient. Many farmers were in debt and there was a severe shortage of crops. Hence, Green Evolution was launched in 1960. The main goal of the Green Revolution was to reduce poverty and malnutrition. Cereals with high yields were introduced as a part of the green revolution. The white revolution in the 1970s aimed to increase milk production using improved animal husbandry practices.

Agriculture policy reformation was done in 1990 and economic liberalization brought changes in agriculture. Reduction in subsidies and trade barriers affected the agriculture sector.

Due to excessive use of fertilizers and pesticides, the chemical residue in food becomes a major concern. Farmers started adopting organic farming and exporting organic products.

Digital technology adoption in farming helped farmers to get information related to weather, crop management, market prices, and pest alerts. For farmers, Interactive Voice Response (IVR) systems, portals, and mobile apps were developed.

Precision agriculture helps farmers to make decisions about crop management, yield optimization, and resource utilization. The technologies for optimizing farming practices include GPS, drones, data analytics, and remote sensing. To support farmers, the Indian government launched several initiatives like Pradhan Mantri Kisan Sanman Nidhi (PM-Kisan) and National Agriculture Marketing (e-NAM). Sustainable agriculture focuses on technology use, such as mechanization, precision farming, and biotechnology to diversify agriculture and improve output.

Out of the two main departments of the Ministry of Agriculture and Farmers' Welfare, (i) The Department of Agriculture, Cooperation, and Farmers' Welfare

DOI: 10.1201/9781032664828-3

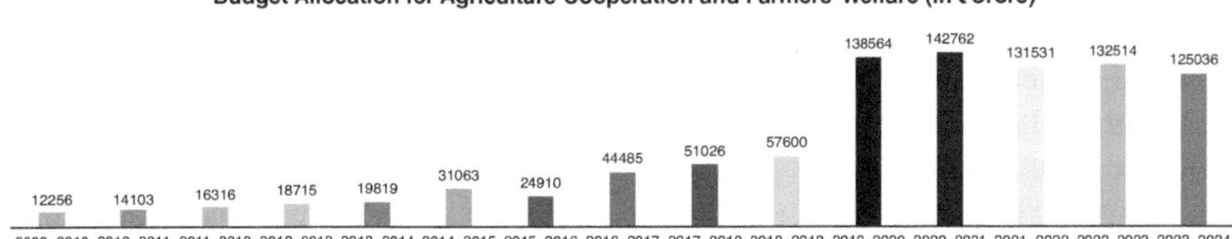

FIGURE 2.1 Budget allocation for agriculture cooperation and farmers' welfare (in ₹ crores).

oversees agriculture inputs and carries out activities and policies about farmer welfare, and (ii) The Agricultural Research and Education Department coordinates and advances agricultural research and education.

In 2019–2020, the ministry has been allotted ₹138,564 crores which is 5% of the central government budget, as shown in Figure 2.1 [1,2]. The allocation exceeds 82.9% as compared to the previous year 2018–2019. This rise in the amount is due to the allocation of ₹75,000 crores to the PM-Kisan scheme, an income assistance program for farmers [3,4].

The agriculture waves improved the lives of millions of farmers in India and the agriculture industry. But still, there are challenges like climate change, water shortages, and poor farm income.

2.2 LITERATURE REVIEW

There is an economic significance of the agriculture industry in India to reduce poverty and foster development. The majority of the labor force in India is still employed in this industry, although it does not significantly make up the majority of the GDP. The Indian government's 2016 vow to "double farmers' income by 2022" remained a dream because the attention of the Government was focused on the causes, trends, effects, and remedies of farmer suicides. The major share of the agriculture funding is allocated to problems associated with the suicides of farmers, with particular attention paid to areas like Maharashtra, Karnataka, and Andhra Pradesh [5].

Among the most promising methods for raising farm incomes is the adoption of contemporary technologies. The most notable obstacles to the adoption of technology are credit and information gaps. In India, formal finance is unavailable to over half of farming households. The implementation of the Pradhan Mantri Kisan Samman Nidhi (PM-Kisan) cash transfer system in December 2018 to alleviate farmers' cash restrictions for input purchases is noteworthy in this regard. Even though the program is marketed as a broad cash distribution program for farmers, it still plays a significant role in encouraging the uptake of contemporary technologies. In the first three months of its launch, the PM-Kisan program reached one-third of farmers, greatly assisting those who are comparatively more reliant on agriculture and have limited access to financing. Additionally, the program has greatly increased Krishi Vigyan Kendra's influence on the uptake of contemporary cultivars [6].

The information on mandi prices and farming tips in ten distinct Indian languages is available on the Nifco Kisan app. The Agri Media Video app offers videos about farming, agricultural programs run by the government, and information about the latest crop cultivation and fertilizer equipment. The Farm Bee-RML Farmers app tells users whether the weather will be suitable to seed any crop, including vegetables, shortly. The Kheti-Badi app encourages organic farming and educates farmers about the benefits of not practicing chemical farming. The Pusa Krishi app, available for Android platforms, provides information on several crop categories originated by the Indian Council of Agriculture Research (ICAR). The IFFCO Kishan Agriculture app offers farmers personalized text and audio–visual information according to their needs, including weather reports, market prices, recommended herbicides, and information on using farming equipment. Kishan Call Centre Services are also provided by it. The Crop Insurance app computes insurance costs, notifies farmers about their crop insurance policy status, and shields farmers from natural calamities like droughts, floods, or declines in agricultural commodity prices. The Kishan Yojana offers details on every government program. The Meghdoot app offers data on precipitation, moisture content, wind direction, and also on animal husbandry [7].

Mobile applications represent feasible digital tools that can be efficiently employed to disseminate agricultural information to a multitude of farmers in a short amount of time. These apps provide accurate information availability, improved farm management, efficient way of agricultural marketing, and connections with government agencies for policy support to farmers thus promoting an increase in farm revenue and productivity. However, there exist certain obstacles such as the low rate of adoption of smartphones in rural India, inconsistent internet connectivity, low levels of digital literacy among farmers, and restricted access to agricultural information in regional languages. But still, the mobile app is considered one of the most significant smart farming technologies. This paper also categorizes and lists various mobile apps as per their utility in Indian agriculture as shown in Table 2.1 [8].

TABLE 2.1

Mobile Apps as per Functionality [8]

Functionality	Name of the Apps
Fields and crop management	RiceXpert, Rice IPM, ICAR-IIOR, IFFCO Kisan, Sugarcane-Mustard-Wheat-IFC, Plantrix, Bharat Agri, Agri-central, Kisan Network, Farmkey, Agriscience Krishi, Mandi Central, KISSAN, Apni Kheti, ICAR-DGR, Krish-e, Agrowon
Horticultural crops management	Horticultural crop-IIHR, Sabji Gyan, Mango-IIHR, Tomato-IFC, Chili-IIHR, Potato-IFC, Okra-IIHR, Brinjal-IIHR, Arka Bagwani, ICAR-DOGR
Animal husbandry	Disease Control IVRI, Veterinary Clinical Care IVRI, Dairy Manager, Organic Livestock Farming, Pig Farming, Bhains Poshahar, Poshupalan
Fishery	Daily Fish India, Ornamental fish disease management, Complete fish farming, Biofloc Aap, Complete freshwater fish farming
Poultry	ICAR-DPR, ICAR-CARI, Poultry App, Poultrybaba, Poultry feed App
remote sensing/GPS	Agri bus NAVI, Plant sat, Bharat Rohan Kisan
Farm machinery	FARMS, Krishi Yantra Mitra, Tractor Junction
Marketing	Agri aap, Agricultural Businesses, Farmers e-market, Bijak-Agri App, Agrimart, eMonda, Marketyard, Bheselo
Others	AgMobile, AGRIPLEX India, Kisan Suvidha, ICAR Technologies

Indian agriculture represents the largest user base globally, they have access to smartphone applications developed by research institutes and nonprofit organizations. The organized evaluation of these smartphone apps is done in this paper [9]. The study examined 25 smartphone apps created for the agricultural and related fields that the Indian farming population mostly uses. Smartphone application usability, accessibility, update frequency, user ratings, and download volume are considered parameters for the assessment. This paper also evaluates how current smartphone applications lack in fulfilling agricultural stakeholders' requirements and provides recommendations for future smartphone applications' features and potential guidelines for researchers, policymakers, and application developers.

Artificial intelligence (AI) and big data analytics are being used to modernize agriculture to solve issues like population increase, dietary changes, and environmental concerns. It highlighted how conventional methods have given way to cutting-edge technologies, underscoring the necessity of paradigm changes in agricultural research. Big data analytics is essential to solving the complexity of agricultural sciences and improving crop production efficiency. AI is emphasized as a tool for precision farming and the solution to agricultural issues. The research highlights the significance of high-quality data for efficient agronomic and environmental decision-making and warns against misconceptions regarding these technologies [10].

Enhancing agricultural productivity and the value chain has been made possible with the usage of sensors and Information and Communication Technology (ICT) in agriculture. The pertinent use of data analytics in agriculture has transformed it from input-intensive to knowledge-intensive by storing, sharing, and analyzing large amounts of agricultural produce. The study is based on the review of the sensors and data analytics methods that are now in use in several agricultural domains. In a case study, the research compared the Korean situation with other industrialized countries. Challenges and opportunities are described in the paper which covers sustainable agriculture, technological constraints, intelligent data processing, investment cost, multidisciplinary collaboration, etc. [11].

Social distancing norms during COVID-19 impacted the direct sales in the market. The paper focuses on the challenges farmers face due to price mismatch and the need to locate optimal markets for their produce. The system comprises an input device for Geographic Information System (GIS) and market price to forecast the price. Probabilistic Neural Network (PNN) is used for agriculture commodity price prediction. A high accuracy of 98.3% is achieved by this system. Farmers can view market locations and expected pricing and the system makes recommendations for the market. The paper suggests further research on implementing big data and other AI methods for diverse agricultural commodities [12].

Currently, the agriculture production growth is not sufficient to fulfill the growing demands of the increasing population. Hence increasing production is an essential task for all countries with specific resources and land. Efficient crop yield prediction is possible with the help of past data, cultivation areas, and environmental conditions. Bangladesh study is included, whose economy is based on agriculture. The data are collected and preprocessed, in this ensemble learning technique K-nearest Neighbor Random Forest Ridge Regression (KRR) is applied for crops like rice, potato, and wheat. The evaluation metrics used were MSE and R2.To check the robustness of the KRR model the Diebold–Mariano test was used. A recommender system was designed that suggests suitable crop cultivation in the next season for a specific type of land area [13].

The AgriNation portal offers comprehensive information about Indian agriculture, including market pricing, news, weather reports, and farming advice. AgriGold is designed to keep Indian farmers in mind, by providing knowledge on crop management, pest control, and agricultural techniques. Krishi Jagran provides the articles, news, and agricultural updates in online mode. AgriTech India is a platform devoted to the latest developments, inventions, and farm equipment in agriculture. FarmNest is a well-known online forum where farmers may exchange stories, talk about a range of agricultural subjects, and ask other farmers for help. Indian agriculture is an all-inclusive portal that addresses a multitude of subjects, such as animal rearing, crop production, and agricultural policy made by the government. The AgriWatch portal provides information on commodities trade, market pricing, and agricultural news. Farmers' Forum India is an online forum for farmers to communicate, exchange queries, and exchange agricultural information. Indian Council of Agricultural Research (ICAR) is an official website that offers a lot of knowledge on agricultural research, education, and extension activities. Department of Agriculture and Farmers Welfare's official website provides details on the programs, policies, and campaigns related to agriculture that the Indian government has launched to assist farmers.

With the help of the m-Kisan portal, farmers can access many databases without internet connectivity and receive information and alerts in the form of text or audio messages, tailored to their preferences. According to May 2014 TRAI data, the percentage of rural areas with internet penetration is still in the single digits, although there are over 38 crore mobile phone connections in the rural area. As a result, mobile messaging has been chosen by the m-Kisan portal as the most efficient way to reach roughly 8.93 crore agricultural families. One hundred and twenty respondents from the state of Telangana were selected for the current study to examine how well the m-Kisan portal provides agricultural information. Six variables, including timeliness of communications, application, ease of understanding, specificity, comprehensiveness, and quality of information, served as the foundation for the development of the index. The findings indicated that over 50% of the participants thought the m-Kisan portal played an important role in disseminating agricultural information [14].

In July 2008, the Madhya Pradesh government launched the Krishinet portal, which is the only 24 · 7 Hindi portal that is specifically designed to meet the information and advisory needs of over 70 lakh agricultural families (75% of the state's entire population) in Madhya Pradesh. The portal can be viewed from any location with internet access. The goal of the ICT-based project Krishnet is to give farmers access to information in their native Hindi so they can raise their standard of life by raising their farming income. The farming community, agriculture-related organizations and societies, NGOs, training facilities, students, researchers, agricultural universities, Krishi Vigyan Kendras, traders, the processing sector, government agencies, extension agents, and policymakers are the principal stakeholders of the Krishinet project. Keeping this in mind, the study was conducted to determine how the Krishinet portal has affected Madhya Pradesh farmers. After performing a paired t-test, the impact of the Krishinet portal was examined in terms of five capitals: human, natural, physical, social, and financial capital. The study's findings also showed a substantial correlation between education, yearly income, operational landholding, information-seeking behavior, attitudes toward ICTs, and portal impact. The primary challenges faced by the farmers were the inaccessibility of the Krishinet portal, internet speed, and shortage of expert trainers which led to insufficient training for using the website. The technical assistance lacuna is also mentioned by the researcher [15].

2.3 ROLE OF INTERACTIVE MEDIA IN INDIAN AGRICULTURE

Farmers, researchers, and other stakeholders use interactive media for information. Figure 2.2 illustrates the salient characteristics of interactive media in agriculture.

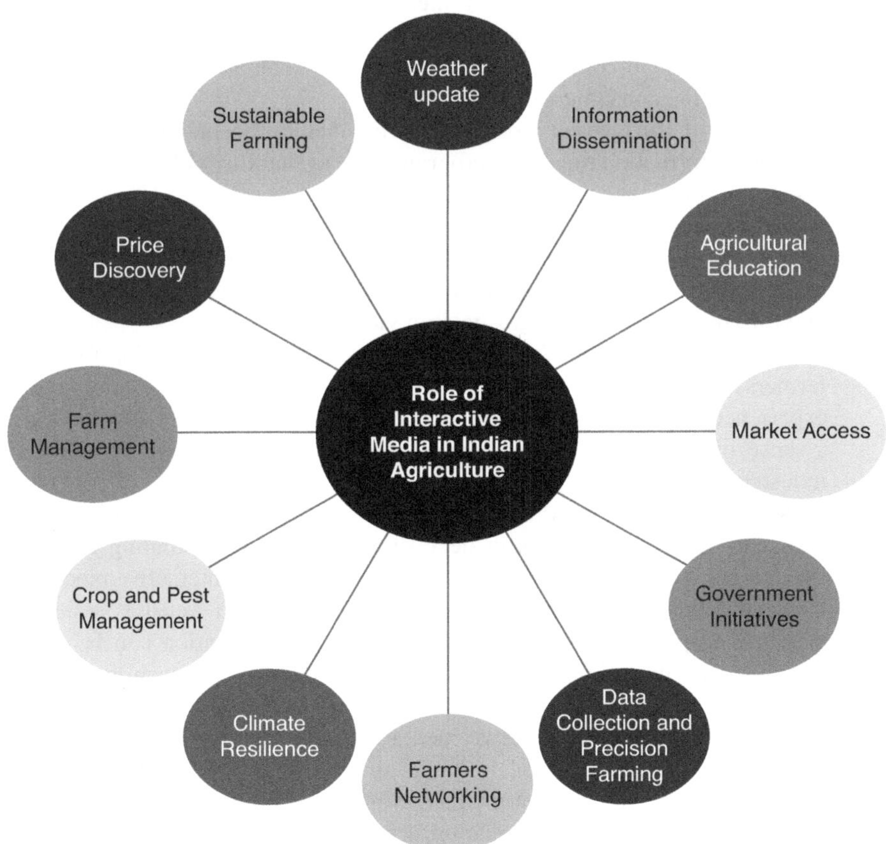

FIGURE 2.2 Key features of interactive media in Indian agriculture.

- **Weather Update:** Weather information is a critical factor in agriculture planning. The real-time updates on weather through interactive media help the farmer in decision-making related to harvest, crop selection, and irrigation. The possibility of crop loss can be reduced with precise weather condition information [16].
- **Information Dissemination:** Farmers are using various interactive media like websites, mobile apps, and social media platforms to acquire up-to-date information on agriculture. These platforms help farmers to access the latest updates related to crops, pests, government initiatives, market rates, and weather conditions.
- **Agricultural Education:** The training and extension services in agriculture are provided through interactive media to farmers. To learn modern farming and practice it, farmers can access a variety of educational resources, webinars, and tutorials. Modern techniques like pest disease management, precision agriculture, sustainable agriculture, and best agriculture practices are accessible through agriculture education.
- **Market Access:** With the help of interactive media platforms, farmers can connect with the wider customer base. The dependency on intermediaries is reduced by using e-commerce platforms, farmers can produce and sell their products to the consumers directly [17].
- **Government Initiatives:** Using interactive media, the Indian government provides information regarding programs like agriculture policies, financial assistance, insurance coverage, and subsidies.
- **Data Collection and Precision Farming:** Farm data can be collected using interactive media. The collected data can be analyzed to improve farming techniques. The collected data are utilized for precision agriculture by maximizing resources and raising crop yield.
- **Farmers Networking:** With the help of social media and online forums, farmers can communicate with other farmers, ask for advice, and share experiences. Virtual networking is a critical factor in the cultivation of a sense of community and solidarity among members of the agriculture industry.
- **Climate Resilience:** Information related to climate change, and mitigation strategies assist the farmers in decision-making with fluctuating climate conditions. Weather changes can be accessible through interactive media.
- **Crop and Pest Management:** Mobile applications and websites assist in pest and crop disease management. The timely information can be used to solve agriculture issues by farmers. This proactive approach leads to a significant reduction in crop losses.
- **Farm Management:** Interactive media helps farmers in record keeping, financial management, and planning for the next season's plantation.
- **Price Discovery:** Interactive media can be utilized by farmers to monitor the market prices of their agricultural products. These data help them to make well-informed and strategic choices regarding the time and location to sell their produce by maximizing their revenue.

- **Sustainable Farming:** Interactive media has the potential to encourage sustainable farming methods by sharing knowledge on organic farming, crop rotation, and conservation techniques. These practices can improve soil health and mitigate environmental consequences.

2.4 SWOC ANALYSIS OF INTERACTIVE MEDIA IN THE INDIAN AGRICULTURE SECTOR

Though agriculture is a primary source of income in most of India still it faces challenges like technology awareness, digital literacy, internet connectivity, etc. With the help of interactive media like mobile apps, and portals, farmers can enhance their productivity up to a certain extent. Figure 2.3 indicates the strengths, weaknesses, opportunities, and challenges of interactive media usage in agriculture.

2.4.1 STRENGTHS

Many smartphone apps, social media sites, and portals are available for exchanging agricultural knowledge for Indian farmers. The government websites also provide agricultural information like PMKissan GoI, e-NAM (National Agriculture Market), Krishi Network, IFFCO (Indian Farmers Fertiliser Cooperative) Kisan Agriculture App, Rashtriya Krishi Vikas Yojana (RKVY) portal, PM-Kisan (Pradhan Mantri Kisan Samman Nidhi), Pradhan Mantri Fasal Bima Yojana (PMFBY), Kisan Credit Card (KCC) Scheme, Krishi Vigyan Kendras (KVKs), and National Food Security Mission (NFSM).

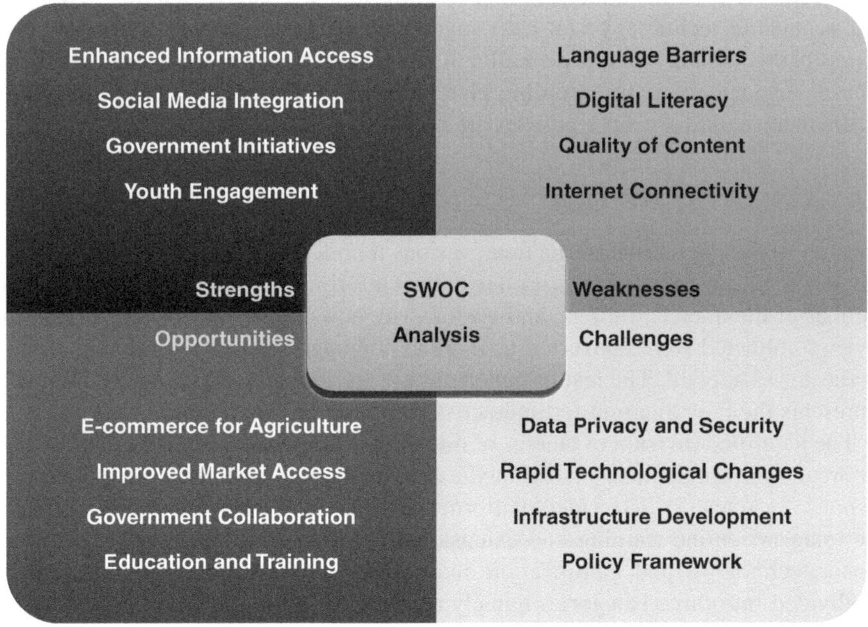

FIGURE 2.3 SWOC analysis of interactive media in the Indian agriculture sector.

2.4.2 WEAKNESSES

Many farm management apps, real-time monitoring apps, and educational content apps require user's personal information. To provide services, websites, software, and social media platforms frequently gather and retain personal data about users. Users may not realize how much privacy they have while using certain websites and apps. Insufficient security measures taken by platforms and apps may lead to a data breach that compromises user privacy.

In rural areas, a lack of digital literacy among farmers is observed. Hence, the available interactive media and services are not efficiently used by the farmers.

2.4.3 OPPORTUNITIES

Increasing use of smartphones, government programs promoting digital agriculture, or innovative ways to reach the market and educate farmers are opportunities in the agriculture sector. The online platforms for agricultural goods can increase market accessibility. Public-private collaborations can be useful for the creation of platforms and farmer-centric apps. Interactive media can be used to train the farmers in contemporary farming methods.

2.4.4 CHALLENGES

For broader adoption of interactive media usage, it is imperative to provide digital literacy and high-speed internet connectivity. Local language and content will be helpful for a better understanding of information content creation. Affordable and accessible technology can help small and marginal farmers. Affordable and accessible technology can help small and marginal farmers by ensuring privacy and security of farmers' data on online platforms. To encourage the use of interactive media in agriculture, proper policies are required.

2.5 RESEARCH METHOD

To analyze the interactive media data, various mobile apps from different categories are shortlisted. Web scraping is used to collect information on the number of reviews, number of downloads, ratings, and customer reviews about mobile apps. Customer reviews collected were converted to English language, and then sentiment analysis on those was applied. The results generated are presented in visual forms. Figure 2.4 represents the flow diagram for interactive media data preparation.

The identified division of classes of interactive media used in the agriculture sector are mobile applications, social media and messaging platforms, interactive voice response systems (IVRS), video platforms, E-Governance portals, drones and satellite imagery, online training and extension services, digital marketplaces, IoT and sensor technology, and gamification and simulation. The mobile applications are subdivided into three categories namely agricultural advice, disease diagnosis, and supply chain management. Social media and messaging platforms are classified as community building and marketplace information. IVRS is used for information dissemination, feedback, and surveys. Video platforms are subdivided into educational

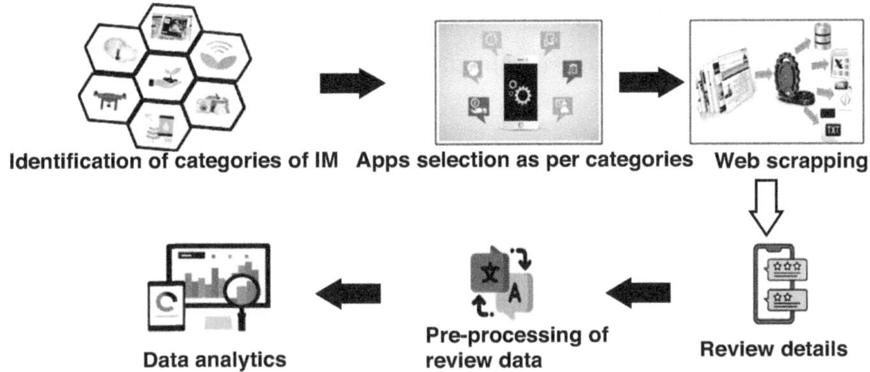

FIGURE 2.4 Flow diagram for interactive media data preparation.

content and live demonstrations. E-Governance portals are used for subsidiary and scheme access as well as license and permit applications. Drone and satellite imagery are used for crop monitoring and precision agriculture. Online training and extension services are provided using virtual training and expert consultation. Digital marketplaces cover online trading and agri-fintech services (credit, insurance, and investment services for farmers). IoT and sensor technology are used in real-time monitoring and smart irrigation. Gamification and simulation include education games and scenario planning. Based on these categories and their subcategories, mobile apps and portals are shortlisted. The criteria used for shortlisting are based on ratings and downloads.

Using google_play_scraper, the app information such as title, developer, score, reviews, and number of installations are identified. It was observed that some of the scrapped reviews were in local languages (Hindi, Marathi, Gujarati). Hence, the review data are preprocessed and translated into English for language standardization. Using the Vader sentiment analyzer, the sentiment score is identified for each review and based on the polarity score, sentiment classification is done.

2.6 DATA ANALYSIS OF INTERACTIVE MEDIA USAGE

Mobile apps from ten different categories are shortlisted. The categories are further subdivided into subcategories, as shown in Table 2.2.

Mobile apps are segregated into private and government apps, and their subcategorywise distribution is depicted in Figure 2.5.

Figure 2.6 represents the number of reviews and downloads of selected mobile applications. It has been observed that government apps have the highest number of reviews and downloads.

Mobile app users' reviews were analyzed using Textblob and Vader sentiment. The reviews are classified as positive, negative, and neutral. Figure 2.7 shows the sentiment analysis of mobile app users' reviews. Plantix app has received the highest number of positive reviews.

TABLE 2.2

Categorywise Subcategories of Apps

Category	Subcategory
Mobile applications	Agricultural advice
	Disease diagnosis
	Supply chain management
Social media and messaging platforms	Community building
	Marketplace information
Interactive Voice Response Systems (IVRS)	Information dissemination
	Feedback and surveys
Video platforms	Educational content
	Live demonstrations
E-Governance portals	Subsidy and scheme access
Drones and satellite imagery	Crop monitoring
	Precision agriculture
Online training and extension services	Virtual training
	Expert consultations
Digital marketplaces	Online trading
	Agri-fintech services
IoT and sensor technology	Real-time monitoring
	Smart irrigation
Gamification and simulation	Educational games
	Scenario planning

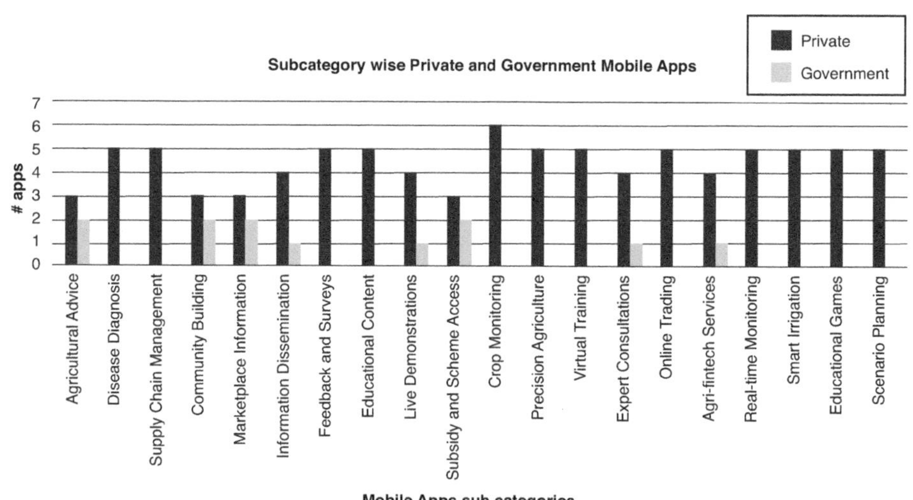

FIGURE 2.5   Subcategorywise private and government apps.

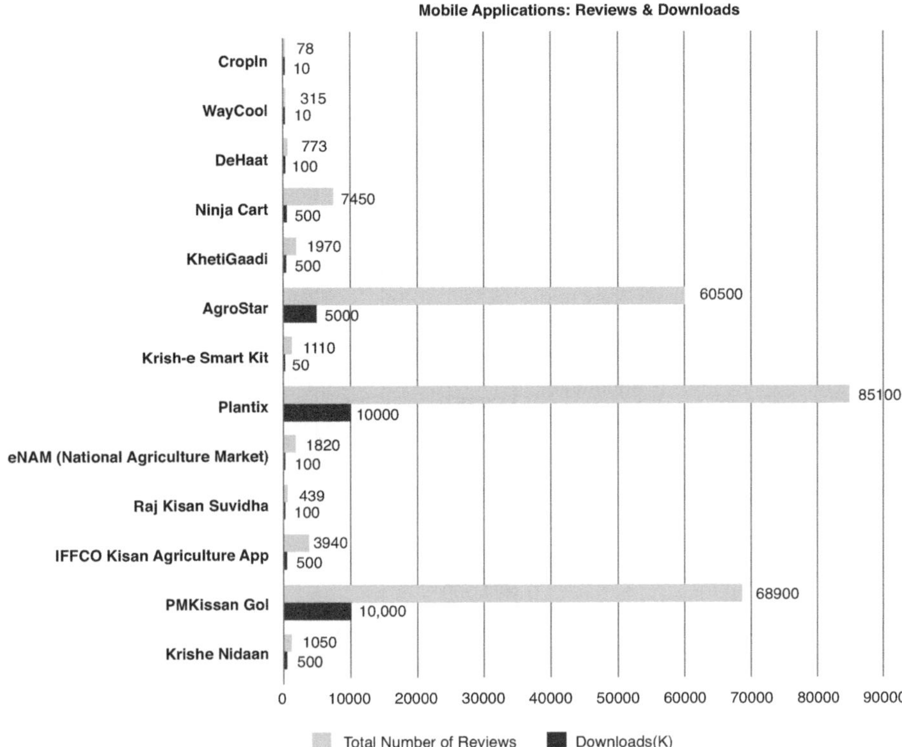

FIGURE 2.6 Mobile applications—reviews and downloads.

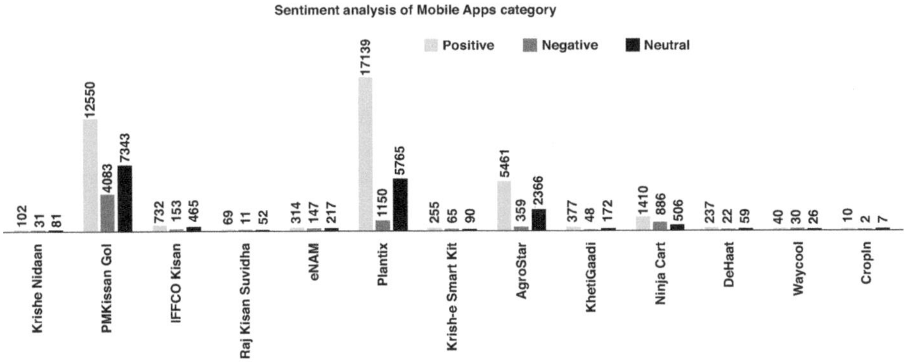

FIGURE 2.7 Mobile apps and positive, negative, and neutral sentiments.

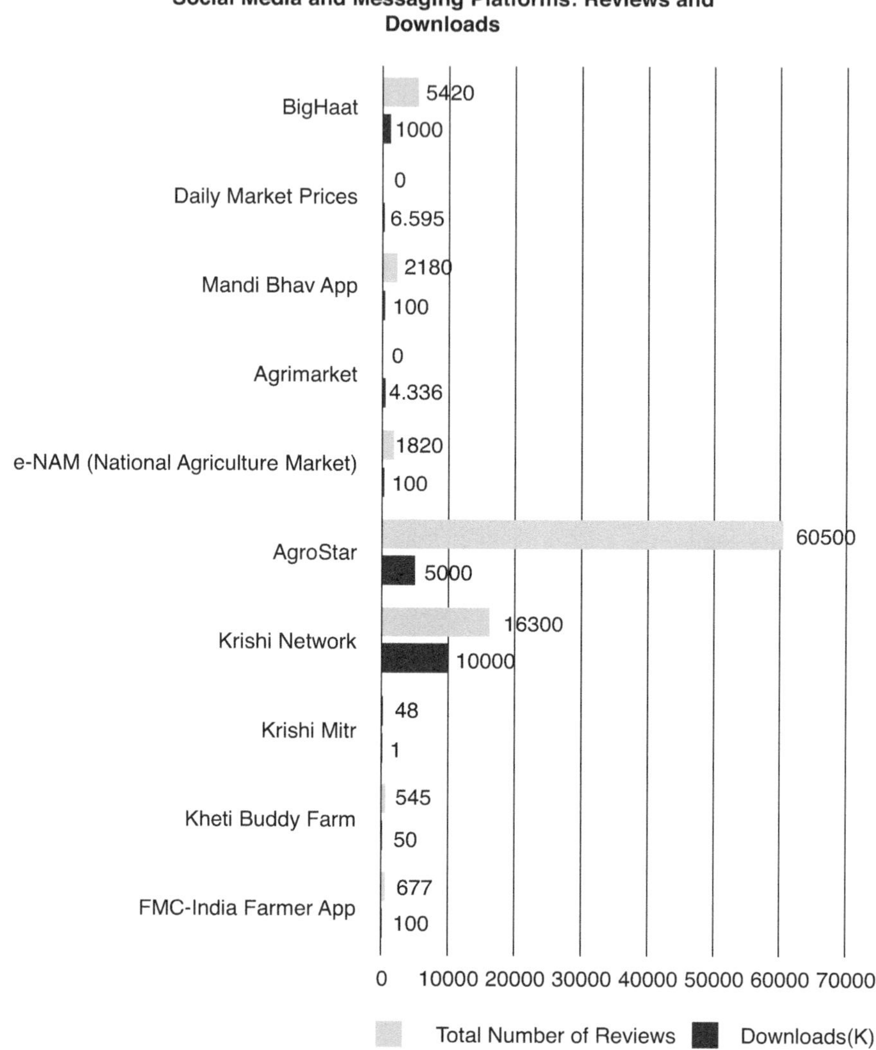

FIGURE 2.8 Social media and messaging platforms—reviews and downloads.

For the category of social media and messaging platforms, the selected mobile app reviews and downloads are depicted in Figure 2.8. Agrostar has got the highest number of reviews as compared to other social media apps.

For the category IVRS, the selected mobile app reviews and downloads are shown in Figure 2.9. Krishify has the highest number of downloads.

To refer to the latest farming information, video platforms are widely used. YouTube is widely used by farmers as depicted in Figure 2.10.

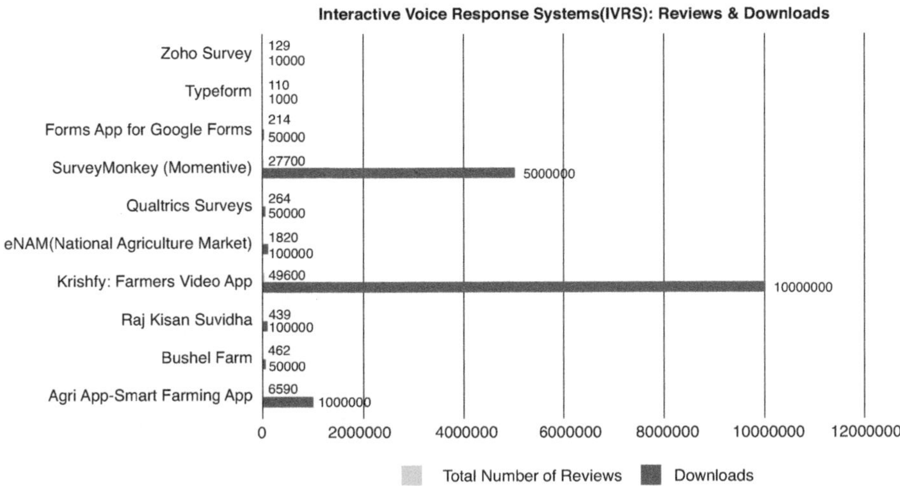

FIGURE 2.9 Interactive voice response systems (IVRS)—reviews and downloads.

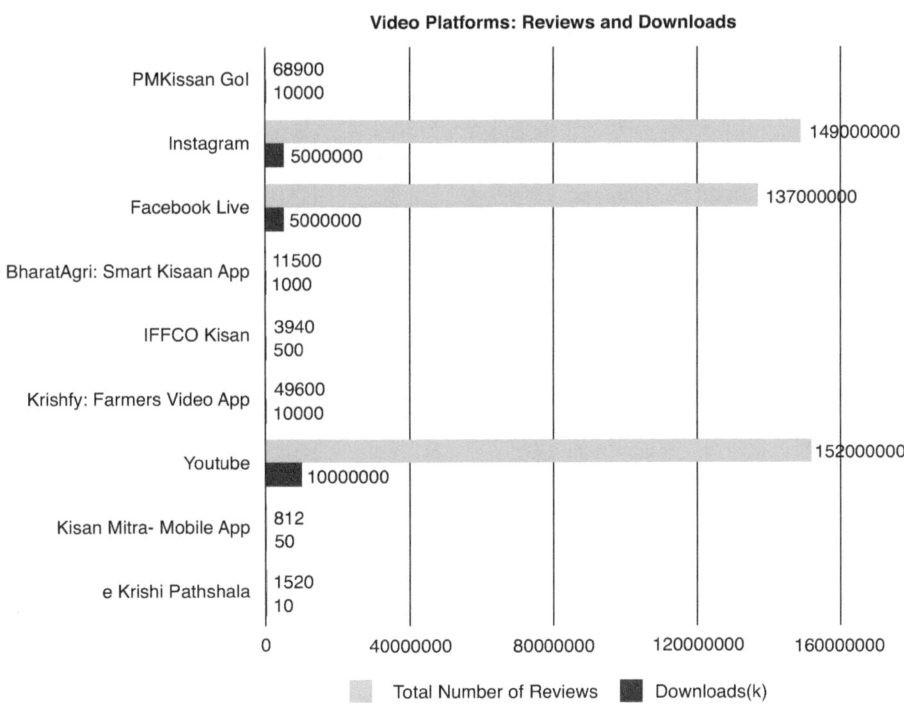

FIGURE 2.10 Video platforms—reviews and downloads.

For precision farming, drones and satellite imagery are used. The reviews and download information of mobile apps is shown in Figure 2.11 where AgriApp and BharatAgri have the highest number of downloads. PMKisan GoI is the most used e-governance portal by farmers in India. The screenshot of the mobile app is shown in Figure 2.12. PM-Kisan is a new Central Sector Scheme introduced by the Indian government, and it aims to increase the income of small and marginal farmers (SMFs).

Mobile applications are used for multiple purposes. The categorywise utilization of mobile apps is depicted in Figure 2.13 [19].

Farmers are using web portals and apps for smart farming. Various portals/apps used by farmers and the features present in the apps are listed in Figure 2.14. The mobile app's popularity is based on ratings and downloads. The shortlisted apps have an average rating of 3.95 out of 5. The reviews collected using web scrapping are processed further to check the sentiments. Using standard sentiment analysis libraries, the sentiments are identified and it was observed that popular apps have maximum positive reviews.

Mostly farmers prefer government apps as compared to private apps. To enhance productivity, farmers are using smart technologies for farming. The various agriculture practices and their technology implementations are depicted in Figure 2.15.

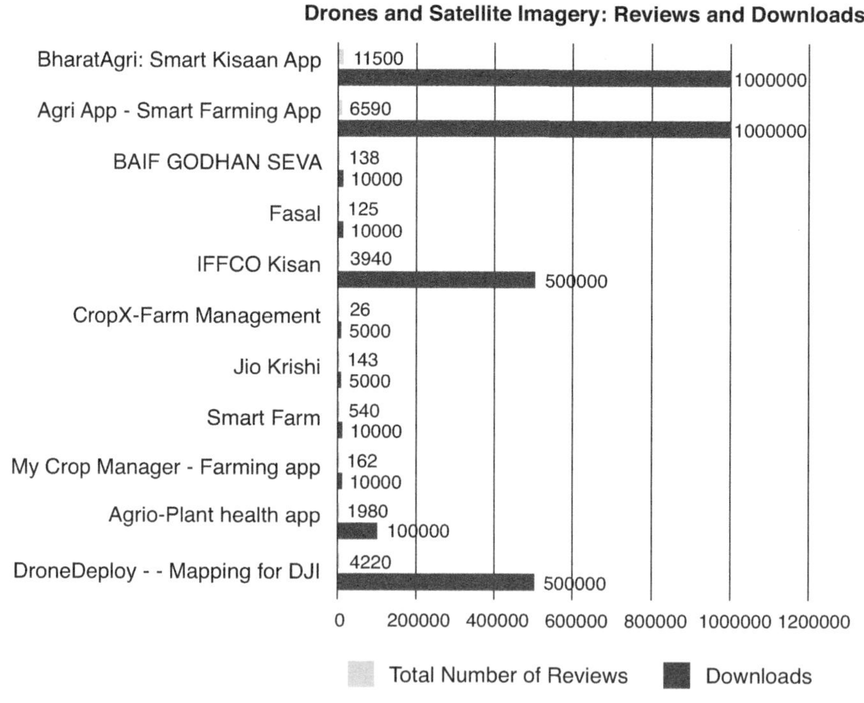

FIGURE 2.11 Drones and satellite imagery—reviews and downloads.

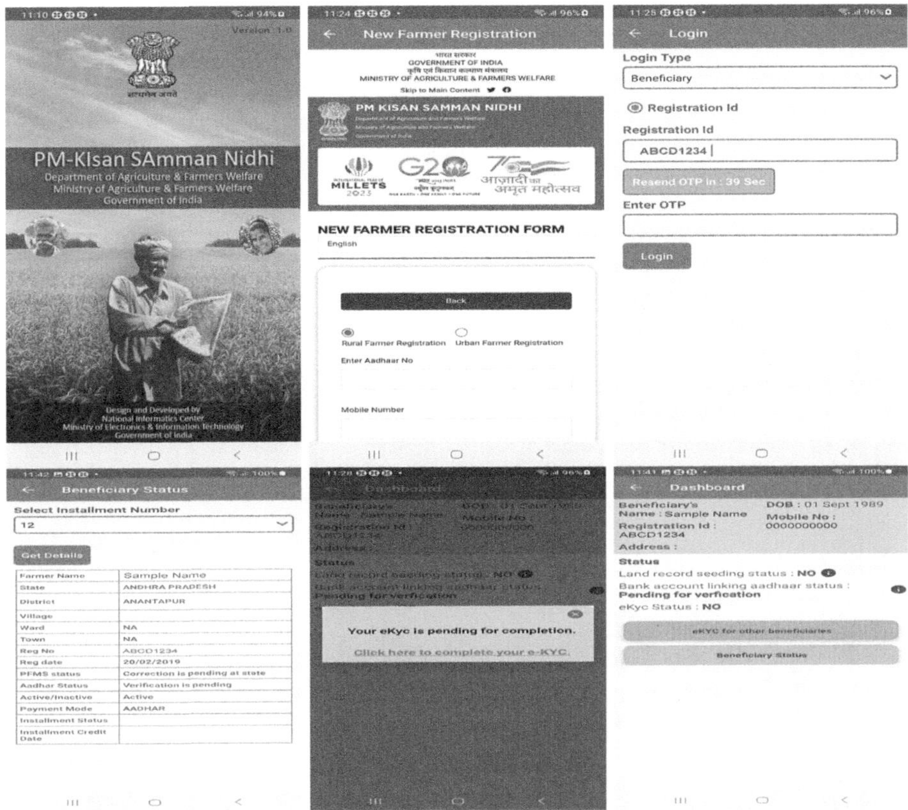

FIGURE 2.12 PMKissan GoI app interface [18].

2.7 DISCUSSION

Agriculture sector outcomes will flourish when farmers start taking advantage of technology-based solutions for farming. To enhance the yield, suggestions from experts and insights from historical data can be identified.

2.7.1 RECOMMENDATIONS FOR IMPROVING THE OUTCOME OF AGRICULTURE IN INDIA

Water Management and Irrigation: Implementing efficient irrigation systems like drip and sprinkler irrigation can help conserve water. Rainwater harvesting and rejuvenating traditional water bodies should also be emphasized to improve water availability, especially in drought-prone areas.

Use of Technology: Crop yields can be increased by utilizing contemporary technologies, such as precision agriculture, which can maximize resource use. To manage and monitor crops, technology like drones, AI, and satellite photography can be used.

FIGURE 2.13 Utility of mobile applications.

	Commodity Prices	Weather Updates	Crop Advisory and Crop Disease Advisory	Market Analysis	Government Policies	News and Events	Mobile Accessibility	Training and Education	SMS Alerts	Language Support	Forums and Community	Tools and Applications
AgriWatch				Yes								
NCDEX (National Commodity & Derivatives Exchange Limited)				Yes			Yes					
Kissan Deals:Mandi Bhav App	Yes			Yes			Yes					
Kisan Call Centre (KCC)	Yes	Yes		Yes			Yes					
Shetkari - Make Business	Yes			Yes				Yes		Yes		
Kisan 2.0		Yes	Yes					Yes		Yes		
AgriMedia:Hi-Tech Agriculture	Yes	Yes		Yes			Yes	Yes			Yes	
agMOOCs by IIT Kanpur			Yes					Yes		Yes	Yes	
Krishi Vigyan Kendras (KVKs)			Yes				Yes	Yes				
Krishi Jagran	Yes			Yes	Yes	Yes	Yes			Yes	Yes	
Pradhan Mantri Fasal Bima Yojana(PMFBY)			Yes		Yes							
Rashtriya Krishi Vikas Yojana (RKVY)	Yes	Yes	Yes	Yes	Yes	Yes	Yes	Yes	Yes	Yes	Yes	Yes
National Food Security Mission (NFSM)					Yes	Yes		Yes				
DBT Agriculture Portal (DBT-Agri)			Yes		Yes	Yes	Yes					
PM-KISAN(Pradhan Mantri Kisan Samman Nidhi)							Yes				Yes	Yes
Soil Health Card Scheme			Yes		Yes	Yes	Yes	Yes			Yes	Yes
Farm Mechanization -MKisan	Yes		Yes	Yes	Yes		Yes	Yes	Yes	Yes	Yes	Yes
iFMS			Yes		Yes	Yes	Yes				Yes	Yes
AgriLMS					Yes		Yes	Yes				
PlantSat-Satellite Precision		Yes	Yes			Yes	Yes	Yes	Yes			Yes
Map My Crop		Yes	Yes			Yes	Yes					
Samunnati		Yes					Yes				Yes	
Soufflet	Yes		Yes			Yes	Yes				Yes	
Rain Machine		Yes				Yes	Yes				Yes	Yes
Open Sprinkler		Yes					Yes					Yes
Smart Irrigation App						Yes	Yes					Yes

FIGURE 2.14 Web applications and feature mapping.

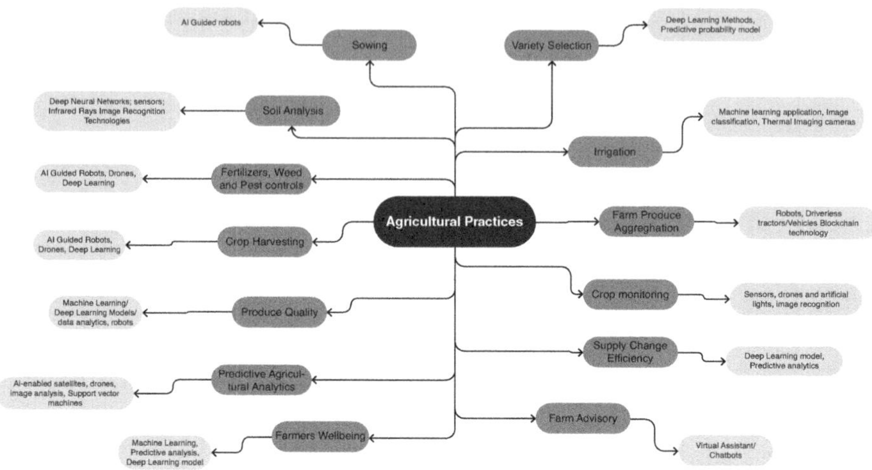

FIGURE 2.15   AI mechanism in agricultural practices.

Soil Health Management: Soil fertility can be preserved with the use of organic farming practices and routine soil health checks. Soil degradation can be avoided by enlightening farmers about the proper use of fertilizers.

Climate-Resilient Crops: By promoting the growth of climate-resilient crop varieties, the climate change effect can be reduced. This includes crops that are pest-resistant and drought-resistant.

Crop Diversification: Crop diversification can help to minimize risk by reducing dependence on a single crop. It can also support maintaining soil health and increasing farm revenue.

Infrastructure and Market Access: To facilitate greater market access, improvement in rural areas is mandatory. This includes lower post-harvest losses, improved roads, storage facilities, and cold chains.

Financial Assistance and Insurance: Sufficient financial assistance to farmers like lower-interest loans and crop insurance can make investments in improved inputs and technologies.

Education and Training: Knowledge dissemination can be possible by holding frequent training sessions on post-harvest technologies, pest management, and new farming methods.

Government Initiatives and Grants: Farmers can benefit greatly by using farmer-friendly policies and subsidies in areas like marketing, equipment purchases, and seed procurement.

Integration with Agri-tech Startups: Traditional farming issues can be addressed by working together with agri-tech startups. This includes apps that provide weather forecasts, market data, and expert connections.

Sustainable Practices: Promoting sustainable farming methods like integrated pest management, agroforestry, and organic farming can help in long-term agriculture results.

Women in Agriculture: Women are a big workforce in agriculture. Agriculture productivity can be achieved by providing education, resources, and land ownership rights to women.

2.7.2 CROP RECOMMENDATION BASED ON SOIL HEALTH AND WEATHER PARAMETERS

Precision agriculture is a technique that suggests the crop to be grown based on information such as climate and soil characteristics. New technologies can assist with this task, including IoT and AI. IoT sensors record weather and soil properties. Using these data, a neural network model can be constructed to suggest which crops would be best suited to grow on a specific farm. The model can help farmers to decide which crop to grow on the farm. The dataset used is secondary data from Kaggle (Crop Recommendation Dataset). The dataset has seven features N, P, K, temperature, humidity, pH, and rainfall. A total of 22 different classes of crops are covered. For the input $N=60$, $P=50$, $K=40$, temperature $=50$, humidity $=50$, pH $=2$, Rainfall $=100$, the recommendation system suggests crops such as mango and pigeon peas [20]. Figure 2.16 shows crop recommendation for input $N=54$, $P=77$, $K=85$, temperature $=17$, humidity $=17$, pH $=8$, rainfall $=83$ as chickpeas. For the set of values 65, 89, 85, 25, 35, 10, and 102, the recommendation system suggests crops such as chickpeas and mothbeans. Different machine learning (ML)

```
test1=sc.transform(np.array([[54,77,85,17,17,8,83]]))
pre1=np.argmax(model.predict(test1))
print(labels[pre1])
img_path=str(labels[pre1])+'.jpg'
x=plt.imread(file_path+img_path)
plt.imshow(x)
plt.show()

1/1 [==============================] - 1s 1s/step
chickpea
```

FIGURE 2.16 Crop recommendation.

methods can be used for prediction of crop yield. Multivariate adaptive regression spline (MARS) is a better approach for lentil crop yield prediction [21].

2.7.3 FERTILIZER AND PESTICIDES RECOMMENDATION FOR CROPS

The dataset used for fertilizer and pesticide recommendation for crops features temperature, humidity, moisture, soil type, crop type, nitrogen, potassium, phosphorus, and fertilizer name [20]. For the input [26, 52, 59, 2, 8, 11, 0, 9], the fertilizer recommendation is "20–2." Label encoding preprocessing technique is used on features like soil type and crop type. The fertilizer and pesticide recommendation samples are shown in Table 2.3.

2.7.4 COMMON OBSERVATIONS REGARDING MOBILE APPLICATIONS USED BY FARMERS

After analyzing various mobile apps and after going through farmers' reviews, some pitfalls in the system are observed:

- **Regional Language:** Most current mobile applications are only available in Hindi and English. For a nation like India, it is crucial to provide agricultural sector information in all regional languages. For instance, riceXpert is a feature-rich program that offers professional guidance, fertilizer and weed control, and disease identification for rice crops. On the other hand, riceXpert has received fewer downloads, most likely as a result of the unavailability of mobile apps in some regional Indian languages. India is a multilingual nation with 23 recognized languages. Most farmers lack formal education or they are only conversant in their mother tongue. So, it could be acceptable to add regional languages to the list of languages to increase the usefulness of mobile applications.
- **Lacks in Regular Updates:** Some mobile applications provide common suggestions and practices that farmers should adhere to in day-to-day farm operations, such applications typically don't require any updates frequently. To track developments daily, some applications, like market prices in riceXpert and daily weather information in Havaamaana Krishi, need frequent updates.

TABLE 2.3

Outcomes of Fertilizer and Pesticide Recommendation System

Input Parameters [Temperature, Humidity, Moisture, Soil Type, Crop Type, Nitrogen, Potassium, Phosphorus]	Recommendation
[30, 60, 63, 3, 1, 9, 9, 29]	'14-35-14'
[31, 65, 34, 3, 2, 15, 0, 37]	'DAP'
[33, 64, 50, 2, 10, 41, 0, 0]	'Urea'
[31, 62, 48, 4, 3, 14, 15, 12]	'17-17-17'
[32, 60, 48, 3, 8, 18, 17, 40]	'14-35-14'

- **Inclusion of Minor Crops:** Most of the mobile applications focus on providing suggestions on crops like rice, wheat, and maize which are commonly cultivated, but mobile applications should also guide the cultivation of millets and pulses.
- **Slow Operation:** Many times, the time taken by mobile applications to load is fairly long, and there are also periodic server connection issues. Regular maintenance is necessary to maintain the application's complete functionality.
- **User-Friendly GUI:** Most mobile applications have textual content rather than a pictorial representation of the information. The farmers find it difficult to understand the content as it lacks an illustrative explanation. Again, if the information is distributed in sections, then it allows farmers to use information efficiently and effectively.

To disseminate technology from researchers to the farming community, agriculture extension is essential. The timely distribution of information using ICT tools for an efficient decision-making process depends on the adoption of the most recent technological information. With more people owning smartphones than ever before, smartphone applications have the potential to completely transform extension work by providing farmers with the most recent technical information. Expert advice, guidance on farm management and development, market information, expert demonstrations, and expert training are typical elements of the smartphone applications made available by extension authorities.

The following are typical limitations on smartphone apps that extension officials can use:

- **Popularity**: Among farmers, applications with high features and ratings are not very well-liked. Due to its lack of popularity, the Farm Extension Manager application has only had about 50,000 downloads.
- **Accessibility**: Network connectivity in rural locations should be taken into account by application developers. For improved accessibility, the majority of agricultural applications require high-speed network connectivity. This results in restricted use and upgrades for agriculture apps in rural areas.
- **Language Barrier:** Due to most applications being created in regional tongues, their widespread use and appeal are limited. The number of downloads for many apps created in Hindi and English is lower than that of apps created in regional languages.

The smartphone app for a government official needs to include the following features. Project implementation at different levels, including the neighborhood, district, and state levels, should be made easier by the application. New rules and regulations must be able to be regularly updated on the department's website by the officials. The application should include a comprehensive knowledge package, pertinent statistical data, financial information, and current backup plans.

Research community expects a smartphone application to have the following features. The researchers may be able to complete their survey and other data-capturing tasks more quickly with the help of a straightforward graphical user interface.

Certain applications are required to collect a variety of data in addition to the querying and dynamic update functionalities in the event of complex field trials. Continuous crop monitoring activities can be provided by smartphone applications that address specific research applications, such as weed or disease identification using AI or ML algorithms and high-resolution sensor-embedded systems [9].

2.8 CONCLUSION AND FUTURE SCOPE

The COVID-19 pandemic hit the entire world, and the global economy has been affected. During the lockdown period, the interactive media helped the farmers to sell the farm produce successfully. The mobile apps created a link between the consumer and the producer, which helped farmers to gain substantial benefits from farm produce.

The data collected from the Play Store gives insights into the usage of different mobile apps by Indian farmers. A total of unique 88 mobile apps are studied to understand their usage. The government app PM KISAN has many subscribers as well as reviews. Farmers are positive about the schemes introduced by the government regarding crop insurance, loan facilities, and fertilizer supply. Mobile apps are highly used by farmers to get daily mandi price updates, fertilizer and pesticide recommendations, soil nutrients, irrigation, and crop monitoring. Weather-based decision-making can be possible because of mobile apps. Farmers can create profiles and can get benefits from government schemes as per the profile. Selling and purchasing seeds/crops is easily possible with the help of interactive media. To create awareness among farmers social media platforms like YouTube, Facebook, Instagram, etc. are used. IVRS like Krishfy: Farmers Video App are helping farmers with smart farming.

Indian agriculture might become much more productive, efficient, and sustainable with the use of interactive media. In the future, blockchain technology can provide transparency in the supply chain from farms to consumers. With the help of agricultural huge data, predictive analytics like crop yields, pest infestations, climate impacts, fertilizer recommendations, and crop cycle suggestions could be possible more precisely.

Mobile and web-based IoT applications can solve the problem of production by organizing crop planning and process management platforms. The real-time monitoring of crops helps farmers to reduce crop damage. The big data about crops are stored, which contain information on weather, terrain, rainfall, geographic conditions, and water. The processing of these data helps agricultural companies and farmers in the prediction of product success and the impact of natural conditions.

REFERENCES

1. Rashtriya Krishi Vikas Yojana (Raftaar), "Statewise Allocation and Release of Funds under Normal RKVY and Sub-Schemes for 2019–20" (2019). https://rkvy.nic.in/static/Statements/DAC_2019-20.pdf
2. Ministry of Statistics and Programme Implementation (MOSPI), National Statistical Office, Government of India, "State-Wise and Item-Wise Value of Output from Agriculture, Forestry, and Fishing, Year: 2011–12 to 2019–20" (2022). https://www.mospi.gov.in/sites/default/files/publication_reports/Agr_forestry-fishingBrochure%202022.pdf
3. Tiwari, S., "Demand for Grants 2019–20 Analysis Agriculture and Farmers' Welfare", July 9 (2019). https://prsindia.org/files/budget/budget_parliament/2019/DFG%20Analysis%20-%20Agriculture%20(2019-20).pdf

4. Vipra, T., "Demand for Grants 2023–24 Analysis: Agriculture and Farmers Welfare", PRS Legislative Research (2023). https://prsindia.org/budgets/parliament/demand-for-grants-2023-24-analysis-agriculture-and-farmers-welfare

5. Economic and Political Weekly-Engage, "Revisiting India's Farming and Agricultural Policies: 13 Questions, 99 Articles", September 17, 2023 (2021), https://www.epw.in/engage/debate-kits/farm-and-agriculture-laws-policies-india

6. Deepak, V., P. K. Joshi, D. Roy, and A. Kumar, "PM-KISAN and the adoption of modern agricultural technologies", *Economic and Political Weekly-Engage* 55(23) (2021): 18. https://www.epw.in/journal/2020/23/special-articles/pm-kisan-and-adoption-modern-agricultural.html?destination=node/157042, Accessed on September 18, 2023.

7. Deepika, K., "A study of agriculture-related mobile apps and its impact on farming system", *International Journal of Creative Research Thoughts* 10(2) (2022): 2882.

8. Raman, R., S. Dhiraj, S. Sudip, K. Ujjwal, and K. Rakesh, "Agricultural mobile apps for transformation of Indian farming", *Agriculture World* 7(4) (2021): 1–12.

9. Sivakumar, S., G. Bijoshkumar, A. Rajasekharan, V. Panicker, S. Paramasivam, V. S. Manivasagam, and S. Manalil, "Evaluating the expediency of smartphone applications for indian farmers and other stakeholders", *AgriEngineering* 4 (2022): 656–673. https://doi.org/10.3390/agriengineering4030042

10. Prakasa Rao, E. V. S, V. Rakesh, and K. V. Ramesh, "Big data analytics and artificial intelligence methods for decision making in agriculture", *Indian Journal of Agronomy* 66(5) (2021): 279–287.

11. Basnet, B., and J. Bang, "The state-of-the-art of knowledge-intensive agriculture: A review on applied sensing systems and data analytics", *Journal of Sensors* 2018 (2018): 1–13.

12. Utomo, A. H., M. A. Gumilang, and A. Ahmad, "Agricultural commodity sales recommendation system for farmers based on geographic information systems and price forecasting using probabilistic neural network algorithm", In *IOP Conference Series: Earth and Environmental Science,* vol. 980, no. 1, p. 012061. IOP Publishing, Bristol, UK (2022).

13. Moon, M. H., M. A. Marjan, M. P. Uddin, M. I. Afjal, S. Kadry, S. Ma, and Y. Nam, "Ensemble machine learning-based recommendation system for effective prediction of suitable agricultural crop cultivation", *Frontiers in Plant Science* 14 (2023): 1234555.

14. V. Preethi, K. Aruna, and M. J. M. Reddy, "A study on effectiveness of mKisan portal in providing agricultural information in Telangana State", *Biological Forum – An International Journal* 14(3) (2022): 1568–1570.

15. Singh, V., and V. L. V. Kameswari, "A study on impact of ICT enabled web portal (Krishinet) on farmers", *International Journal of Agriculture, Environment, and Biotechnology, IJAEB* 12(2) (2019): 163–174. https://doi.org/10.30954/0974-1712.06.2019.13

16. Dhulipala, R. K., P. Gogumalla, R. Karuturi, R. Palanisamy, A. Smith, S. Nagaraji, S. A. Rao et al., "Meghdoot: A mobile app to access location-specific weather-based agro-advisories pan India." CGIAR Research Program on Climate Change, Agriculture and Food Security Working Paper (2021).

17. Borkar, S., "How Digital Mandis Are Transforming the Lives of Farmers in India" (2021). https://yourstory.com/socialstory/2021/09/digital-mandis-transforming-lives-farmers-india

18. "PMKISAN GoI - Apps on Google Play." https://play.google.com/store/apps/details?id=com.nic.project.pmkisan, Accessed December 30, 2023 (2019).

19. Balkrishna, A., R. Pathak, S. Kumar, V. Arya, and S. K. Singh, "A comprehensive analysis of the advances in Indian digital agricultural architecture", *Smart Agricultural Technology* 5 (2023): 100318.

20. Gladiator. "GitHub - Gladiator07/Harvestify: A Machine Learning Based Website That Recommends the Best Crop to Grow, Fertilizers to Use, and the Diseases Caught by Your Crops." GitHub. May 16 (2023). https://github.com/Gladiator07/Harvestify

21. Das, P., G. K. Jha, A. Lama, and R. Parsad, "Crop yield prediction using hybrid machine learning approach: A case study of lentil (*Lens culinaris* Medik)", *Agriculture* 13(3) (2023): 596. Accessed December 30 2023. https://doi.org/10.3390/agriculture13030596

3 Addictive Interfaces
How Persuasion, Psychology Principles, and Emotions Shape Engagement

Saniya Atalatti and Urmila Pawar

3.1 INTRODUCTION

This chapter analyzes the design principles of social media platforms, focusing on the intersection between technological features and human psychology. Platforms leverage features like infinite scrolling, personalized content, and gamification, demonstrating a deep understanding of user behavior.

The analysis explores the multifaceted effects of social media. It examines the "Fear of Missing Out" (FoMO) driving compulsive engagement, the influence of social validation on self-esteem, and the rapid content cycles fueled by instant gratification. Parallels are drawn between dopamine release through positive feedback and established addictive mechanisms.

Recognizing the complexities of these digital environments, the chapter highlights the dynamics of social comparison, where users constantly compare their lives to curated online personas. Additionally, it focuses on the potential correlation between social media use and perceived social isolation, suggesting a nuanced issue of loneliness.

This work emphasizes the need to understand the symbiotic relationship between technology and psychology within social media. It advocates for a deliberate and mindful approach to online interactions, urging both users and developers to acknowledge these mechanisms and make informed choices.

3.1.1 DEFINITION OF ADDICTIVE INTERFACES

Addictive interfaces refer to user interfaces, digital applications, or platforms intentionally designed to capture and maintain users' attention for extended periods of time, often at the expense of the users' well-being. These interfaces leverage principles rooted in social psychology and behavioral techniques, which exploit underlying human emotions, leading users to engage with the platform compulsively and repeatedly.

DOI: 10.1201/9781032664828-4

3.1.2 PREVALENCE OF ADDICTIVE INTERFACES IN ENTERTAINMENT

Social media platforms have emerged as prime examples of addictive interfaces due to their extensive user bases and the strategic incorporation of design elements aimed at captivating and retaining users. According to Statista, as of September 2023, statistics illustrate the immense reach and engagement levels of these platforms:

- **Facebook:** With a staggering 3.03 billion monthly active users worldwide, users spend an average of nearly 19.5 hours per month on the platform. This highlights both the widespread prevalence of Facebook and the significant amount of time users invest in it.
- **Netflix:** The streaming giant, Netflix, boasts 238.39 million monthly users. While specific time spent figures are not provided here, it is well documented that Netflix users often binge-watch entire series, indicating high user engagement.
- **Instagram:** With 1.358 billion monthly active users, Instagram has established itself as a key player in the field of addictive interfaces, according to the statistics of 2022, a person spent an average of 12 hours per month on the platform.

These statistics underscore not only the widespread prevalence of these platforms but also the substantial amount of time users allocate to them. Users are drawn in by addictive design elements and features that keep them engaged.

3.1.3 IMPACT OF DIGITIZATION ON USER ENGAGEMENT

The ongoing process of digitization has significantly transformed how individuals interact with technology and engage with digital interfaces. This transformation has given rise to the emergence of addictive interfaces. As technology continues to evolve, more aspects of our daily lives become intertwined with the digital field. Users are naturally exposed to and influenced by these interfaces, which are constantly present, shaping their everyday behaviors and routines.

Digitization has ushered in a new era of loaded accessibility, where digital technologies and platforms enable individuals to interact, connect, and engage with others. This impact extends to the user engagement landscape, with digital platforms being leveraged to connect, communicate, and build relationships with target audiences.

The pervasiveness of addictive interfaces, particularly in entertainment and social media, is a testament to the profound impact of digitization on user engagement. The extensive reach and the time users invest in these platforms underscore the effectiveness of addictive design elements in capturing and retaining user attention.

The addictive nature of social media interfaces is not confined to a single platform. It extends to various apps and devices, as users carry these platforms with them throughout their daily lives. They can access their feeds on smartphones, tablets, desktops, and even smartwatches, making it exceptionally convenient and accessible.

3.2 THEORETICAL FRAMEWORK

The field of social psychology has long been devoted to unraveling the intricacies of human likability and persuasiveness. While a singular, comprehensive theory on interpersonal likability or persuasiveness remains elusive, various mid-range theories and dynamics within social psychology have proven to be robust predictors. In terms of likability, five key predictive factors have emerged from the literature, as highlighted by Cialdini in 1993 and further elaborated upon by Forsyth in 1995. These factors serve as essential building blocks, providing valuable insights into the factors that contribute to people being perceived as likable and persuasive in social interactions.

3.2.1 PREDICTIVE FACTORS FOR LIKABILITY

With respect to user engagement, five main predictive factors have emerged from the literature [1,2]. These factors serve as the foundational framework guiding the art of captivating user experiences in digital interfaces.

3.2.1.1 Compliments

Compliments play a pivotal role in social interactions, influencing behavior and fostering compliance [3]. Compliance researchers have delved into the notion that compliments hold the power to enhance compliance by exerting a significant impact on interpersonal assessments. The principle of liking serves as a bridge between compliments and compliance, highlighting the psychological link between positive affirmations and the likelihood of individuals adhering to requests.

In social media applications, the influence of compliments on user engagement and interaction is profound. Platforms often incorporate features that allow users to express admiration and appreciation for content through likes, comments, and shares. Compliments, whether conveyed through text, emojis, or other interactive elements, contribute to the establishment of a positive online environment. The principle of liking, as identified in compliance research, aligns with the dynamics of social media interactions, where positive feedback and affirmations can enhance the sense of connection and community. For instance, a user receiving compliments on their posts may experience a boost in confidence and a greater inclination to participate actively within the platform, showcasing the real-world applicability of compliments in the digital environment.

3.2.1.2 Similarity

The concept of similarity holds significant weight in interpersonal dynamics, particularly in the way individuals are naturally drawn to those who share similar attitudes and values [4]. The similarity effect is rooted in the idea that when people find common ground with others, it cultivates a sense of connection and understanding. This inclination is deeply tied to the establishment of respect and trust, as the alignment of attitudes and values creates a foundation for perceiving the other person as trustworthy and deserving of respect.

The bedrock of trust and respect formed through similarity is crucial in fostering positive interpersonal relationships. When individuals identify shared beliefs and values, there is a natural tendency to place trust in one another, creating a conducive environment for mutual understanding and collaboration. In social media applications, the principle of similarity plays a pivotal role in shaping user experiences. Algorithms often leverage user preferences, interactions, and shared content to suggest connections and content that align with an individual's interests and values. By facilitating connections based on similarity, these platforms enhance user engagement and satisfaction, illustrating the practical application of the similarity effect in the digital landscape.

3.2.1.3 Attraction

Attraction, particularly in the context of physical appearance, is intricately linked to the Status Characteristics Theory, which posits that individuals who are perceived as physically attractive tend to be positively regarded by others [5]. According to this theory, the positive attributions assigned to attractive individuals not only influence how they are perceived but also result in more favorable treatment from others. This positive treatment, in turn, elicits increased positive social behaviors from attractive individuals, creating a continuous and reinforcing cycle of socially positive interactions.

In social media applications, the impact of physical attraction on user interactions is evident.

Profiles featuring visually appealing content, whether in the form of attractive images or well-curated aesthetics, often attract more positive attention from other users. This initial positive perception can lead to enhanced engagement, such as likes, comments, and shares, creating a digital manifestation of the continuous cycle described by the Status Characteristics Theory. In this way, the dynamics of attraction, as outlined by the theory, find resonance in online environments, shaping social behaviors and interactions within digital communities.

3.2.1.4 Familiarity

Familiarity, within interpersonal attraction, stands as a cornerstone principle influencing how individuals connect with each other [6]. The significance of the familiarity principle is a fundamental driver of attraction. Frequent encounters contribute significantly to the development of attraction.

This effect posits that the more exposure one has to a particular stimulus, the more likely they are to develop a preference or liking for that stimulus. In the context of interpersonal relationships, repeated interactions catalyze familiarity, leading to an increased sense of comfort and liking. This phenomenon is not confined to offline interactions; in the digital age, social media platforms facilitate frequent exposure to individuals through shared content, posts, and interactions. The familiarity principle is amplified by personalized algorithms, engagement metrics, and interactive features that enhance exposure to content and individuals. Algorithmic feeds deliver tailored content, notifications prompt repeated interactions, and real-time features facilitate instant engagement. The sharing and virality of content across networks contribute to widespread exposure, while consistent branding and community building further strengthen connections. Social media's digital landscape adapts and extends the mere exposure effect, playing a pivotal role in shaping online relationships through frequent and varied interactions.

3.2.1.5 Association

The association plays a pivotal role in shaping interpersonal preferences, as individuals tend to develop positive feelings toward those who are psychologically linked with desirable things or events [7]. This phenomenon underscores the impact of cognitive associations on the formation of interpersonal attraction. When individuals are connected in people's minds with positive experiences, traits, or events, it contributes significantly to the likability of those individuals. This principle is deeply rooted in the psychology of perception and cognitive biases. People naturally gravitate toward those associated with positive attributes, creating a positive halo effect around individuals linked with favorable circumstances. In social media applications, this is evident in the way users curate their online personas. Individuals often share content that reflects positive experiences, achievements, or affiliations, strategically aligning themselves with desirable aspects. As a result, these associations contribute to the attractiveness of individuals within the digital space.

The exploration of psychological principles in the context of user engagement within digital interfaces reveals a nuanced understanding of human behavior in both physical and digital fields.

Lastly, the principle of association underscores the power of positive cognitive connections, shaping likability in both online and offline contexts. As we navigate the digital age, understanding and leveraging these psychology principles provide invaluable insights for designing user-centric experiences that resonate with the intricacies of human behavior and interaction.

3.2.2 Contributing Factors for Persuasion

In the context of social media, persuasion plays a central role, emphasizing the importance of understanding fundamental principles. Cialdini's work in 1987 and 1993 explores six key factors guiding compliance in the digital age, from cultivating digital connections through liking to creating urgency with scarcity. These principles intricately shape the dynamics of influence in social media.

3.2.2.1 Liking

Liking, rooted in the inclination to comply with those we admire, is a vital psychological principle in social media. This fundamental aspect gains exceptional importance in the context of virtual interactions and digital connections.

Liking extends beyond personal attraction. Algorithms on social media platforms curate content based on user preferences, creating echo chambers of like-minded individuals. Users consistently encounter and engage with content endorsed by those they admire, fostering a persuasive environment. For instance, followers are more likely to engage with posts from influencers they like, leading to increased trust and influence.

3.2.2.2 Consensus

Consensus, emphasizing the human inclination to be influenced by group opinions, is a significant factor in social media dynamics. This principle highlights the impact of collective viewpoints on individual decision-making. The influence of consensus is particularly pronounced in team dynamics within social media communities. Users are often swayed by the prevailing opinions and behaviors of online groups. Social media platforms amplify consensus through features like comments, shares, and likes, creating a powerful persuasive force. An example is the virality of trending topics or hashtags, where users align with the group's prevailing sentiments, influencing the wider online community.

3.2.2.3 Reciprocation

Reciprocation, centered on the human tendency to feel obligated to repay favors or gifts is a fundamental principle in social media persuasion. This principle forms the basis for understanding the dynamics of digital interactions where users engage with content, expecting a reciprocal response.

The principle of reciprocation is evident in various forms, from liking and sharing content to participating in online challenges. Social media platforms thrive on the reciprocity principle, as users often feel compelled to reciprocate the engagement they receive. This principle underlies the foundations of online collaboration, content sharing, and community building. An example is the reciprocal nature of engagement on platforms like Twitter, where users retweet content to reciprocate the favor.

3.2.2.4 Consistency

Consistency, reflecting the human desire to appear consistent to others, is a significant factor shaping digital personas in social media. This principle unfolds in the construction of online identities.

The pursuit of consistency is evident in various aspects, such as users aligning their content with established themes, maintaining a consistent posting schedule, or upholding certain beliefs. Social media platforms provide users with tools to curate and present a consistent image, fostering predictability and reliability. The perception of consistency plays a pivotal role in building trust and influence. An example is content creators on YouTube who maintain a consistent style, topic, or upload schedule, attracting and retaining a dedicated audience.

3.2.2.5 Authority

Authority, underscoring the human inclination to believe and be influenced by authorities, is a persuasive factor in social media. In this field, this translates to the impact of perceived expertise and influence.

Social media platforms amplify the influence of authority through verified badges, professional affiliations, and expertise-driven content. Users tend to trust and engage with content produced by recognized experts or influencers in specific niches. The credibility of an authoritative figure enhances the persuasive impact of their messages, shaping opinions, and influencing user behavior. An example is a health influencer with relevant qualifications endorsing a wellness product, leading followers to trust and possibly adopt the recommended lifestyle.

3.2.2.6 Scarcity

Scarcity, highlighting the human tendency to find scarce things more desirable, is a key element in social media persuasion. This principle is manifested in the creation of a sense of urgency and exclusivity, triggering the FoMO phenomenon.

The concept of scarcity is strategically employed in digital marketing campaigns, limited-time offers, and exclusive content releases. Social media platforms enable the rapid dissemination of information, making scarcity-driven narratives highly effective. Users are more inclined to engage with and share content that conveys limited availability, contributing to the virality and persuasive impact of digital campaigns. For example, flash sales or exclusive online events create a sense of urgency, driving user engagement and participation.

Navigating the vast expanse of social media reveals the intricate integration of persuasion into every digital interaction. Liking, consensus, reciprocation, consistency, authority, and scarcity transcend theoretical abstraction; they emerge as dynamic forces shaping the dissemination of ideas, endorsement of products, and formation of communities. Recognizing and adeptly utilizing these principles equips digital content creators, marketers, and influencers with the essential tools to not only captivate but also mobilize audiences in the ceaselessly evolving and competitive domain of social media. In an era where connections materialize in the digital sphere, a profound understanding of these principles isn't merely enlightening; it is indispensable for those aspiring to wield influence and effect change in the ever-changing field of social media.

3.3 DESIGN TECHNIQUES

In the field of digital interactions, the design of applications and platforms has evolved into a sophisticated interplay of technology and psychology. This symbiotic relationship not only seeks to captivate user attention but also endeavors to shape user behaviors and foster sustained engagement. With insights from studies by Refs. [8,9], we dissect the deliberate integration of psychological principles in platform interfaces. From endless scrolling and personalized content delivery to gamification, these design strategies aim to capture user attention and shape behaviors. The following examination provides a comprehensive understanding, rendering further citations unnecessary.

3.3.1 ENDLESS SCROLLING/STREAMING

The concept of endless scrolling or streaming in app design is rooted in creating an immersive experience that fosters a positive mental state known as "flow." Flow occurs when the difficulty of a task matches the individual's abilities, resulting in a seamless and captivating user experience. In social media, endless scrolling is prominently featured, aiming to keep users engaged by presenting a continuous stream of content. Platforms strategically design interfaces to induce a flow state, encouraging users to lose track of time and space while navigating through posts and updates.

Social media platforms leverage endless scrolling to keep users immersed in their feeds, by minimizing interruptions and seamlessly presenting content, platforms like Instagram and Facebook aim to extend usage time and maintain user attention. The continuous flow of content is carefully curated to align with individual preferences, further enhancing the immersive experience (Figure 3.1).

FIGURE 3.1   Infinite scroll.

The strategic implementation involves optimizing the app interface for fluid and uninterrupted scrolling. However, the immersive nature of endless scrolling can pose dangers, such as distraction and potential risks in real-world scenarios. Developers strategically aim for users to lose track of time within social media apps, contributing to extended usage. While this engagement is valuable, it also raises concerns about well-being. Monetization occurs indirectly as prolonged user engagement increases the likelihood of exposure to advertisements and sponsored content.

3.3.2 ENDOWMENT, MERE EXPOSURE, AND GAME PROGRESSION

Endowment and mere exposure effects are observed in the slow progression of game elements without monetary investment. The endowment effect refers to users feeling attached to their virtual possessions, while the mere exposure effect enhances user affinity through repeated interactions. Game progression signifies the advancement and achievements within a game.

In social media, the endowment effect is evident as users invest time and effort in curating their profiles, creating a sense of attachment to their virtual identity. The mere exposure effect is harnessed as repeated interactions with posts and content increase user liking and engagement.

Social media platforms leverage endowment and mere exposure effects by encouraging users to invest time in creating and curating their digital presence. Features like "memories" and "highlights," on platforms such as Facebook and Instagram, respectively, capitalize on the mere exposure effect, resurfacing past content to reinforce positive interactions and memories (Figure 3.2).

The strategic implementation involves designing features that promote user investment in their virtual presence. Game progression, evident in achievements and milestones, serves as a psychological reward. Monetization occurs through targeted advertising, as platforms use user-generated content and engagement data to deliver personalized ads.

3.3.3 SOCIAL PRESSURE AND COMMUNICATION DYNAMICS

Social pressure, exemplified by features like WhatsApp's double tick function, creates an expectation for quick responses in communication dynamics. The double tick indicates message delivery and reading, inducing a sense of urgency.

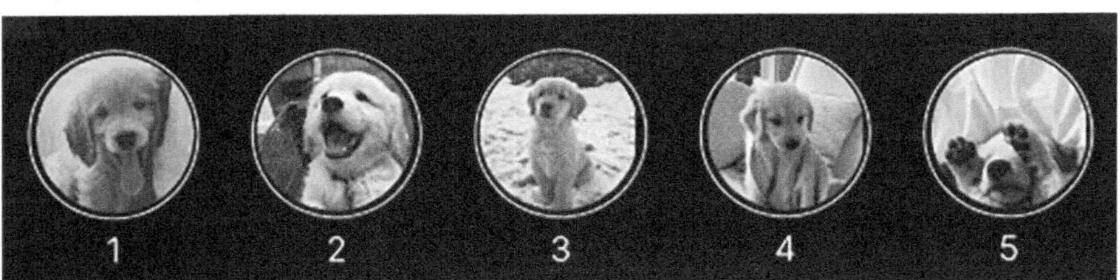

FIGURE 3.2   Highlights on Instagram profile.

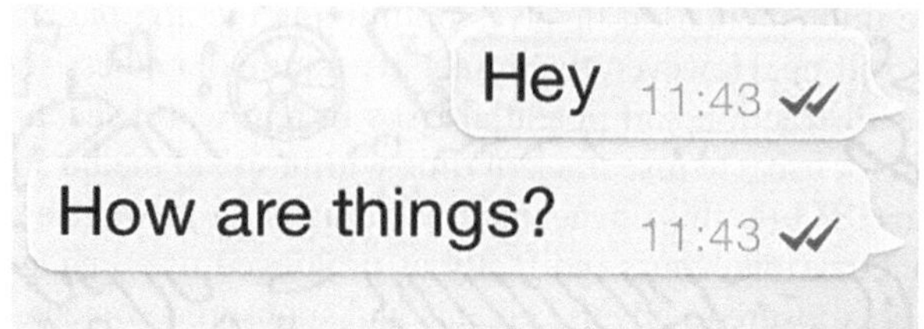

FIGURE 3.3 WhatsApp double tick.

In social media communication, platforms employ features that create social pressure. For instance, read receipts and online status indicators heighten expectations for quick responses, influencing communication dynamics.

Social media platforms strategically leverage the double tick function to create social pressure for quick responses, the immediate visibility of when a message is delivered and read compels users to respond promptly, fostering real-time interaction and engagement (Figure 3.3).

Strategic implementation involves integrating features that enhance real-time communication dynamics. Monetization is indirect, as features, like read receipts, contribute to user engagement and the platform's overall appeal, attracting advertisers looking to reach an actively engaged audience.

3.3.4 Personalization and User-Specific Tailoring

Personalization involves tailoring content to individual user preferences. Social media platforms, like Facebook, employ machine learning to analyze user behavior and present content customized to individual interests.

In social media, personalization is evident in algorithms that curate content based on user interactions, such as likes, shares, and comments. The user's feed is dynamically adjusted to showcase content that aligns with their interests.

Social media platforms strategically leverage personalization to enhance user engagement [10]. The more personalized the content, the higher the likelihood of prolonged user interaction. Default settings, such as the "double tick function" on WhatsApp, play a role in nudging users toward specific behaviors within the personalized experience.

Strategic implementation involves refining algorithms to better understand and cater to individual preferences. Monetization occurs through targeted advertising, as personalized content provides advertisers with a more effective means of reaching specific audience segments.

3.3.5 Social Reward and Comparison Dynamics

Social reward involves the neuroscientific impact of positive social interactions, such as receiving "likes" on social media. Comparison dynamics refer to users gauging their standing within their social networks based on these social rewards. In social media, the "like" button serves as a prominent social reward. Users actively engage

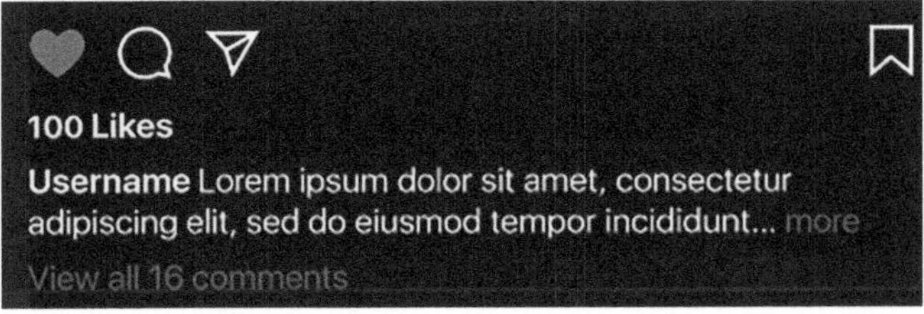

FIGURE 3.4 Likes feature on Instagram.

with content, seeking positive reinforcement in the form of likes, shares, and comments. Social comparison dynamics are at play as individuals measure their popularity and influence against their peers (Figure 3.4).

Social media platforms strategically leverage the power of "likes" to reinforce positive user behavior. The visibility of these social rewards creates a dynamic where users actively seek validation through increased engagement and content creation.

Strategic implementation involves designing features that amplify the visibility and impact of social rewards. Monetization occurs through sponsored content and partnerships with influencers, as platforms capitalize on the influence and engagement of users who accumulate significant social rewards.

3.3.6 Variable Reward and Intermittent Conditioning

Variable reward mechanisms involve providing unpredictable rewards to users and maintaining engagement through intermittent conditioning principles. Examples include endless scrolling and auto-play features.

Social media platforms incorporate variable reward mechanisms in features like endless scrolling and auto-play. These features keep users engaged by offering unpredictable rewards, such as interesting content, at intermittent intervals.

Social media platforms strategically leverage variable reward mechanisms to prolong usage time. The unpredictability of rewards, coupled with the continuous flow of content, creates an environment where users find it challenging to disengage.

Strategic implementation involves optimizing features like endless scrolling and auto-play to maximize the impact of variable reward mechanisms. Monetization occurs through increased ad exposure, as users spend more time on the platform, encountering a higher volume of advertisements.

3.3.7 Personalization and Behavioral Nudging

Behavioral nudging involves subtly guiding user behavior. In the context of social media, personalization strategies serve as a form of nudging by tailoring content to influence user engagement and interaction.

Personalization and behavioral nudging are intertwined in social media. Algorithms analyze user behavior to tailor content, nudging users toward specific actions, such as increased interaction and time spent on the platform.

Social media platforms strategically leverage personalization to nudge user behavior positively. By creating a sense of social comparison and tailoring content to individual preferences, platforms influence users to engage more actively.

Strategic implementation involves refining personalization algorithms to effectively nudge user behavior. Monetization occurs through increased user engagement, attracting advertisers seeking to reach a more actively involved and responsive audience.

3.3.8 Zeĭgarnik Effect

The Zeĭgarnik effect suggests that individuals remember and return to interrupted tasks more effectively than completed ones, driven by the cognitive tension created by unresolved tasks.

In social media, the Zeĭgarnik effect is strategically harnessed by leaving certain actions or content incomplete. This creates a psychological trigger, compelling users to return and seek closure by completing the interrupted tasks, such as viewing Stories or scrolling through posts.

Social media platforms intentionally leverage the Zeĭgarnik effect to maintain user engagement, features like incomplete story markers or partially viewed content capitalize on the cognitive tension, prompting users to return regularly (Figure 3.5).

FIGURE 3.5 Incomplete story markers.

Strategic implementation involves designing features that deliberately leverage the Zeĭgarnik effect. Monetization occurs indirectly through increased user engagement, as the effect contributes to prolonged usage and the potential for exposure to advertisements and sponsored content.

3.3.9 Gamification: Elements and Achievement Dynamics

Gamification involves integrating game-like elements into non-game contexts to enhance user engagement. Elements include points, badges, leaderboards, and challenges, contributing to achievement dynamics within the platform.

In social media, gamification is evident in features like points for interactions, badges for accomplishments, leaderboards for rankings, and challenges for users to participate in, creating a sense of competition and achievement.

Social media platforms strategically leverage gamification to enhance user engagement. Elements like badges for active participation or leaderboards for content popularity motivate users to actively engage and strive for recognition within the community.

Strategic implementation involves seamlessly integrating gamified elements to maximize their impact on user behavior. Monetization occurs through sponsored challenges or partnerships, as platforms capitalize on the competitive spirit and engagement of users participating in gamified activities.

The field of social media and digital platforms demands a nuanced understanding of the psychological underpinnings that define our online interactions. The seamless integration of immersive features, personalized content, variable rewards, social dynamics, and gamification creates a combination that not only seizes attention but also molds habits and engagement patterns. Recognizing these mechanisms empowers users to make judicious choices in their digital interactions. As the digital panorama continues to evolve, comprehending the subtle interplay between technology and psychology emerges as a pivotal consideration for both users and developers. It advocates for a deliberate and thoughtful approach to digital engagement, where awareness and choice take precedence in the user experience.

3.4 SOCIAL MEDIA'S DUAL IMPACT: EFFECTS AND THE CYCLE OF RETURN

In the field of contemporary digital connectivity, the omnipresence of social media unveils a nuanced interplay of advantages and drawbacks. It serves as a conduit for immediate interpersonal connection and gratification, instigating neurochemical responses that engage users. Conversely, these platforms cast shadows, cultivating sentiments of profound isolation, heightening the apprehension associated with the FoMO and ensnaring individuals within enduring cycles of seeking social validation and engaging in comparative scrutiny. Paradoxically, the very platforms accentuating these challenges become magnets for users, perpetuating a complex oscillation between the virtual interactions.

3.4.1 FEAR OF MISSING OUT

FoMO is a contemporary psychological phenomenon fueled by the prevalence of social media platforms (SMP) [11]. It encapsulates the pervasive and often anxious desire individuals have to stay connected with the experiences and activities of others. Users experiencing FoMO fear being excluded from rewarding social interactions and events, and this fear is intensified by the constant stream of updates on social media platforms. FoMO drives individuals to compulsively check social media feeds, engage in online discussions, and prioritize digital connection over other activities, reflecting a broader societal concern of missing out on shared experiences.

Social media platforms contribute significantly to the amplification of FoMO. The constant flow of curated content showcasing the accomplishments and social interactions of peers creates a virtual environment where users feel compelled to stay digitally connected. The FoMO on the latest updates and experiences drives users to intensively use social media as a means of alleviating the anxiety associated with potential exclusion. This continuous engagement, however, is suggested to lead to Social Media Fatigue (SMF), emphasizing the intricate relationship between FoMO and the potential negative impact of intensive social media use on users' psychological well-being.

3.4.2 SOCIAL VALIDATION

Social validation is a pivotal force shaping online experiences, with users actively seeking validation through likes, comments, and shares to bolster self-esteem, establish social status, and foster a sense of belonging [12]. The emotional intricacies are tied to social validation, emphasizing the role of positive feedback in generating happiness and satisfaction, while negative feedback can evoke emotions such as envy and disappointment.

Users curate their online personas to align with social norms, navigating a delicate balance between authenticity and conformity. The platform's influence on self-presentation and identity construction is evident, as individuals seek approval from others while striving to maintain their true selves. Users' desire for validation sometimes does have a potential negative impact on well-being.

3.4.3 INSTANT GRATIFICATION

Instant gratification refers to the immediate satisfaction or fulfillment of a desire, often without delay or the need for prolonged effort. It involves seeking and obtaining rewards, pleasures, or results promptly, providing a swift and often fleeting sense of pleasure or contentment. In the context of social media, instant gratification may manifest through the quick responses, likes, comments, or other forms of immediate validation that users receive when sharing content or seeking social engagement [13].

Users engage in updating Stories on social media platforms with a multifaceted aim that encompasses elements of instant gratification [14]. The motivation of socially rewarding self-promotion is evident as individuals seek to promote themselves, and the swift feedback in the form of likes and comments provides an immediate sense of validation and reward. Similarly, the act of social sharing, where users quickly disseminate personal experiences and achievements, serves as a means to swiftly connect with their social circle, fostering immediate reactions and contributing to a sense of instant gratification. The use of Stories for entertainment or escape further reflects a desire for immediate enjoyment or distraction, offering users a quick reprieve from their current reality. Moreover, updating Stories to showcase trendy fashion not only aligns with a personal sense of style but also underscores the immediate recognition and approval sought, emphasizing the role of instant gratification in the social media storytelling experience.

3.4.4 DOPAMINE RELEASE

Dopamine release refers to the discharge of the neurotransmitter dopamine between nerve cells in the brain. It plays a vital role in regulating mood, reward-motivated behavior, and pleasure. In the context of the text, positive feedback on social media triggers dopamine release, creating a pleasurable experience. Excessive and frequent release, however, may lead to potential health issues, including addiction-like behaviors and mood effects over time [15].

Following a positive social interaction, dopamine is released upon receiving favorable feedback on social networks. In summary, these social media platforms exploit the same neural pathways described as being "utilized by slot machines and cocaine to maintain our engagement," as outlined by Trevor Haynes, a research technician at Harvard Medical School, in his article "Dopamine, Smartphones & You: A battle for your time" [16]. Over time, the excessive release of dopamine (in some cases, in unusually high amounts) leads to a deficiency in the brain: users find less enjoyment in non-social media activities because their dopamine levels drop below the baseline [17]. Dr. Anna Lembke, a professor of psychiatry, elaborates, "We enter a state of dopamine deficit. This is the brain's way of restoring balance: if there's a significant surge upward, there will be a corresponding drop." Thereby creating an urge to stay online, watch another video, or connect with one more person . In the short term, the dopamine deficit presents itself similarly to depression and anxiety, replicating the same symptoms and emotions.

3.4.5 SOCIAL COMPARISON

Social comparison is a psychological process wherein individuals evaluate their own abilities, opinions, or social standing by comparing themselves to others. This can involve both upward comparison, where individuals assess themselves against those perceived as superior, and downward comparison, where individuals gauge themselves against those considered less fortunate. Social media platforms contribute significantly to social comparison dynamics. The constant exposure to curated content on these platforms, such as Facebook, can lead individuals to engage in upward social comparison, as users often present their highlight reels, showcasing achievements, appearances, and experiences. This curated presentation can foster feelings of inadequacy or inferiority in individuals who perceive themselves as falling short in comparison to their online peers [18].

Frequent use of social media, particularly platforms like Facebook, has been associated with lower self-esteem, primarily due to the prevalence of upward social comparison [19]. The nature of the content on these platforms, emphasizing positive aspects of users' lives, intensifies the comparison process. Exposure to profiles with certain characteristics, such as high social network activity or upward-healthy comparisons, can further contribute to diminished self-esteem. Thus, social media platforms not only facilitate social comparison but also amplify its impact, potentially influencing individuals' overall well-being.

3.4.6 LONELINESS

Loneliness is a subjective emotional state characterized by a perceived gap between one's desired and actual social connections. It goes beyond mere physical isolation and is more about the quality and depth of social interactions. Individuals experiencing loneliness often feel a sense of disconnection, isolation, and a lack of understanding or meaningful relationships. It can arise from various factors, such as social isolation, a lack of close connections, or a feeling of being undervalued [18]. Loneliness is associated with negative impacts on mental health and overall well-being, emphasizing the importance of meaningful social connections for emotional fulfillment and psychological health.

There is an intricate association between social media use and loneliness, revealing a robust link between higher levels of social media engagement and heightened feelings of loneliness [20]. This correlation persisted even after adjusting for various factors like age, living arrangements, employment, and health concerns, echoing findings from previous COVID-19 pandemic surveys, thereby implying a broader connection to mental distress beyond loneliness. As the above-stated source highlights nuances in this relationship are based on individuals' motives for using social media. Participants leveraging social media to evade difficult emotions showed no significant link to loneliness, while those prioritizing maintaining contact displayed a weak but statistically significant association between increased daily usage and elevated loneliness levels. The above-stated source also explains the potential mechanisms behind this association, citing the addictive nature of social media and its potential impact on self-regulation, as well as the emotional consequences of consuming content portraying others' positive experiences or challenges without offering direct support. While overall findings support a consistent negative correlation between social media use and mental health, particularly loneliness, caution is advised in generalizing these results, acknowledging the limitations inherent in cross-sectional studies and the potential for reverse causation.

3.5 ANALYSIS OF SURVEY

While encompassing the age range of 18–40, the survey primarily attracted participants aged 18–25, indicating a significant usage of social media among young adults. This survey, regardless of gender or profession, aims to comprehend individuals' engagement with social media and its impacts. Stringent adherence to ethical guidelines ensured the absence of personal or intrusive inquiries, safeguarding

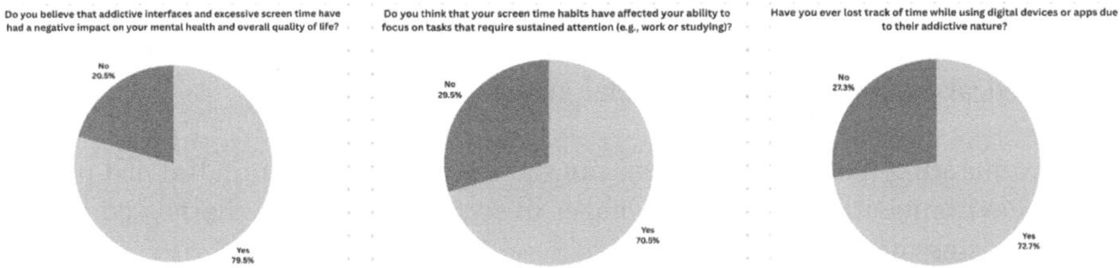

FIGURE 3.6   Results of analysis of the survey.

respondents' privacy for research purposes exclusively. Distribution of the survey was conducted within the student community via a Google Form.

Indicates the results for, Do you believe that addictive interfaces and excessive screen time have had a negative impact on your mental health and overall quality of life?

A staggering 79.5% answered yes, while 20.5% responded no, underscoring the alarming prevalence of adverse effects stemming from social media usage.

Figure 3.6 indicates the results for, Do you believe that your screen time habits have impacted your capacity to concentrate on tasks requiring sustained attention, such as work or studying? A significant 70.5% answered yes, while 29.5% responded no, emphasizing the disruptive influence of social media.

Despite acknowledging earlier that social media can adversely affect mental health and concentration, according to Figure 3.6, a striking 72.7% of respondents admitted to losing track of time while using digital devices or apps due to their addictive nature. This finding underscores the addictive tendencies inherent in social media, corroborating the discussion above regarding the design techniques that contribute to its addictive nature.

3.6   STRATEGIC SOLUTIONS FOR USERS AND DEVELOPERS

As social media continues to evolve and becomes more persuasive day by day, it is crucial to not only understand but also address the challenges highlighted in this study. Here are some specific recommendations for users and developers to navigate the complexities of social media more mindfully:

1. **Promote Digital Well-being:** Users can set intentional limits on their social media usage, schedule digital detox periods, and prioritize in-person interactions to maintain a healthy balance.
2. **Enhance Algorithmic Transparency:** Developers should strive for transparency in algorithmic processes, ensuring users have a clearer understanding of how content is curated and presented on their feeds.
3. **Implement User Empowerment Features:** Social media platforms can introduce features that empower users to have more control over their digital experience, such as customizable algorithms, content filters, and privacy settings.

4. **Educational Initiatives:** Both users and developers can benefit from educational initiatives that promote digital literacy and awareness of the psychological dynamics at play on social media.

By incorporating these strategies, we can move toward a more mindful and positive digital environment, where the advantages of social media can be harnessed without compromising mental well-being.

3.7 CONCLUSION

In our in-depth examination of the intersection between technology and human psychology within the sphere of social media, this research sheds light on the intricate design techniques that define the captivating yet, at times, contentious nature of digital interactions. From the intentional construction of endless scrolling to induce a state of flow, to the strategic utilization of endowment effects, social pressure dynamics, and gamification, social media platforms exhibit a sophisticated understanding and integration of psychological principles into their interfaces.

As of 2024, users are deeply engaged in tailored content, actively seeking social validation, and succumbing to the allure of instant gratification. However, it is crucial to recognize the rapidly evolving nature of technology and social media. Regular updates to this research or consideration of potential future developments, such as emerging platforms or shifts in user behavior, would enhance the chapter's long-term relevance. As we anticipate the evolution of social media, a proactive approach to updating this information will ensure its continued applicability to the ever-changing dynamics of social media.

The FoMO propels compulsive digital connections, social validation becomes a potent influence on self-esteem, and instant gratification fosters swift cycles of content creation and consumption. Amid this, the dopamine release triggered by positive feedback draws parallels to the addictive mechanisms observed in other domains.

The magnification of social comparison dynamics accentuates these effects, as users frequently find themselves evaluating their own lives against curated online personas. Loneliness emerges as a nuanced consequence, revealing the intricate relationship between social media use and the perceived gap in social connections.

It also becomes crucial to acknowledge in the digital world the impact of social media can vary significantly across cultures. Cultural nuances play a vital role in shaping user behaviors and responses to digital interactions. Therefore, both users and developers are encouraged to adopt a culturally sensitive approach, recognizing and respecting diverse perspectives. This nuanced understanding fosters an environment where judicious choices prevail in navigating the evolving digital sphere, ensuring the relevance and applicability of insights to a global audience.

This paper serves as an invitation to deliberate and comprehend the profound implications of the symbiotic relationship between technology and psychology in the world of social media and digital platforms, advocating for a conscientious and thoughtful approach to digital interactions.

REFERENCES

1. Cialdini, R. B. 1984. *Influence: The Psychology of Persuasion.* New York, NY: William Morrow and Company.
2. Fogg, B. J. 1998. *Charismatic Computers: Creating More Likable and Persuasive Interactive Technologies by Leveraging Principles from Social Psychology.* Stanford, CA: Stanford University.
3. Grant, N., Fabrigar, L., and Lim, H. 2010. "Exploring the efficacy of compliments as a tactic for securing compliance.*" Basic and Applied Social Psychology* 32(3): 226–233.
4. Singh, R., Goh, A., Sankaran, K., and Bhullar, N. 2016. "Similarity and liking effects on interpersonal attraction: Test of the two-dimensional trust-respect model." *Psychologia* 59: 1–18.
5. Gordon, R. A., Crosnoe, R., and Wang, X. 2013. "Physical attractiveness and the accumulation of social and human capital in adolescence and young adulthood: Assets and distractions." *Monographs of the Society for Research in Child Development* 78(6): 1–137.
6. Reis, H. T., Maniaci, M. R., Caprariello, P. A., Eastwick, P. W., and Finkel, E. J. 2011. "Familiarity does indeed promote attraction in live interaction." *Journal of Personality and Social Psychology* 101(3): 557–570.
7. Cialdini, R. B., and de Nicholas, M. E. 1989. "Self-presentation by association." *Journal of Personality and Social Psychology* 57(4): 626–631.
8. Montag, C., Lachmann, B., Herrlich, M., and Zweig, K. 2019. "Addictive features of social media/messenger platforms and freemium games against the background of psychological and economic theories." *International Journal of Environmental Research and Public Health* 16(14): 2612.
9. Garcia, J., Rodrigues, P., Simoes, J., and Fonseca, M. 2022. "Gamification strategies for social media." In Oscar Bernardes, Vanessa Amorim, António Carrizo Moreira (eds), *Handbook of Research on Gamification Strategies for Business and IT, Chapter 7.* Hershey, PA: IGI Global, doi:10.4018/978-1-6684-5538-8.ch007
10. Eg, R., Demirkol Tønnesen, Ö., and Tennfjord, M. K. 2023. "A scoping review of personalized user experiences on social media: The interplay between algorithms and human factors." *Computers in Human Behavior Reports* 9: 100253.
11. Jabeen, F., Tandon, A., Sithipolvanichgul, J., Srivastava, S., and Dhir, A. 2023. "Social media-induced fear of missing out (FoMO) and social media fatigue: The role of narcissism, comparison, and disclosure." *Journal of Business Research* 159: 113693.
12. Ballara, N. 2023. "The power of social validation: A literature review on how likes, comments, and shares shape user behavior on social media." *International Journal of Research Publication and Reviews* 4: 3355–3367.
13. Eyal, N., and Hoover, R. 2014. *Hooked: How to Build Habit-Forming Products.* New York: Portfolio/Penguin.
14. Menon, D. 2022. "Updating 'stories' on social media and its relationships to contextual age and narcissism: A tale of three platforms - whatsapp, instagram, and facebook." *Heliyon* 8(5): e09412.
15. Fernandez, V. 2022. "Social media, dopamine, and stress: Converging pathways." *Dartmouth Undergraduate Journal of Science.* Retrieved from https://sites.dartmouth.edu/dujs/2022/08/20/social-media-dopamine-and-stress-converging-pathways/#:~:text=Over%20time%2C%20the%20abundant%20release,baseline%20(McNamara%2C%202021).
16. Haynes, T. 2018. "Dopamine, smartphones & You: A Battle for Your Time." *Science in the News.* Retrieved from https://sitn.hms.harvard.edu/flash/2018/dopamine-smartphones-battle-time/.

17. McNamara, B. 2021. "The Science behind Social Media's Hold on Our Mental Health." Teen Vogue. Retrieved from https://www.teenvogue.com/story/the-science-behind-social-medias-hold-on-our-mental-health.
18. Orlowski, J. 2020. The Social Dilemma. Directed by Jeff Orlowski. Exposure Labs.
19. Vogel, E. A., Rose, J. P., Roberts, L. R., and Eckles, K. 2014. "Social comparison, social media, and self-esteem." *Psychology of Popular Media Culture* 3(4): 206–222.
20. Bonsaksen, T., Ruffolo, M., Price, D., Leung, J., Thygesen, H., Lamph, G., and Geirdal, A. Ø. 2023. "Associations between social media use and loneliness in a cross-national population: Do motives for social media use matter?" *Health Psychology and Behavioral Medicine* 11(1): 2158089.

4 Overview of Dark Patterns

Akshay Manchekar, Ganesh Bhutkar,
and Yohannes Kurniawan

4.1 INTRODUCTION

Digital trust is a fundamental concept in the realm of cybersecurity, representing the confidence that individuals and organizations have in the integrity, reliability, and security of the digital environment. When users trust the various digital platforms, they are more likely to engage in interaction, share information, make transactions, and continue using digital platforms over time. The enforcement of digital trust becomes even more crucial as users navigate the internet, where the use of dark patterns by developers poses a threat to the very foundation of digital trust.

Dark patterns, a pervasive issue in user experience design, refer to deceptive tactical interface, intentionally crafted to manipulate a user behavior. These Graphical User Interfaces (GUIs) can divert users from their intended actions, may coerce them into providing personal information or may encourage unintended purchases. A significant body of evidence highlights the prevalence of dark patterns in GUIs. For instance, In 2022, a report by the European Commission found that about 97% of the most popular websites and apps deployed at least one dark pattern and affected the interaction of European Union (EU) consumers [1].

Figure 4.1 is a brief example of a dark pattern in QuillBot—an English writing website. Here, a user wishes to unsubscribe from a mailing list. But, the "Unsubscribe" button is tiny and buried at the bottom of an email, as indicated by a red arrow in Figure 4.1. It's a clear indication that QuillBot is throwing up subtle hurdles in the user's way [2]. This is an instance, where the user is at a disadvantage and such an instance is termed as a dark pattern.

This in-depth overview takes a closer look at dark patterns in online websites. The goal is to understand these tricky design choices—how dark patterns work, what they mean for users, and how common they are. By breaking down the details of dark patterns, the aim is to categorize them for both developers and users. The paper also analyzes each type of dark pattern in detail along with explanations of several real world examples; offering insights into their different forms, functions, and consequences. This knowledge serves as the foundation for the subsequent direction of this research work. Finally, it aims to provide actionable guidelines for both developers and users, fostering ethical design practices that prioritize user well-being and digital trust.

DOI: 10.1201/9781032664828-5

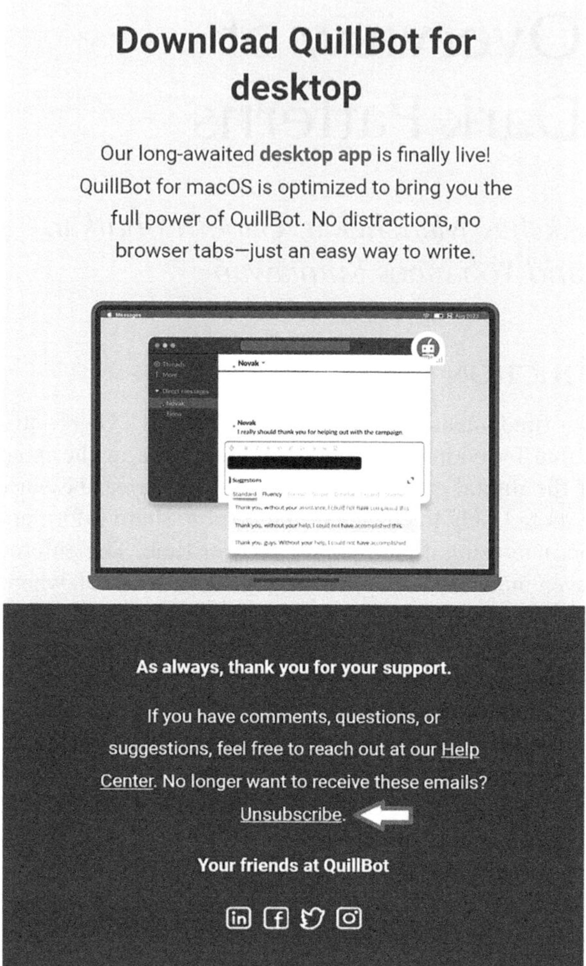

FIGURE 4.1 Dark pattern in QuillBot—an English writing website.

4.2 LITERATURE SURVEY

The literature survey on dark patterns has been conducted using Google Scholar. This scholarly search consists of all English-language research papers. Search criteria of either "dark patterns" or "black patterns" (the entire terms) or "dark," "black," and "patterns," separately; being present in either the title, abstract, or keywords are used.

The first paper examines dark patterns in shopping websites. The research work in this paper discovers that dark patterns are extremely widespread, with at least one dark pattern appearing on 11.1% of the 11,000 websites examined. The most common sorts of dark patterns this paper explores are Fake Social Proof and Sneaking [3]. The focus of this literature has been on analyzing dark patterns in e-Commerce websites which is crucial for safeguarding customer's digital security.

The second paper claims that dark patterns represent a complicated psychological phenomenon rather than merely immoral design practices. The paper highlights the impact of dark patterns on users' psychology. The paper explores Trust Erosion among users when they encounter dark patterns [4]. This paper helps readers see dark patterns in a more detailed way and makes readers think about their impact on user's psychology and ethical design.

The next paper takes an in-depth look at dark patterns, covering their definition, history, various kinds, and impact on consumers. The types of dark patterns identified are Obstruction and Forced Action. The paper also provides a guideline to avoid dark patterns by promoting awareness of dark patterns [5]. The paper conveys to the readers that more research on dark patterns is needed to protect users from tricky design tricks which tend to confuse the human mind.

Another paper does a detailed assessment of the literature on dark patterns in mass media or communication channels. Dark patterns were discovered to be used across a number of media channels. The paper highlights the influence of dark patterns on users' psychology by exploring the Emotional Responses given by users after encountering dark patterns [6]. This research work shows that users need to be more aware and the government should have guidelines in place to protect users when using digital media.

The subsequent paper performs a user study to determine how people perceive dark patterns in mobile applications. The research paper also sheds light on the negative effect of dark patterns on users' psychology. The paper explores behavioral changes in users after they are exposed to dark patterns. The paper also gives a guideline to report a dark pattern when users encounter it [7]. This literature seems to urge users to be more aware of dark patterns in mobile applications.

This paper presents website adaptations, especially scraping the consent pop-ups, which subtly manipulate user decisions through web content design. It also explains how dark patterns challenge the effectiveness of Digital Privacy Laws [8]. This research reveals hidden tactics for obtaining user consent, during interaction with dark patterns.

Another paper explores dark patterns in games. This research work discusses manipulative design in gaming GUIs. It also dissects various game aspects, exposing how dark patterns influence player decisions. The paper also explores user satisfaction before and after exposure to dark patterns [9]. Ethical concerns about player autonomy and informed choices are raised in the paper.

The next paper explores the negative aspects of how websites and mobile applications are designed. The type of dark pattern identified is Misdirection. The research paper also explores the psychological effect like Trust Erosion on users [10]. The paper explains how dark patterns can be sneaky; making readers think about how developers should design websites and mobile applications in a fair way.

Table 4.1 shows a comparative analysis of various research papers related to dark patterns. The research papers have been analyzed by using key parameters like the Type of Dark Patterns identified, the Impact of Dark Patterns on User's Psychology, Guidelines Provided for Users, Emphasis on Digital Trust, and Application Domain.

TABLE 4.1

Key Parameters of Dark Patterns in Literature Review

Article/ Authors	Type of Dark Patterns Identified	Impact of Dark Patterns on Users' Psychology	Guidelines Provided for Users	Emphasis on Digital Trust	Application Domain
Mathur et al. [3]	Yes (fake social proof, sneaking)	No	No	Yes	e-Commerce
Mathur et al. [4]	No	Yes (trust erosion)	No	Yes	General
Narayanan [5]	Yes (obstruction, forced action)	No	Yes (awareness)	Yes	General
Araujo et al. [6]	No	Yes (emotional responses)	No	Yes	Communication channels
Di Geronimo et al. [7]	No	Yes (behavioral changes)	Yes (report)	Yes	Mobile applications
Nouwens et al. [8]	No	No	No	Yes	Digital privacy law
Zagal et al. [9]	No	Yes (user satisfaction)	No	No	Gaming
Gray et al. [10]	Yes (misdirection)	Yes (trust erosion)	No	Yes	General

The vital observations from the literature survey related to dark patterns as depicted in Table 4.1, are as follows:

- Most of the (seven out of eight) research papers have emphasized the **need for digital trust**.
- Five out of eight research papers have explored the impact of dark patterns on user's psychology. **The impacts identified and explored are Trust Erosion, Emotional Responses, Behavioral Changes, and User Satisfaction**.
- Three out of eight research papers have identified common dark patterns. The dark patterns identified are **Fake Social Proof, Sneaking, Obstruction, Forced Action**, and **Misdirection**. Each paper has identified different types of dark patterns.
- The research papers focus on diversified domains like **e-Commerce, Communication Channels, Mobile Applications, Digital Privacy Law**, and **Gaming**.
- Very few (two out of eight) research papers have explored **guidelines for users to avoid dark patterns**.

4.3   TYPES OF DARK PATTERNS

Dark patterns is an emerging topic in the field of Cyber Security and Human-Computer Interaction. Dark patterns are typically categorized based on the specific techniques and tactics used to manipulate the user behavior. The categorization considers various parameters related to the design and interaction patterns on websites and related Information and Communication Technology (ICT) applications. The key parameters taken into consideration are User Interface Design, User Interaction, Information Presentation, Communication, and Language utilized. The following are the major types of dark patterns:

1. **Untrustworthy Behavior**: Websites/Applications engaging in untrust-worthy behavior violate user trust. This includes hidden fees, unclear terms and conditions, or actions that go against privacy expectations, eroding the user's confidence in the applications. The following are the key features of the Untrustworthy Behavior patterns:
   - **Parameter Used**: Communication and Language
   - **Severity**: High
   - **Most Frequent Domain**: e-Commerce
   - **Example**: Unclear terms and conditions of Amazon Prime

2. **Misdirection**: Misdirection patterns involve diverting a user's attention away from crucial information or actions. This could be achieved through clever design that leads users to click on options that they don't intend to, resulting in unintended consequences. The following are the key features of the Untrustworthy Behavior patterns:
   - **Parameter Used**: Communication and Language
   - **Severity**: High
   - **Most Frequent Domain**: Travel and Tourism
   - **Example**: Booking.com website misdirects users to pay more for a hotel room.

3. **Forced Action**: Users may be forced into actions that they don't intend to take. This could involve automatically adding items to a shopping cart, enabling features without user consent, or making it challenging to opt out certain services [5]. The following are the key features of the Forced Action patterns:
   - **Parameter Used**: User Interaction
   - **Severity**: High
   - **Most Frequent Domain**: e-Commerce
   - **Example**: Forced acceptance of Audible subscription.

4. **Sneaking**: Sneaking involves hiding certain details or options from users. This could be pre-checked boxes during sign-ups, where users unknowingly subscribe to additional services, or tricky opt-out mechanisms that lead to unintentional actions [3]. The following are the key features of the Sneaking patterns:
   - **Parameter Used**: User Interface Design
   - **Severity**: High
   - **Most Frequent Domain**: e-Commerce
   - **Example**: Unknowingly adding an extra item in the cart while checkout.

5. **Manipulative Language**: Dark patterns use language that exploits emotions or misrepresents information to influence user behavior. This could include using misleading button labels, exaggerated claims, or ambiguous terms. The following are the key features of the Manipulative Language patterns:

- **Parameter Used**: Communication and Language
- **Severity**: Medium
- **Most Frequent Domain**: Healthcare
- **Example**: Guilt-ridden messages on healthcare websites

6. **Obstruction**: Obstruction dark patterns intentionally create hurdles for users to complete a desired action. For example, making it complicated to unsubscribe from a service, intentionally hiding the "Cancel" button, or making it difficult to find essential information [5]. The following are the key features of the Obstruction patterns:
   - **Parameter Used**: User Interface Design
   - **Severity**: Medium
   - **Most Frequent Domain**: Social Media/Communication Channel
   - **Example**: Obscured Facebook settings

7. **Interface Interference**: Designers may use confusing GUIs to mislead users. This includes intentionally placing buttons in locations, where users would not expect them or using deceptive visual elements to guide users toward unintended actions. The following are the key features of the Interface Interference patterns:
   - **Parameter Used**: User Interface Design
   - **Severity**: Medium
   - **Most Frequent Domain**: Software
   - **Example**: Adding disguised advertisements on a website

8. **Nagging**: This type involves persistent and intrusive prompts that pressure users into taking action. This can include pop-ups that repeatedly ask users to subscribe to newsletters, enable notifications, or upgrade to premium services. The following are the key features of the Nagging patterns:
   - **Parameter Used**: User Interaction
   - **Severity**: Low
   - **Most Frequent Domain**: Social Media/Communication Channel
   - **Example**: Constant Notifications from Instagram

9. **Fake Social Proof**: Websites employ this by displaying false endorsements or testimonials to create a sense of trust and popularity. This could involve fake user reviews, fake follower counts, or fake endorsements from non-existent influencers [3]. The following are the key features of the Fake Social Proof patterns:
   - **Parameter Used**: Information Presentation
   - **Severity**: Low
   - **Most Frequent Domain**: e-Commerce
   - **Example**: Fake reviews on Amazon

10. **Fake Scarcity**: This tactic involves falsely presenting a limited quantity of a product or service to create a sense of urgency. Countdown timers or notifications suggesting limited stock are examples of this dark pattern. The following are the key features of the Fake Scarcity patterns:
    - **Parameter Used**: Information Presentation
    - **Severity**: Low
    - **Most Frequent Domain**: e-Commerce
    - **Example**: Fake "Items sold" numbers on Amazon and Flipkart

Awareness of these dark patterns is essential for users to navigate online spaces more effectively and for designers to create ethical, user-friendly GUIs that prioritize transparency and user consent.

4.4   REVIEW OF WEBSITES PRESENTING DARK PATTERNS

A systematic review has been conducted by analyzing various websites, which are in the English language. These include about 32 unique websites belonging to various domains that have dark patterns of diverse types. The domains of the websites analyzed are e-Commerce, Travel and Tourism, Software, Communication Channels, Retail, and Healthcare. Figure 4.2 shows the websites investigated for dark patterns. In Figure 4.2, the bigger font size of a website indicates a great amount of dark patterns in it.

The research conducted on these websites has been compiled in a comprehensive table as a database of dark patterns. The parameters used for the comparative analysis of these websites are name of dark pattern, website name, type of dark pattern, severity of dark pattern, type of loss for users due to dark pattern, information of dark pattern, the link of the website and domain of the website. Figure 4.3 shows an example of the Bait and Switch dark pattern on the Booking.com website. This dark pattern first shows a cheaper price for booking a flight and after clicking on the 'Show Flights' button, it shows an increased price, thereby deceiving the customer.

The dark patterns encountered in these websites have been categorized into the following types: **Nagging**, **Fake Social Proof**, **Obstruction**, **Sneaking**, **Interface Interference**, **Forced Action**, **Fake Scarcity**, **Manipulative Language**,

FIGURE 4.2   Word cluster of websites examined.

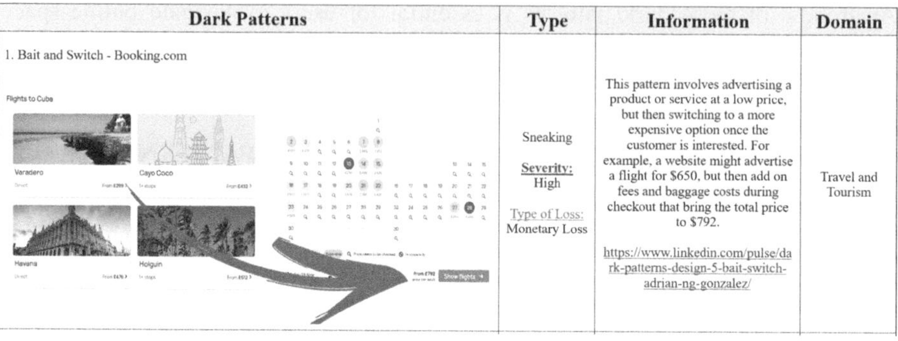

FIGURE 4.3   Sample entry in the database of dark patterns.

TABLE 4.2
Statistics by the Type of Dark Pattern

Type of Dark Pattern	Number of Occurrences	Most Frequent Domain
Sneaking	12	e-Commerce
Obstruction	7	Communication channel
Forced action	6	e-Commerce
Manipulative language	5	Healthcare
Misdirection	5	Travel and tourism
Fake social proof	4	e-Commerce
Fake scarcity	4	e-Commerce
Interface interference	3	Software
Untrustworthy behavior	3	e-Commerce
Nagging	1	Communication channel

Misdirection, and **Untrustworthy Behavior** as discussed in Section 4.3. The most common types of dark patterns are Sneaking, Obstruction, and Forced Action, while the least common ones are Fake Scarcity and Nagging.

The frequency of dark patterns by type is displayed in Table 4.2 and Figure 4.4. Out of 50 dark patterns, Sneaking, Obstruction, and Forced Action make up the majority (25). Fourteen out of fifty dark patterns are of the following types: deceptive language, fake social proof, and manipulative language. Only a small percentage (11 out of 50) of the dark patterns are of the following types: nagging, untrustworthy behaviour, interface interference, and fake scarcity.

Along with the types, each dark pattern is categorized based on severity as shown in Figure 4.5. The tokens used for the severity of a dark pattern are **High, Medium,** and **Low.** High severity is given to a dark pattern, which induces a monetary loss to the user. Medium severity is given to a dark pattern, which causes a data loss/privacy violation to the user. Low severity is given to a dark pattern, which induces a negative

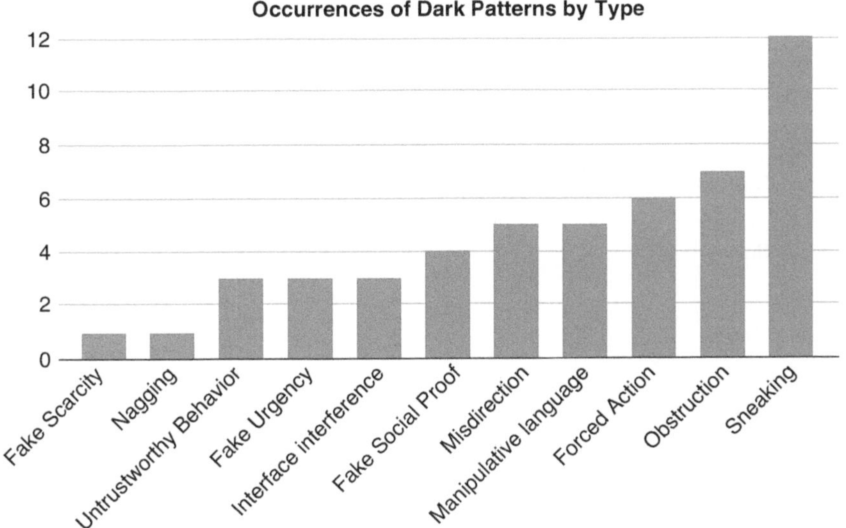

FIGURE 4.4 Occurrences of dark patterns by type.

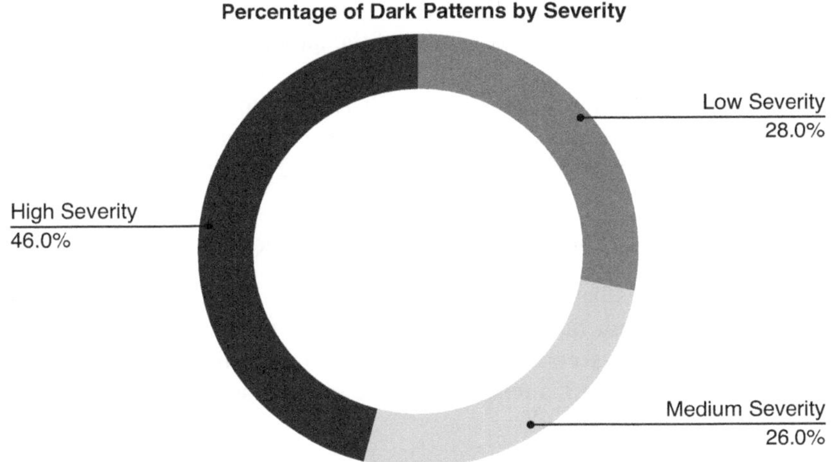

FIGURE 4.5 Percentage of dark patterns by severity.

psychological impact on the user. The majority of the high severity of dark patterns are from the e-Commerce domain due to the possibility of monetary loss incurred to the user.

The research also statistically analyzes the occurrences of dark patterns in websites of certain domains like e-Commerce, Communication Channels, Software, Travel & Tourism, and Healthcare. The research found out that the majority of the dark patterns are found on e-Commerce websites as shown in Figure 4.6.

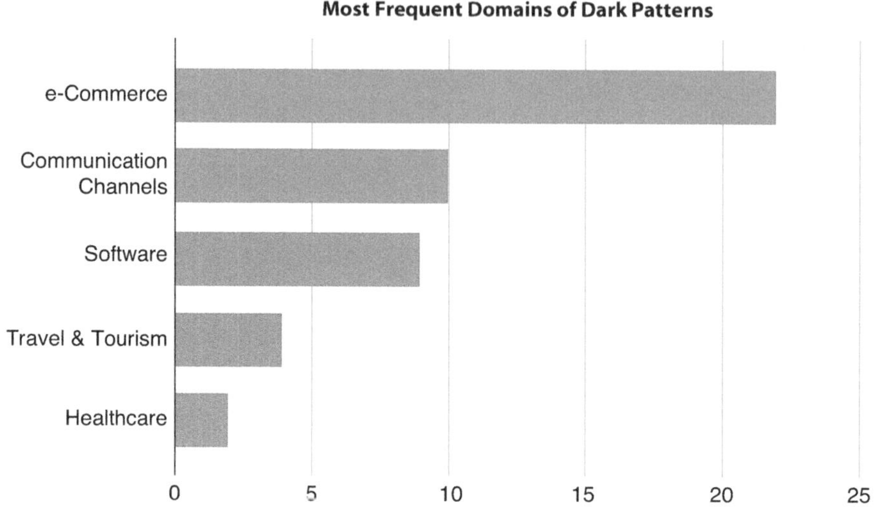

FIGURE 4.6   Most frequent domains of dark patterns.

The majority of the dark pattern types found in e-Commerce websites were Sneaking and Forced Action. The majority of the dark pattern types found in Travel and Tourism websites was Misdirection. The majority of the dark pattern type found in Social Media websites was Obstruction, whereas the majority of the dark pattern type found in healthcare websites was Manipulative Language. Table 4.2 showcases the number of occurrences of dark patterns in selected 50 websites.

4.5   GUIDELINES FOR DEALING WITH DARK PATTERNS

Dark patterns in GUIs have become a major concern. To help users and developers counter unethical practices, comprehensive guidelines for dealing with dark patterns are required. This section is focused on proposing guidelines for dealing with dark patterns. In this regard, a related literature survey has been pursued in the next subsection.

4.5.1   LITERATURE SURVEY ON GUIDELINES

The literature survey on guidelines for dealing with dark patterns has been conducted using Google Scholar. Search criteria of either "guidelines," "suggestions," "advice," "dark patterns," or "dark" and "patterns" separately; being present in either the title, abstract, or keywords are used.

The first paper highlights the value of transparency in digital interactions and promotes **user empowerment** via well-informed choices. It gives ideas such as **promote clarity, regulate the digital domain**, and **stay aware of manipulative practices** [11]. It also proposes that there should be digital literacy programs to equip users with the skills to identify and deal with dark patterns.

The second paper proposes a **framework for ethical design** that encourages developers to prioritize user well-being over short-term gains. The paper introduces the principle of **Respect user autonomy**, which means that developers should design user interfaces that allow users to make their own choices without being manipulated or coerced [12]. This principle is particularly relevant to the issue of dark patterns, as these techniques are often used to trick users into taking actions that they may not want or understand.

Another paper proposes a set of design principles that support the inclusion of moral considerations into the development of a website or an application, ensuring the end product aligns with the user interests. The principles include **transparency, user autonomy, privacy**, and **fairness** [13]. These principles align with the growing recognition of the ethical implications of technology and the need to design products that are not only functional but also beneficial to society.

Another article explains **Ben Shneiderman's Eight Golden Rules of User Interface Design**. It gives rules for developers to develop ethical GUIs. The eight key rules include Strive for Consistency, Seek Universal Usability, Offer Informative Feedback, Design Dialogs to Yield Closure, Prevent Errors, Permit Easy Reversal of Actions, Keep Users in Control, and Reduce Short-Term Memory Load [14]. These rules form the foundation for developing guidelines for dealing with dark patterns.

The subsequent article is a notification from the Government of India, which proposes draft guidelines for prevention and regulation of dark patterns. Although the aim of the notification is to provide guidelines, there are no guidelines provided. Instead, each type of dark pattern has been defined, and few illustrations for respective types are given. The document specifies 10 types of dark patterns such as false urgency, basket sneaking, confirm shaming, forced action, subscription trap, interface interference, bait and switch, drip pricing, disguised advertising, and nagging [15]. The information about dark patterns provided in the notification acts as a progressive step by the Government of India for countering dark patterns.

Despite the growing body of research on guidelines for dealing with dark patterns, a concrete set of guidelines has yet to emerge. The literature survey revealed various valuable insights, including the emphasis on clarity, transparency, user empowerment, ethical design principles, and the incorporation of existing frameworks such as Ben Shneiderman's Eight Golden Rules of User Interface Design. However, the absence of concrete, standardized guidelines is apparent in the examined literature. This gap highlights the need for an actionable set of guidelines. Recognizing this issue, this research paper seeks to contribute by formulating comprehensive guidelines based on the principles identified in the literature, aiming to provide a practical way to deal with dark patterns.

4.5.2 KEY PRINCIPLES CONSIDERED IN THE DESIGN OF GUIDELINES FOR DARK PATTERNS

Based on the above literature and Ben Shneiderman's Eight Golden Rules of Interface Design, this research paper has proposed some principles for designing guidelines for users and developers to deal with dark patterns. Each guideline adheres to the following key principles:

1. **Clarity of Understanding**: A guideline must be expressed in clear and straightforward language, avoiding ambiguity. Users and developers should easily understand the intended action or behavior.
2. **Alignment with Ethical Design Principles**: A guideline must follow ethical design principles such as transparency, user autonomy, privacy, and fairness. Each guideline should reflect a commitment to create a digital experience that prioritizes user well-being and ethical considerations.
3. **User Empowerment and Awareness**: A guideline must promote digital literacy and awareness among users. Users should feel equipped to navigate digital interfaces with confidence, identifying and dealing with dark patterns effectively.
4. **Integration of Existing Frameworks**: A guideline can use established frameworks, such as Ben Shneiderman's Eight Golden Rules of User Interface Design, as a foundation. This ensures that the guidelines are built on well-established principles that contribute to ethical and user-friendly digital interactions.

4.6   DESIGN OF GUIDELINES FOR DARK PATTERNS

Effective guidelines play a pivotal role in empowering users and guiding developers toward practices that prioritize user well-being. While formulating guidelines for users, the guideline should empower users to navigate digital spaces with confidence by providing clear, actionable guidelines. Whereas, while formulating guidelines for developers, the guideline should prioritize user well-being over short-term gains, promoting ethical design through principles of transparency, user autonomy, privacy, and fairness [16]. While creating guidelines for dealing with dark patterns, the following rules have been followed:

1. **Start with a Verb**: A guideline should start with an actionable verb to convey a specific action or behavior.
2. **Be Concise**: A guideline should be brief and to the point, ensuring that it is easy to understand and implement.
3. **One Idea per Guideline**: A guideline should focus on a single idea or concept to avoid confusion and maintain clarity.
4. **Use Simple Language**: A guideline should use uncomplicated language to enhance accessibility for a diverse audience.
5. **Ensure Actionability**: A guideline should be such that it can be easily translated into practical actions.
6. **Length**: An ideal guideline should be not more than thirteen words long.

The following are the guidelines for **users** to deal with dark patterns:

1. Stay informed and get educated about common dark patterns.
2. Pay close attention to GUI for signs of manipulation.
3. Check regularly and update the privacy settings on websites and apps.
4. Report instances of dark patterns to the relevant authorities.

5. Use ad blockers to minimize exposure to manipulative advertisements.
6. Verify information when asked for personal details or financial information.

The following are the guidelines for **developers** to deal with dark patterns:

1. Create GUIs that are straightforward and easy to understand.
2. Provide users with the ability to control their preferences, settings, and data.
3. Avoid hiding important details or using manipulative language.
4. Listen actively to user feedback regarding their experiences with the website.
5. Ensure that users understand the terms and conditions before completing an action.
6. Conduct regular reviews of GUIs to identify any unintentional dark patterns.

4.7 CONCLUSION

The research work on dark patterns has been explored in depth, revealing their presence in the realm of digital trust. The literature survey emphasizes the critical importance of digital trust in the face of dark patterns, with a majority of research work emphasizing its significance. Despite the breadth of exploration across various domains, there remains a notable gap in research, with only a few papers offering practical guidelines for users to navigate and mitigate the effects of dark patterns. The categorization of dark patterns on parameters like User Interface Design, User Interaction, Information Presentation, Communication, and Language utilized has provided a comprehensive understanding of their diverse types which include Sneaking, Obstruction, Forced Action, Manipulative language, Misdirection, Fake Social Proof, Fake Scarcity, Interface Interference, Untrustworthy Behavior, and Nagging.

The systematic review of websites across domains like e-Commerce, Travel and Tourism, Software, Communication Channels, Retail, and Healthcare further explores the prevalence of dark patterns in different sectors, reinforcing the urgency of addressing these issues. Guidelines for users and developers offer actionable insights to mitigate the impact of dark patterns. Users are advised to stay informed, scrutinize GUIs, and actively manage privacy settings, while developers are encouraged to prioritize transparency, user empowerment, and ethical design.

In the future, the guidelines proposed in this research can be reviewed and tested by researchers and modifications can be made. Ongoing research will contribute to the evolution of strategies in mitigating dark patterns, ensuring sustained progress in promoting ethical design practices, and safeguarding digital trust.

REFERENCES

1. Lupianez, F., Boluda, A., Bogliacino, F., Liva, G., Lechardoy, L. and Rodríguez, T.: Behavioural Study on Unfair Commercial Practices in the Digital Environment. Publications Office of the EU, 001-303 (2022).
2. Mazzella, F., Sundararajan, A., Espous, V. and Molmann, M.: How digital trust powers the sharing economy: The digitization of trust. *IESE Insight*, 2887, 24–31 (2016). https://doi.org/10.15581/002.art-2887

3. Mathur, A., Mayer, J. and Weiser, M.: Dark patterns at scale: Findings from a Crawl of 11K shopping websites. *Proceedings of the ACM on Human-Computer Interaction, Association for Computing Machinery,* New York, pp. 01–32 (2019). https://doi.org/10.1145/3359183

4. Mathur, A., Mayer, J. and Kshirsagar, M.: What makes a dark pattern... dark? *Proceedings of the ACM on Human-Computer Interaction, Association for Computing Machinery,* New York, pp. 01–48 (2021). https://doi.org/10.1145/3411764.3445610

5. Narayanan, A.: *Shining a Light on Dark Patterns.* Oxford University Press, Oxford, pp. 01–67 (2021). https://doi.org/10.1093/jla/laaa006

6. Araujo, A., Matias, V., and Pontes, R.: Dark patterns in the media: A systematic review. *Media & Communication,* 2019, pp. 105–113 (2019). https://doaj.org/article/e3d388b79eba4d66a92b12b0e0e2dc78

7. Di Geronimo, L., Braz, L., Fregnan, E., Palomba, F. and Bachelli, A.: UI dark patterns and where to find them: A study on mobile applications and user perception. *Proceedings of the 2020 ACM Conference on Human Factors in Computing Systems, Association for Computing Machinery,* New York, pp. 01–13 (2020). https://doi.org/10.1145/3313831.3376600

8. Nouwens, M., Liccardi, I., Veale, M., Karger, D. and Kagal, L.: Dark patterns after the GDPR: Scraping consent pop-ups and demonstrating their influence. *Proceedings of the 2020 CHI Conference on Human Factors in Computing Systems,* Association for Computing Machinery, New York, pp. 01–13 (2020). https://doi.org/10.1145/3313831.3376321

9. Zagal, J., Bojork, S. and Lewis, C.: Dark patterns in the design of games. *Foundations of Digital Games Conference, FDG 2013,* Chania, Greece, pp. 01–08 (2013).

10. Gray, C., Kou, Y., Battles, B., Hoggatt, J. and Toombs, A.: The dark (patterns) side of UX design. *Proceedings of the 2018 CHI Conference on Human Factors in Computing Systems,* Association for Computing Machinery, New York, pp. 01–14 (2018). https://doi.org/10.1145/3173574.3174108

11. Anderson, J. and Rainie, L.: *The Future of Well-Being in a Tech-Saturated World.* Pew Research Center, Washington, DC (2018)

12. Narayanan, A., Mathur, A., Chetty, M., and Kshirsagar, M.: Dark patterns: Past, present, and future. *ACM Queue,* 63, 67–92 (2020). https://doi.org/10.1145/3400899.3400901

13. Mathur, A., Acar, G., Friedman, M. J., Lucherini, E., Mayer, J., Chetty, M., and Narayanan, A.: Dark patterns at scale. *Proceedings of the ACM on Human-computer Interaction, Association for Computing Machinery, New York,* pp. 1–32 (2019). https://doi.org/10.1145/3359183

14. Teja, V.: Ben Shneiderman Eight Golden Rules of Interface Design. Geeksforgeeks [https://capian.co/shneiderman-eight-golden-rules-interface-design], (Accessed on August 2023).

15. Department of Consumer Affairs: Draft Guidelines for Prevention and Regulation of Dark Patterns. Consumer Affairs - Government of India [https://consumeraffairs.nic.in/sites/default/files/file-uploads/latestnews/Draft%20Guidelines%20for%20Prevention%20and%20Regulation%20of%20Dark%20Patterns%202023.pdf] (Accessed on August 2023).

16. Patil, A., Siriah, H., Shah, A., and Bhutkar, G.: Review and prioritization of accessibility guidelines for video game development. *Special Proceedings of 2021 Asian CHI Symposium,* Yokohama, Japan, pp. 43–52 (2021).

Section B

Next-Gen Technologies

5 Review of VR Technologies for Visually Impaired

Rahul Kanade, Ganesh Bhutkar, Kshitij Magare, Adarsh Londhe, and Shubham Kasar

5.1 INTRODUCTION

The biggest challenge for visually impaired people is to navigate around places. Obviously, they roam easily in their homes without any help, because they are very well aware of those home surroundings. But, when it comes to exploring unknown or new places, they have to heavily rely on other people, thus making them dependent. For many years, a lot of experiments have been carried out for assisting visually impaired people, aiming at improving the way they perceive their surrounding space. People with visual impairment may be born with vision loss or develop a visual impairment later in life caused by some accident or eye diseases. There are a number of terms used to describe the different degrees of vision loss of a person. The terminologies "visually impaired" as well as "visual impairment" are used to include all people or users with low vision, regardless of the degree of vision loss or blindness.

Major types of visual impairments include Early blind, Late blind, Congenital blind, and/or Total blind. The early blind category includes people who have lost their vision in their early stages of life (age < 18) or by birth. Congenital blindness is a group of diseases and conditions in childhood or early adolescence (under 16 years), which, if left untreated, can lead to blindness or severe visual impairment that is likely to cause permanent blindness later in life [1]. Total blindness is a term used to describe those who have a complete lack of perception of light. In 2020, around 43 million people were facing some kind of visual impairment. Out of that, almost 24 million were women and 19 million were men [1]. The Global Burden of Disease study shows that South Asia is home to the majority of the blind (11.7 million), accounting for 32.5% of the total world blind population. Similarly, in South Asia, around 61.2 million people have moderate to severe vision impairment, which is around 28.2% of the global dimension [2]. India has the second largest population in the world, and it accounts for more than 20% of the world's blind population. Unfortunately, India also has the greatest number of blind children in any one country [3].

Nowadays, Virtual Reality (VR) along with spatial audio is playing a major role in solving various problems specifically focusing on navigational issues faced by people with visual impairments. This article is a literature-review-based study of the various device prototypes developed for assisting visually impaired people mainly focusing on navigation.

5.2 BACKGROUND

Spatial sound is defined as sound positioned in a three-dimensional physical or virtual space around the listener. VR is an artificial environment that is experienced through sensory stimuli (such as sights and sounds) provided by a computer and in which one's actions partially determine what happens in the environment [4]. Spatial audio in VR involves handling audio signals that mimic acoustic behavior in the real world. A precise sonic representation of a virtual world is a very powerful way to create a fascinating and immersive experience.

In today's world, VR has spanned across a wide range of fields from the entertainment industry to creating complex training environments such as flight simulation, battle simulation, vehicle simulation, etc. Although using VR is very effective, when combined with spatial/3D audio it can provide us a highly realistic experience that produces far better results when compared to plain old VR (with normal/2D audio or no audio). For visually impaired individuals haptic (touch) and auditory senses play a major role as far as perception of the surroundings is concerned. In the early stages of experimentation, haptic sense along with 2D audio were used. Over the period of time 2D audio was replaced with 3D audio providing better results in real time. This 3D audio is now integrated with VR which allows us to create/simulate any environment that a visually impaired person needs to face in real life, thus improving the effectiveness of the training by breaking all physical barriers/constraints [5].

5.3 METHODOLOGY

A literature review was conducted using Google Scholar wherein all English language articles/papers from journals published between January 1999 to January 2021 with the keywords "Blind Virtual Reality," "Acoustic Virtual Reality," "VRWorks Audio," and "Audio Virtual Reality" were taken into consideration. Articles/papers on the development of different devices for assisting visually impaired people (with and without the use of virtual environments) were taken into consideration. Few papers based on experiments showing the impact of audio assistance on how visually impaired people perceive their surroundings are also taken into consideration.

Figure 5.1 represents the Preferred Reporting Items for Systematic Reviews and Meta-Analyses (PRISMA) workflow of this review with details of each step in the process [6]. Initially, thirty (30) review papers were identified based on titles that comprised the recent activities in VR technology. Out of these, 24 were screened with their abstracts, and four were removed due to their non-specific nature. These 20 papers were further assessed for eligibility. Criteria for eligibility were determined

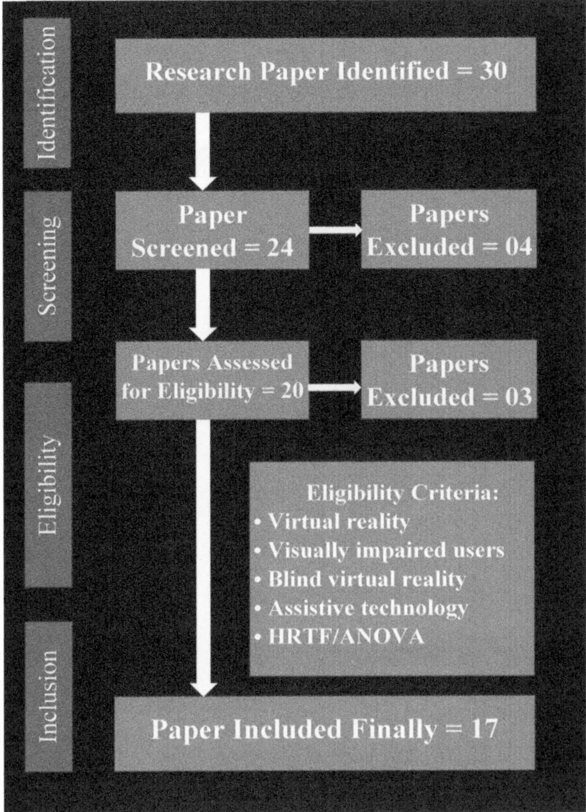

FIGURE 5.1 PRISMA workflow depicting the selection of research papers.

based on VR, visually impaired users, blind VR, and assistive technology. The presence of ANOVA/HRTF in the research along with its interaction with technology is a necessary measure. As a result, a total of 17 journal papers were taken into consideration and reviewed in this research study. The papers were organized into two groups. The first group had 15 papers based on experimentations and device prototype development. The second group consisted of the remaining two papers based on theoretical discussions on specific subtopics in VR audio.

A synthesized summary of the papers reviewed is provided in Table 5.1. The following parameters have been taken into consideration:

1. **Number of Participants**: The number of individuals considered for the experiment.
2. **Age Group:** The age range of participants included in the experiment.
3. **Types of Visually Impaired Users**: Categories of visually impaired users participated in the experiment.
4. **Evaluation/Environment Settings**: The settings utilized for conducting the study.

TABLE 5.1
Summary of the Experiments/Devices/Protypes Studied

Authors	Study Objectives	Number of Participants	Age Group (Years)	Types of Visually Impaired Users	Evaluation Environments/ Settings	Software/Hardware Tools	Sensors	Tracking System	Evaluation Techniques for Conclusion
Jaco, Jacquemin et. al. [9] (France)	To study spatial cognition in an immersive virtual audio environment	54	25–59	1. Congenitally blind 2. Late blind 3. Blindfolded	Physical and virtual room with six virtual sound objects	Head tracker device, stereophoric headphones, handheld tracked pointing device, IRCAMreverberation, SPAT	Ultrasonic sensor and infrared sensor	6 DOF Max/MSP	Visualizing Trajectory
Amir Amedi et al. [10] (Israel)	Demonstrate several games that were developed in the media lab as part of the training programs for the EyeCane, which augments the traditional White Cane with additional distance and angles	43	Not stated	Blindfolded	Computerized virtual environment	Headphone, keyboard (navigation in-game)	IR sensors that capture the distance to the object it is pointed at	NA	fMRI (Functional magnetic resonance imaging) technique for getting the image
Brian F. et al. [11] (France)	Immersive audio virtual reality system, ability to generate precise spatial representations from locomotor experience, building a metrically accurate representation of a spatial environment	72	21–63 years	24 congenitally blind, 24 people with late blindness (including male and female)	A room-type virtual immersive environment	Open circumaural headphones and stereophonic headphones	Handheld tracked pointing device for navigation purposes	Aerial navigation tracking system	ANOVA(statistical method)

(Continued)

TABLE 5.1 (*Continued*)

Summary of the Experiments/Devices/Protypes Studied

Authors	Study Objectives	Number of Participants	Age Group (Years)	Types of Visually Impaired Users	Evaluation Environments/ Settings	Software/Hardware Tools	Sensors	Tracking System	Evaluation Techniques for Conclusion
F. Grani et al. [12] (France)	Analyze how user attention is diverted while physically walking in a virtual environment when audio and/or visual attractors are present	13(4 female, 9 male)	21–59	Not stated	Four-sided CAVE-like VR systems	Unity3D (for creating a virtual environment), headphone	IR sensors that capture the distance to the object it is pointed at	NA	ANOVA (statistical method)
Freeman et al. [13] (London)	Comparing levels of presence for the same film clip with no AD, standard AD and cinematic AD	54(29 men, 25 women)	21–83	Three groups (no vision, partial vision, full vision)	Small theater room	7.5 minutes film clip, standard AD, cinematic AD for the clip, ITC_SOPI presence subscales	NA	NA	The ITC-sense of presence inventory (questionnaire)
Gonzál ez-Mora et al. [6] (Spain)	A portable electronic prototype that helps blind people with perception of space using sound only	5	People falling into different age groups are taken into consideration	Blindness onset, duration, and severity taken into consideration	A room with a table and a window and other unstated environments	Headphones, micro cameras, processors, artificial vision algorithms (geometric feature detection, stereo-vision, etc.)	NA	NA	Visual representations drawn by the subjects, verbal feedback
José Luis González-Mora et al. [5] (Spain)	Developing an interactive device based on virtual acoustic reality oriented to blind rehabilitation	58	Not stated	19-totally blind 20-partially blind 19-sighted	Not stated	Headphone, VFX3D virtual reality helmet IIS, Interactive Imaging System, micro cameras mounted in a virtual reality helmet	Two micro cameras mounted over a virtual reality helmet	Three space fastrak	WAIS-R (block design test with textures)

(*Continued*)

TABLE 5.1 (*Continued*)

Summary of the Experiments/Devices/Protypes Studied

Authors	Study Objectives	Number of Participants	Age Group (Years)	Types of Visually Impaired Users	Evaluation Environments/ Settings	Software/Hardware Tools	Sensors	Tracking System	Evaluation Techniques for Conclusion
Lorenzo Picinalli et al. [7] (France)	Exploration of architectural spaces by blind people using auditory virtual reality for the construction of spatial knowledge	10 Real 2-CB 3-LBVirtual 4-EB1-LB	24–57	Congenitally Blind (CB), Early Blind (EB), Late Blind (LB)	Allow them to navigate into the environment, along the corridor containing rooms on two sides using different sounds like computer keyboard clicking, men's voice, etc.	Second order-Ambisonic sound rendering engine along with nine channels, Joystick, speakers	Xsens	Not required	Using Lego kits
R. Iglesias et al. [14] (Spain)	Development of three applications for visually impaired people an adventure game, a city map explorer, and a chart explorer	Not stated	Not stated	Not stated	City-map type audio virtual environment	Headphones for sound feedback	Speech recognition and voice synthesis sensors	Not required	Not stated
Rodriguez-Hernandez et al. [8] (Spain)	Testing a device that captures the form and the volume of the space in front of the blind people and sends this information in the form of a sound map through headphones in real time	12	16–52	Six blind subjects, six sighted volunteers	A room with a table and a window and other unstated environments	stereoscopic machine vision system (two colored cameras), Pentium two processor, HRTF-based headphones	NA	NA	ANOVA (statistical method), Bonferroni's Multiple Comparison Test (statistical analysis)

(Continued)

TABLE 5.1 (*Continued*)

Summary of the Experiments/Devices/Protypes Studied

Authors	Study Objectives	Number of Participants	Age Group (Years)	Types of Visually Impaired Users	Evaluation Environments/ Settings	Software/Hardware Tools	Sensors	Tracking System	Evaluation Techniques for Conclusion
Sami Abboud et al. [15] (Israel)	Testing an EyeMusic SSD, which conveys shape and color information, to blind people	15 (13 female, 2 male)	Not stated	Five blind, ten blindfolded	Virtual environment	Blender 2.49 and Python 2.6.2, Eyemusic (SSD), standard keyboard, headphone	IR sensors (that capture the distance to the object it is pointed at)	NA	Verbal feedback
Shachar Maidenbaum et al. [16] (Israel)	Visual-to-audio Sensory Substitution Devices (SSDs) can potentially increase their accessibility by sonifying the on-screen content regardless of the specific environment while allowing the user to capitalize upon his experience from other uses of the device such as in the real world	15	Not stated	Five blind and ten blindfolded	Computerized virtual environment	Camera, headphone	IR sensors (that capture the distance to the object it is pointed at)	NA	Verbal feedback
Tiponut et al. [17] (Romania)	A man-machine interface included in an integrated environment that improves the mobility of blind persons into a limited area	3	Not stated	Visually impaired people	Not stated	AVR board, software module, headphones	Ultrasonic system, a 2-Axis accelerometer, a 3-Axis magnetic sensor	GSM/ GPRS/ GPS module	Not stated

(*Continued*)

TABLE 5.1 (*Continued*)

Summary of the Experiments/Devices/Protypes Studied

Authors	Study Objectives	Number of Participants	Age Group (Years)	Types of Visually Impaired Users	Evaluation Environments/ Settings	Software/Hardware Tools	Sensors	Tracking System	Evaluation Techniques for Conclusion
Torres-Gill et al. [18] (Spain)	Applications of virtual reality for visually impaired people	Not stated	Not stated	Visually impaired people	Acoustically isolated room of length 5.5 m, 4 m wide and 3 m height covered with panels at top	3D tracking system, tracker processor, PEGASO (sound rendered)	3D distance sensor (glasses), time of flight for enhancement for real-time scene capturing	Fastrak (magnetic field-based tracker), long ranger antenna	Visualizing trajectory
Yoshikazu Seki et al. [19] (Japan)	The present training system can reproduce a virtual training environment for orientation n and mobility, The virtual training environment is described in extensible markup language (XML), and the O&M instructor can edit it easily according to the training	5	Not stated	Not stated	Three-dimensionally a virtual training environment	3-D processors, mixer, headphone amplifier	Position sensor	Audio GPS	ANOVA (statistical method)

5. **Software/Hardware Tools**: The software and hardware tools employed in the studies.
6. **Sensors**: Sensors used for sensory computations.
7. **Tracking System**: Tracking systems used during the experiment.
8. **Evaluation Technique**: Evaluation methods for post-experiment understanding.

A summary of the theoretical papers reviewed has been provided in Table 5.2. The summary includes study objectives and concise conclusions. These papers provide insights into the utilization of sound formats and rendering in VR. Tables 5.3 and 5.4 provide a summary of the frequency of tools used for assistance and evaluation techniques used during the studies/experiments conducted mentioned in Table 5.1.

TABLE 5.2

Summary of the Theoretical Papers Studied

Authors	Agenda	Study Objectives	Conclusion
Keating [20] (England)	Sound formats for VR audio	Theoretical paper on different sound formats that can be used in AVR, representation and manipulation of the direction of sound in AVR	The ambisonic B-format representation of the direction of acoustic sources has been proposed as one of the most effective formats to use in a VR System
Mirza Beig et al. [21] (Canada)	Spatial sound in VE	Introduce spatial sound rendering in virtual environments, explore the aspects required to recreate spatial sounds, and provide a brief overview of the spatial sounds and common tools used in today's major game development engines to create spatial sounds in VEs	As new technologies evolve rapidly with much interest and investment in virtual and augmented reality, there will be new competing APIs that offer performance benefits rather than the necessary auditory benefits for the listener

TABLE 5.3

Tools Used for Assistance

Tool Used for Assistance	Number of Studies
EyeCane	2
EyeMusic	2
Processor	3
Headphones	12
Guided dog	2

TABLE 5.4

Frequency of the Evaluation Methods Used

Evaluation Method	Frequency
Visualizing trajectory	2
ANOVA	4
Questionnaires/verbal questioning	4
WAIS-R block design test	1
Lego kits	1

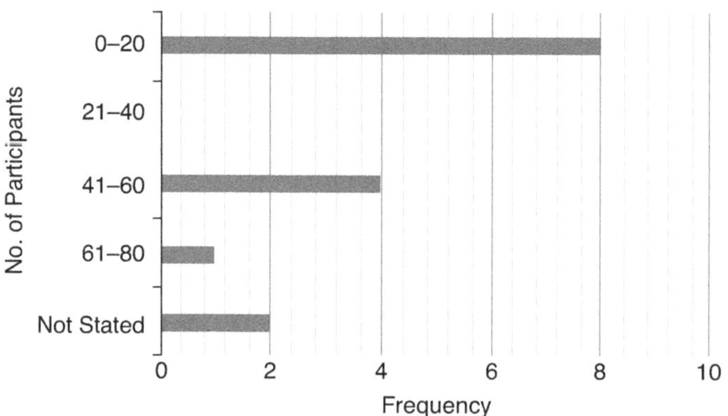

FIGURE 5.2 Frequency of participant's count considered across experiments.

5.4 FINDINGS

5.4.1 NUMBER OF PARTICIPANTS

The number of participants in the studies ranges from a minimum of 3 to a maximum of 72. Most of the studies have included both male and female research participants, where a number of male participants are most likely to be overweight in most of the cases (a similar trend is seen in many fields of research). In two of the studies, the data regarding the number of participants have not been stated. Out of the remaining studies, about 62% (8/13) had small groups, with less than 20 participants, and there were only 4 studies with a group of more than 50 participants (refer Figure 5.2). This shows that smaller groups with participants less than 20 are usually preferred.

5.4.2 AGE GROUP

As far as the age group is concerned, none of the studies have participants with age less than 16, which indicates that no such experiments have been carried out on children with visual impairments. Similarly, most of the studies have not included research participants having age above 60 (late adulthood). The preferred age group taken into consideration ranges from 25 to 52.

5.4.3 Evaluation Environments

Evaluation of experiments is carried out in both physical (real-world) as well as in virtual environments where 7 out of 15 papers use virtual environments for evaluation. After performing the experiments on different types of Visually Impaired Users, it is observed that virtual environments are very useful to train the user to understand how to handle different tools (like EyeCane and EyeMusic) for navigation and visualization. Out of all studies, 46% (7/15) of studies are making use of VR technology, whereas the remaining studies 54% (8/15) make use of real environments. This shows a gradual shift from the use of real environments to the use of VR and is expected to grow in the future.

5.4.4 Hardware

Among the different tools used for assistance, the use of headphones is most prominent (refer to Table 5.3). Around 80% (12/15) of the experiments are based on the use of headphones to achieve better results (refer to Table 5.4). In order to provide the 3D audio rendering in a virtual environment, most of the experiments have made use of head-related transfer function (HRTF) 67% (8/12), which will transform the audio by attenuating and boosting the frequencies according to location in virtual space.

5.4.5 Evaluation Methods Used

Table 5.4 shows the most commonly used evaluation methods used for evaluating the participants after performing the experiments. Among these, ANOVA (analysis of variance) has a frequency of 4/12 (34%), and the use of Questionnaires/Verbal Questioning also has a frequency of 4/12 (34%). Also, other techniques like WAIS-R (block test) with some modification, that is, instead of colors, the textures are used. Also, the graphs of the path followed by the participant (visualizing trajectory) are also taken into consideration, and in some of the cases, the Lego kits are used for building the scene of the virtual environment.

5.4.6 Regional Findings

Most of the studies reviewed in this paper have been carried out in European countries Spain, France, and England. Out of 17 papers studied, five of the papers are from Spain, four are from France, two are from England, and one from Romania. This indicates that a substantial amount of research in assistive technology for visual impairments is been carried out in European countries 71% (12/17) (Figure 5.3).

5.5 DISCUSSION

5.5.1 Tracking System

Tracking system plays a very important role for the movement of visually impaired individuals, as it will ensure that they move in right the direction. As far as outdoor environments are concerned tracking systems like GPS (Global Positioning System) have been used. For indoor environments, we find different systems have been used.

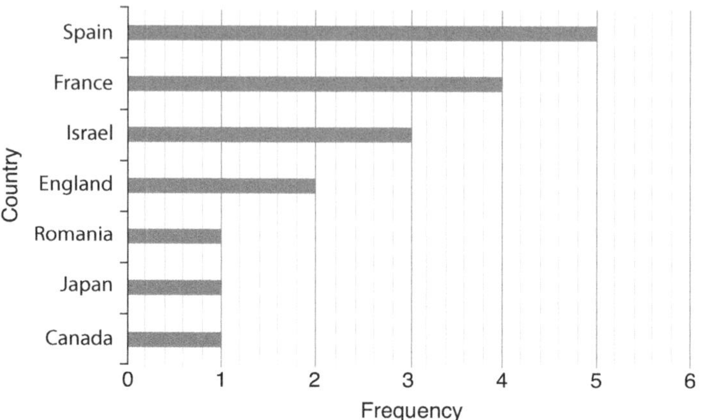

FIGURE 5.3 Regional frequency.

This includes 6DOF Max/MSP which helps to locate the user in the 3-dimensional environment (6 Degree of Freedom), 3 Space fastrak is used to locate the user and render the virtual image at that location in order to calculate the depth maps of the scene and for generating 3D audio in the virtual environment, the coordinates of the user in a virtual environment are given to acoustic sound renderer. Also as compared to standard antennae, the fastrak and LongRanger antennae have greater space to interact and therefore are able to cover the entire environment (room).

5.5.2 Sensor

A sensor is a device that detects and responds to a particular type of input from the physical environment. Ultrasonic sensors (measure the distance to an object using ultrasonic sound waves) and Infrared sensors (used in motion detectors) are the most common types of sensors used in these experiments. Apart from these micro cameras (used to collect image data), mobile navigation systems using spatial audio (Spatial Audio enables you to hear three-dimensional audio) have also been used in some experiments.

5.5.3 Types of Visually Impaired Participants

From the experiments, it can be seen that there are four types of visual impairments that have been taken into consideration—early blind or congenital blind, late blind, partial blind and total blind. Participants with partial blindness have limited vision, whereas participants with total blindness have complete loss of vision. Congenital blindness refers to a group of diseases partial blindness and total blindness [7]. Early blindness refers to loss of vision in the early stages of life (childhood), whereas late blindness refers to loss of vision in later stages of life (adulthood) and conditions occurring in childhood or early adolescence (below age 16), which, if left untreated, result in blindness or severe visual impairment that is likely to cause permanent blindness in later stages of life. Apart from this, some experiments have also included a group of blindfolded participants (having normal vision). Such experiments should include all types of groups mentioned above as significant differences in the results can be seen in most of the cases.

5.5.4 EVOCATION

Evocation is the appearance of the evoked visual stimuli that are found when visually impaired people are exposed to spatialized sounds [8]. This phenomenon refers to the generation/appearance of visual information in the form of phosphenes [8]. Among the 15 papers studied which were based on real-time experimentation and testing, the phenomenon of evocation was observed in only two experiments/studies. The exact reason for the occurrence of this phenomenon is not known yet. As this phenomenon is seen in very rare cases, it cannot be used as a measure to come up with a conclusion.

5.6 CONCLUSION

Out of the many challenges, one of the practical challenges faced during the use of new assistive technology for the blind is the training process. It is mostly faced while learning to master the device and more importantly while learning to use it in a specific environment. External help is usually required for such training which is not always available, can be expensive, and trying to navigate without prior training/ preparation can be dangerous [22]. Our review of experimentations carried out for visually impaired individuals for better environment perception is a step in the direction of understanding and improving the techniques used in this field. The results of this review lead us to the following conclusions:

- Smaller groups with participants less than 20 are usually preferred.
- Preferred age group taken into consideration ranges from 25 to 52.
- Experiments studied in this paper have used either virtual or real environments. With advancements in technology, there is a gradual shift from the use of real to virtual environments.
- Use of headphones with HRTF for sound rendering is a prominent trend that is observed.
- The ANOVA and Questionnaire (Verbal and Written) are the most commonly used evaluation methods to check the effectiveness of the proposed systems mentioned above.
- Most of the research in assistive technology for the visually impaired has been carried out in European countries.
- Use of VR for training/assisting visually impaired users has a great number of advantages as the use of VR allows us to generate any complex environment as per the requirements which makes the process smooth and safe.
- In situations where the setup cost for training is very high, the use of VR technology is scalable and can be profitable over a period of time.

In future, the researchers can develop some guidelines for the usage and development of assistive technology for visually impaired users.

REFERENCES

1. World Health Organisation, https://www.who.int/news-room/fact-sheets/detail/blind-ness-and-visual-impairment, last accessed 2021/12/10.
2. NCBI (National Center for Biotechnology Information) Homepage, https://www.ncbi.nlm.nih.gov/pmc/articles/PMC5820638/#:~:text=Blindness%20and%20visual%20impairment,in%20South%20Asia%20 was%200.7%25, last accessed 2021/12/26.
3. Express Health Care, https://www.expresshealthcare.in/publichealth/india-is-home-to-over-20-per-cent-of-the- worlds-blind-population/417470/, last accessed 2021/12/26.
4. Marxentlabs, https://www.marxentlabs.com/what-is-virtual-reality/, last accessed 2021/12/26.
5. González-Mora, J.: VASIII: Development of an interactive device based on virtual acoustic reality oriented to blind rehabilitation, *Jornadas de Seguimiento de Proyectos en Tecnologías Informáticas* 2(1), 2001–3976 (2003).
6. Gonzalez-Mora, J., Rodriguez-Hernandez, A.,Burunat, E., Martin, F., Castellano, M.: Seeing the world by hearing: Virtual acoustic space (VAS) a new space perception system for blind people, In: *2nd International Conference on Information & Communication Technologies,* pp. 837–842, Damascus, Syria (2006).
7. Picinali, L., Afonso, A., Denis, M., Katz, B.: Exploration of architectural spaces by blind people using auditory virtual reality for the construction of spatial knowledge, *International Journal of Human-Computer Studies* 72(4), 393–407 (2014).
8. González-Mora, J., Rodríguez-Hernández, A., Rodríguez-Ramos, L., Díaz-Saco, L., Sosa, N.: Development of a new space perception system for blind people, based on the creation of a virtual acoustic space. In: Mira, J., Sánchez- Andrés, J. (eds) *Engineering Applications of Bio-Inspired Artificial Neural Networks.* IWANN 1999. Lecture Notes in Computer Science, vol. 1607, pp. 321–330. Springer, Berlin, Heidelberg (1999).
9. Jaco, A., Katz, B., Blum, A., Jacquemin, C., Denis, M.: A study of spatial cognition in an immersive virtual audio environment: comparing blind and blindfolded individuals, In: *International Conference on Auditory Display,* pp. 228–235, Limerick, Ireland (2005).
10. Maidenbaum, S., Hanassy, S., Abboud, S., Buchs, G., Chebat, D., Levy-Tzedek, S., Amedi, A.: The "EyeCane", a new electronic travel aid for the blind: Technology, behavior and swift learning, *Restorative Neurology and Neuroscience* 32(6), 813–824 (2014).
11. Afonso, A., Blum, A., Katz, B.F.G.: Structural properties of spatial representations in blind people: Scanning images constructed from haptic exploration or from locomotion in a 3-D audio virtual environment, *Memory & Cognition* 38, 591–604 (2010).
12. Grani, F., Serafin, S., Argelaguet, F., Gouranton, V., Badawi, M., Gaugne, R., Lecuyer, A.: Audio-visual attractors for capturing attention to the screens when walking in CAVE systems, In: *IEEE VR Workshop: Sonic Interaction in Virtual Environments (SIVE),* pp. 3–6, Minneapolis, USA (2014).
13. Louise, F., Freeman, J.: Presence in those with and without sight: Audio description and its potential for virtual reality applications, *Journal of Cyber Therapy and Rehabilitation* 5(1), 15–23 (2012).
14. Iglesias, R., Casado, S.: Computer graphics access for blind people through a haptic and audio virtual environment, In: *The 3rd IEEE International Workshop on Haptic, Audio and Visual Environments and Their Applications,* pp. 13–18, Ottawa, Canada (2004).
15. Maidenbaum, S., Abboud, S., Buchs, G., Amedi, A.: Blind in a virtual world: Using sensory substitution for generically increasing the accessibility of graphical virtual environments, In: *2015 IEEE Virtual Reality (VR),* pp. 233–234, Arles, France (2015).

16. Maidenbaum, S., Amedi, A.: Non-visual virtual interaction: Can Sensory Substitution generically increase the accessibility of graphical virtual reality to the blind? In: 2015 *3rd IEEE VR International Workshop on Virtual and Augmented Assistive Technology (VAAT)*, pp. 15–17, Arles, France (2015).

17. Tiponut, V., Haraszy, Z., Ianchis, D., Lie, I.: Acoustic virtual reality performing man-machine interfacing of the blind, In: *Proceedings of the 12th WSEAS International Conference on Systems (ICS'08)*, pp. 345–349, Heraklion, Greece (2008).

18. Torres-Gil, M., Casanova-Gonzalez, O., Gonzalez-Mora, J.: Applications of virtual reality for visually impaired people, *WSEAS Transactions on Computers* 9(2), 184–193 (2010).

19. Seki, Y., Sato, T.: A training system of orientation and mobility for blind people using acoustic virtual reality, *IEEE Transactions on Neural Systems and Rehabilitation Engineering* 19(1), 95–104 (2011).

20. Keating, D.: The generation of virtual acoustic environments for blind people, In: *Proceedings of the 1st European Conference on Disability,* pp. 201–208, Maidenhead, UK (1996).

21. Beig, M., Kapralos, B., Collins, K.: An introduction to spatial sound rendering in virtual environments and games, *Computer Game* 8, 199–214 (2019).

22. Maidenbaum, S., Amedi, A.: Blind in a virtual world: Mobility-training virtual reality games for users who are blind, In: *IEEE Virtual Reality (VR)*, pp. 341–342, Arles, France (2015).

6 Virtual Reality-Based Experience of AICTE Idea Development, Evaluation, and Application Lab (KKWIEER)

Kushal Birla, Snehal Kamalapur, and Shirish Sane

6.1 INTRODUCTION: BACKGROUND AND DRIVING FORCES

Human–computer interaction (HCI) and Virtual Reality (VR) are closely related fields that can be used to enhance user experiences, improve interactions, and create immersive environments for various applications. VR serves as a valuable instrument for education and training in HCI, providing various advantages such as elevating engagement, motivation, collaboration, feedback, and the transfer of skills. VR has ushered in a new era of engaging with digital environments. The transformative power of VR technology has revolutionized various industries, and one area where it holds exceptional promise is education and scientific research. VR offers a level of immersion that was previously unimaginable, making it an ideal candidate for reshaping the landscape of learning and exploration. In this chapter, we explore an innovative project that harnesses VR's capabilities to recreate the AICTE IDEA Lab of the K. K. Wagh Institute of Engineering Education and Research (a.k.a. KKWIEER), which is located in Nashik city, of the state of Maharashtra in India. AICTE has granted funds with the objective of developing a physical AICTEIdea Lab at KKWIEER to (i) foster creativity and innovation in graduates and undergraduate students; (ii) equip students with essential 21st-century skills; (iii) enhance teachers' proficiency in pedagogical techniques, research, and project-based learning; (iv) revolutionize engineering education; and (v) promote student engagement in hands-on projects and internships.

6.1.1 THE VR REVOLUTION

VR technology, often characterized by its ability to immerse users in artificial, yet highly realistic, 3D environments, stands as a testament to the power of human imagination and innovation. VR offers an unprecedented level of immersion,

DOI: 10.1201/9781032664828-8

enabling users to transcend the boundaries of the physical world and step into a digital world that engages all their senses.

This capability to create immersive digital worlds has found applications in diverse fields, from gaming and entertainment to healthcare. However, one of the most exciting and promising domains for VR's transformative potential is education and innovation.

6.1.2 THE PROMISE OF VR IN EDUCATION

VR has the potential to redefine the educational experience. By immersing learners in dynamic, interactive, and realistic virtual environments, it can make learning more engaging and effective. Di Natale et al. presented insights through a systematic review that showed interactive VR is a technology that can influence a student's motivation to learn [1]. In the ever-evolving landscape of education and technology, the integration of VR has emerged as a promising frontier. It serves as a beacon of inspiration for engineers, educators, practitioners, and instructional designers. When creating virtual content, it is important to set up specific conditions to make it engaging. Saif Alatrash et al. used techniques like gamification and storytelling to prove this. These techniques make it more interactive and fun to explore heritage artifacts and information [2]. The work of Adurangba V. Oje et al. shows how important it is to create VR content and research agendas that are firmly rooted in strong educational theories and multimedia learning principles. This work shows how important it is for educators and researchers to use VR's transformative potential in education, which will lead to a more immersive and effective learning experience [3]. Kavanagh et al. describe a thematic analysis of 90 papers that describe the use of VR in education [4].

A systematic literature review conducted by Muhammad et al. showed that VR is used as a pedagogical tool for subject areas, including engineering, medical, language, and social learning, to encourage involvement and provide first-hand experience of the environment [5]. It explored the effect of VR as a pedagogical tool for enhancing students' experiential learning. The review was based on an analysis of 26 selected articles from various contexts and methodologies. Nine themes were identified, including VR as an emerging educational technology, digital transformation, the teaching-learning model, architectural and pedagogical tools, communication skills, reading and writing skills, social learning, and experiential learning. Adopting VR as a pedagogical method in education has challenged the conceptual definition of what constitutes a learning environment. VR's essential capability to give users a sense of presence and immersion has unlocked new opportunities in education if implemented appropriately [5]. According to Campos et al., learners reported that the integration of VR aided their understanding of course materials, particularly in areas like 3D visualization, identification, and overall comprehension. The learners viewed VR as a positive influence on their learning experience and a valuable tool for enhancing the understanding and retention of course content [6]. The use of VR in education has been found to increase the intrinsic motivation of students and improve learning outcomes [1,3,5,6].

6.1.3 THE AICTE IDEA LAB AT KKWIEER

The AICTE IDEA Lab serves as the backdrop for our VR project. The All India Council for Technical Education (AICTE) is renowned for its innovative approach to education. The AICTE IDEA Lab, located at KKWIEER (K. K. Wagh Institute of Engineering Education and Research), stands as a testament to this innovative spirit. It is more than just a physical space; it represents a hub of scientific exploration, experimentation, and educational innovation. Here, students and researchers come together to explore new ideas, conduct experiments, and push the boundaries of knowledge. The AICTE IDEA Lab is a space where innovation is nurtured and encouraged. It represents an educational environment that values hands-on learning, problem-solving, and experimentation. Our project aims to leverage VR technology to make this innovative environment accessible to a broader audience.

The lab offers various facilities, including:

- **Work Spaces:** Dedicated workspaces with essential infrastructure, such as computers, workbenches, and tools, to facilitate project development.
- **Prototyping Equipment:** The lab is equipped with 3D printers, drones of varying sizes, CNC machines, electronics prototyping kits, and other fabrication tools.
- **Collaboration Areas:** Specially designed spaces for collaborative efforts and brainstorming sessions, equipped with whiteboards, large displays, and comfortable seating arrangements to facilitate idea discussions and refinement.
- **Networking Possibilities:** The lab allocates spaces for organizing events, workshops, and seminars, fostering connections among students, faculty, industry professionals, and entrepreneurs. These areas also serve as platforms for showcases and exhibitions, allowing students to display their projects and prototypes, thereby encouraging networking and knowledge exchange.

6.2 OBJECTIVES

The objectives of this experiment are as follows:

- To use VR development software and design a visually immersive laboratory environment that faithfully replicates the AICTE IDEA Lab of KKWIEER, encompassing its layout, equipment, and overall ambiance
- To enable users to virtually engage with the lab and manipulate objects and data within the VR environment
- To evaluate the success of the replicated lab by analyzing feedback collected from users of the virtual lab.

The work provides insights into user interactions and feedback analysis, shedding light on the transformative potential across various dimensions of using VR by students and staff of the institute.

6.2.1 Harnessing VR for Lab Replication

The primary objective is to replicate the AICTE IDEA physical laboratory at KKWIEER in a VR environment and give a contextual overview of the transformative potential of VR. A VR environment helps to create an immersive experience that involves recreating the lab's layout, equipment, and overall atmosphere. The emphasis is to ensure the virtual lab experiences a striking resemblance to the physical environment.

- **Lab Layout:** The VR environment meticulously mimics the physical layout of the AICTE IDEA Lab, including the placement of various sections, workstations, and equipment. Users should feel a sense of familiarity as they navigate through the virtual space.
- **Equipment:** An essential aspect of the replication process involves accurately modeling and representing the lab's equipment, tools, and machinery. This includes scientific apparatus, drones, instruments, and specialized tools commonly used within the lab. Each piece of equipment is designed to closely resemble its real-world counterpart, down to the smallest details.
- **Atmosphere:** Creating an authentic atmosphere is key to ensuring that users feel as if they are truly within the AICTE IDEA Lab (KKWIEER). This involves attention to details such as lighting, textures, and ambiance to provide an immersive and realistic environment.

6.2.2 Visual Immersion and Interactivity

In addition to replication, the project places a strong emphasis on creating a visually immersive and interactive experience for users. Visual immersion is central to making the virtual lab environment engaging and compelling.

- **Visual Immersion:** The use of VR technology allows for a heightened level of visual immersion. High-quality graphics and realistic rendering techniques are employed to create a visually convincing virtual environment. Users should feel as though they are physically present in the recreated lab.
- **Interactivity:** The project also aims to make the virtual lab highly interactive. Users are not passive observers; they are active participants. They can engage with the environment, explore various elements, and interact with objects. This interactivity fosters a sense of involvement and engagement, making the learning experience more dynamic.

6.2.3 Feedback Collection and Evaluation

The work by Vlahovic et al. presented a review of several methods and Quality of Experiences (QoEs) for evaluating user acceptance factors of a VR application [7]. Wienrich et al. discussed the implications for a holistic evaluation framework [8]. Beyond replication and immersion, another objective is to conduct a user satisfaction survey to evaluate the VR-based application replicating the AICTE IDEA Lab at KKWIEER by following a structured approach.

6.3   METHODOLOGY

This section gives an outline of the methodology for a project that involves using VR development software, such as Unity and Blender, along with programming languages, to create a virtual laboratory environment. Here's the breakdown of the key points in the methodology.

6.3.1   Designing the Virtual Experience of the Lab and Its Components

The project involves meticulously designing the virtual laboratory environment. This process includes creating 3D models and visual representations of the lab layout, equipment, and surroundings. The aim is to make the virtual environment as real as possible. This means that the virtual lab should closely replicate the physical lab, including its layout, equipment, and overall appearance.

The complete video of a walk-through experience of the VR-based application is captured by wearing a head-mounted VR device called the HTC VIVE Starter Kit, and it is made available on YouTube at the following URL: https://www.youtube. com/watch?v=vSMOJBeUI9M.

Walls, windows, walking areas, roofs, and trees are shown in Figure 6.1. It is an actual view of the lab captured using a camera.

The frame or a scene visible through the VR device in Figure 6.2 shows outer walls, roofs, entrance points of the lab, and security fencing around the lab.

A notice board, the entrance of the lab, and spaces for vehicles are visible as a scene, as shown in Figure 6.3, through the VR device.

6.3.2   Immersive Experience

One of this project's key components is providing an immersive experience to end users. This means that when users interact with the virtual laboratory, they should feel like they are actually in a real lab. The use of VR technology allows for a more engaging and realistic experience.

The actual internal view of the lab and a notice board are shown in Figure 6.4.

An internal view of the lab upon entering the lab and visible through a head-mounted VR device (HTC VIVE Starter Kit) is shown in Figure 6.5. It shows entrances to various compartments (having doors and windows), such as a laser cutting machine

FIGURE 6.1   The actual photographs of the lab.

FIGURE 6.2 Aerial scene (view) of the lab visible through the VR device.

FIGURE 6.3 Entrance of the lab visible through the VR device.

FIGURE 6.4 An actual photograph of the internal view of the lab.

FIGURE 6.5   Internal view of the lab visible through the VR device.

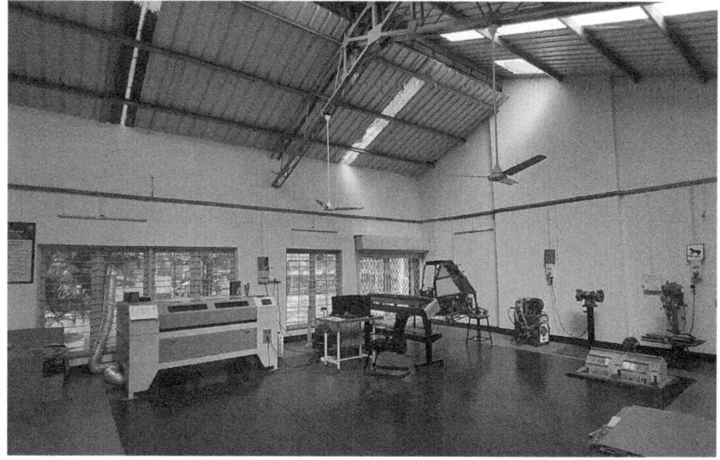

FIGURE 6.6   Internal view of the laser cutting machine section of the lab.

lab, 3D printing labs, and a walking passage with multimedia-assisted audio-video guides, replicated notice boards, and signboards.

An actual photograph depicting the internal view of the laser cutting machine section in Figure 6.6 illustrates equipment, including a laser cutting machine, a computer system, and other displayed apparatus.

A video-assisted guide about the operating instructions for a laser cutting machine for a viewer is shown in Figure 6.7.

An actual internal view shows various elements within the lab are visible, including chairs, walking areas, glass cabins, doors, windows, and display boards as shown in Figure 6.8.

The scene is visible through the VR device, as shown in Figure 6.9, and various elements within the lab are visible, including chairs, walking areas, glass cabins, doors, windows, and display boards.

A photograph in Figure 6.10 captures the interior of the ideation-discussion room within the lab, featuring elements such as a television, book racks, and a plant. This space serves as a dedicated area for conducting brainstorming activities, discussions, and collaborative endeavors.

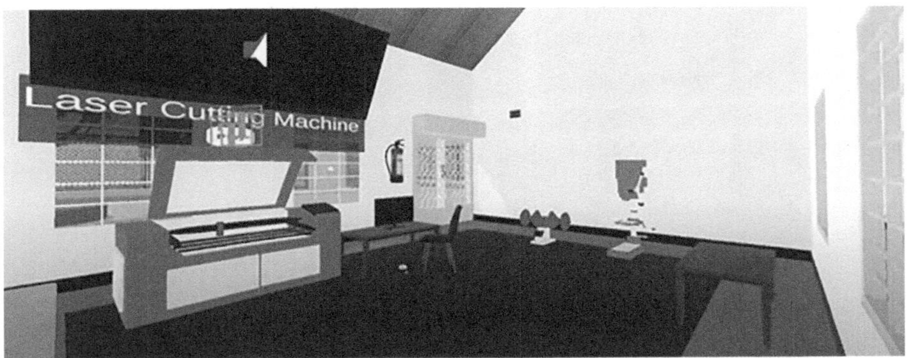

FIGURE 6.7 Internal view of the laser cutting machine compartment of the lab visible in the VR device.

FIGURE 6.8 Actual photograph of the internal view of seating spaces, glass cabins, and display boards of the lab.

FIGURE 6.9 Scene of an internal view of seating spaces, glass cabins, and display boards of the lab visible in VR device.

FIGURE 6.10   Actual photograph of ideation-discussion room.

FIGURE 6.11   Scene of ideation-discussion room visible in the VR device.

Figure 6.11 is a scene of the internal view of the ideation-discussion room visible in a VR device. It shows the television, book rack, table and chairs, whiteboard, and a plant in the lab.

6.3.3   ASSESSMENT AND EVALUATION

The research project employed a convenience sampling method, selecting participants based on factors such as their availability, and their willingness to engage in the study. This approach ensured the inclusion of individuals who were accessible within a specific location and time frame, while also adhering to ethical guidelines for involving users in the research. This involves evaluating how well users can navigate and interact with the VR application and whether they gain a better understanding of the lab infrastructure and equipment through this virtual experience, therefore gathering insights and feedback from users. The feedback mechanisms, as outlined in Table 6.2, were designed to evaluate diverse dimensions of users, including sensory, technical, cognitive, and educational aspects. A total of 64 responses were gathered from willing participants who had either visited the lab physically before or after experiencing the VR presentation of the same lab. Demographic details from the survey are presented in Table 6.1. The assessment and analysis of the responses are delineated in the column labeled 'Count of Responses' in Table 6.2.

The summary and visualization of the questions and respective responses are discussed as follows:

1. **Overall Experience:** 9 respondents rated the experience as excellent, 29 as very good, 14 as good, 12 as fair, and none as poor.
2. **Immersion and Realism:** 15 respondents found the VR replication incredibly real, 49 found it quite immersive, and none found it not very immersive or not immersive at all.
3. **Comfort and Motion Sickness:** 33 respondents experienced very mild discomfort, 29 experienced moderate discomfort, and 2 experienced significant discomfort.
4. **Usability and Navigation:** 39 respondents found the VR replication very user-friendly, 25 found it somewhat user-friendly, and none found it challenging to use.
5. **Visual Quality:** 35 respondents rated the visual quality as excellent, 29 as good, and none as average or poor.
6. **Audio Experience:** 38 respondents rated the audio quality and immersion as very good, 26 as adequate, and none as below average.
7. **Content and Interactivity:** 41 respondents found the activities and interactions very engaging and informative, 23 found them moderately engaging and informative, and none found them slightly engaging and informative.
8. **Technical Issues:** 25 respondents found the VR replication mostly smooth with a few minor hiccups, 39 experienced occasional performance issues, and none found it unusable due to performance problems.
9. **Duration and Comfort:** 38 respondents used the VR experience for less than 15 minutes, 26 used it for 15–30 minutes, and none used it for 30 minutes to 1 hour.

TABLE 6.1

Demographics of the Survey

Sr. No.	Survey Question	Answer Choices and Count
1.	Gender	i. Male (35)
		ii. Female (29)
2.	Age	i. 18–22 years (34)
		ii. 23–25 years (09)
		iii. 26–30 years (03)
		iv. More than 30 years (18)
3.	Occupation	v. Student (38)
		vi. Staff (26)

TABLE 6.2

Questionnaire and Responses

Question Number	Question	Choices of Answers	Count of Responses
1.	**Overall Experience** Please rate your overall experience with the AICTE IDEA Lab VR replication	i. Excellent	09
		ii. Very good	29
		iii. Good	14
		iv. Fair	12
		v. Poor	Nil
2.	**Immersion and Realism** How immersive did you find the VR replication of the IDEA Lab?	i. It felt incredibly real	15
		ii. It was quite immersive	49
		iii. It was not very immersive	Nil
		iv. It didn't feel immersive at all	Nil
3.	**Comfort and Motion Sickness** Did you experience any discomfort or motion sickness while using the VR experience?	i. Very mild discomfort	33
		ii. Moderate discomfort	29
		iii. Significant discomfort	02
4.	**Usability and Navigation** How user-friendly was the VR replication in terms of usability and navigation?	i. Very user-friendly	39
		ii. Somewhat user-friendly	25
		iii. Challenging to use	Nil
5.	**Visual Quality** Please rate the visual quality of the VR replication	i. Excellent	35
		ii. Good	29
		iii. Average	Nil
		iv. Poor	Nil
6.	**Audio Experience** How would you rate the audio quality and immersion in the VR experience?	i. Very good audio	38
		ii. Adequate audio	26
		iii. Below-average audio	Nil
7.	**Content and Interactivity** Were the activities and interactions in the VR replication engaging and informative?	i. Very engaging and informative	41
		ii. Moderately engaging and informative	23
		iii. Slightly engaging and informative	Nil
8.	Did the VR replication run smoothly without technical issues?	i. Mostly smooth, with a few minor hiccups	25
		ii. Occasional performance issues	39
		iii. Unusable due to performance problems	Nil
9.	**Duration and Comfort** How long did you use the VR experience in one session?	i. Less than 15 minutes	38
		ii. 15–30 minutes	26
		iii. 30 minutes to 1 hour	Nil

(*Continued*)

TABLE 6.2 (*Continued*)
Questionnaire and Responses

Question Number	Question	Choices of Answers	Count of Responses
10.	**Loading Times** How would you rate the loading times when starting the VR experience and transitioning between scenes or activities?	i. Fast (less than 1 s) ii. Average (between 1 and 2 s) iii. Slow (more than 2 s)	29 35 Nil
11.	**Educational Value** How intellectually stimulating and educational did you find the VR experience?	i. Highly educational and intellectually stimulating ii. Moderately educational and intellectually stimulating iii. Not educational or intellectually stimulating	26 38 Nil
12.	**Physical Involvement** Did the VR experience require you to be physically active or engaged in physical tasks?	i. It required some physical movement ii. It was mostly sedentary iii. It was entirely sedentary	24 40 Nil
13.	**Age** of participant (respondent)	i. 18–22 years ii. 23–25 years iii. 26–30 years iv. More than 30 years	34 09 03 18

10. **Loading Times:** 29 respondents rated the loading times as fast (less than 1 second), 35 as average (between 1 and 2 seconds), and none as slow (more than 2 seconds).
11. **Educational Value:** Figure 6.12 shows a pie chart indicating that 26 respondents found the VR experience highly educational and intellectually stimulating, 38 found it moderately educational and intellectually stimulating, and none found it not educational or intellectually stimulating.
12. **Physical Involvement:** 24 respondents found the VR experience required some physical movement, 40 found it mostly sedentary, and none found it entirely sedentary.
13. **Age of Participant:** The pie chart as shown in Figure 6.13 shows the ages of respondents. 34 respondents were between 18 and 22 years old, 9 were between 23 and 25 years old, 3 were between 26 and 30 years old, and 18 were more than 30 years old.

6.3.4 RESULTS AND DISCUSSIONS

We performed a chi-squared analysis to test for an association between the "Age of participant" and "Educational Value" variables. Comparing the observed counts (O) with the expected counts (E), we can calculate the chi-squared statistic for each cell:

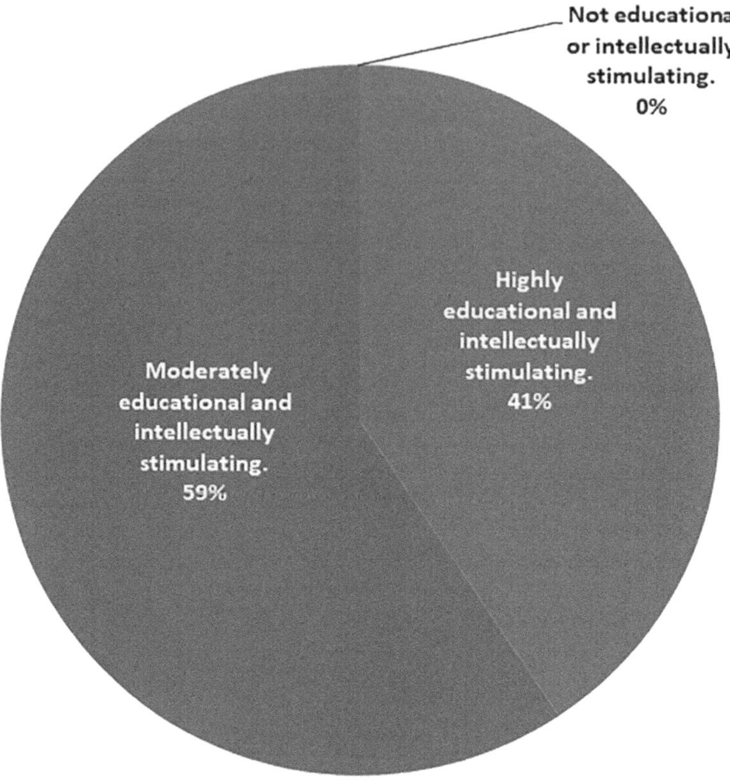

FIGURE 6.12   Educational value.

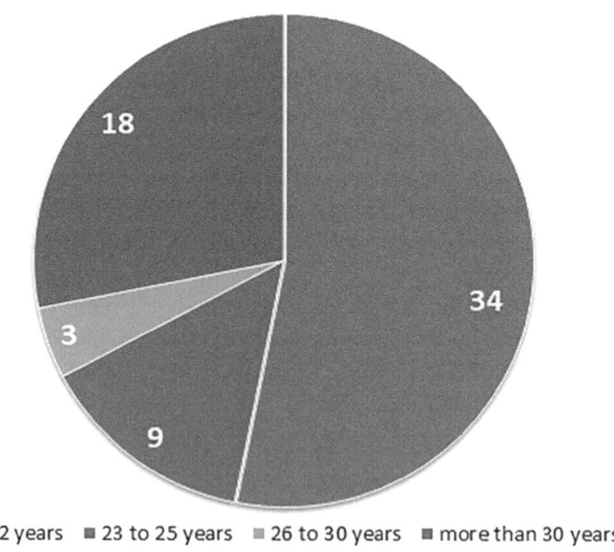

FIGURE 6.13   Age of participants (respondent).

$$x^2 = \frac{(O-E)^2}{E}$$

The chi-squared statistic for the entire table is the sum of the chi-squared values for each cell:

$$\chi^2 = \sum x^2$$

With 1 degree of freedom, the p-value for the chi-squared test is calculated as follows:

$$\text{p-value} = \text{Probability}\left(\chi^2, df\right)$$

Based on the p-value, it can be determined if there is a significant association between the "Age of participant" and "Educational Value" variables. If the p-value is less than the chosen significance level (e.g., 0.05), we can reject the null hypothesis and conclude that there is a significant association between the variables. In this case, the p-value for the chi-squared test is approximately 0.047. Since the p-value is less than the significance level of 0.05, we can reject the null hypothesis and conclude that there is a significant association between the "Age of participant" and "Educational Value" variables. This means that the educational value of the VR experience varies depending on the age of the participant, with younger participants (18–22 years) finding it more intellectually stimulating than older participants (23–25 years and above 30 years).

Similarly, performing the chi-squared test on "Immersion and Realism" and "Usability and Navigation," we get a chi-squared test statistic of 64.64 with 2 degrees of freedom and a p-value of less than 0.001. This indicates that there is a significant association between "Immersion and Realism" and "Usability and Navigation." We can reject the null hypothesis and conclude that the two variables are dependent. The results suggest that participants who found the VR replication to be more immersive were more likely to find it user-friendly, and those who found it less immersive were more likely to find it challenging to use.

In the context of the HCI, the responses of 'Immersion and Realism' and 'Comfort and Motion Sickness' indicate a successful HCI design in terms of immersion. The participants' responses can help identify areas for improvement in terms of 'Content and Interactivity' and 'Overall Experience', suggesting that the HCI design could be improved to minimize these issues.

6.4 CONCLUSION AND FUTURE DIRECTIONS

This study gives detailed insights into the design, development, assessment, and evaluation of the VR experience of AICTE IDEA Lab at KKWIEER.

From a technical analysis perspective, the study reveals a generally positive VR experience with notable strengths and areas for improvement. The majority of users reported smooth VR replication with minor hiccups, indicating satisfactory overall performance. However, some users faced occasional performance issues, highlighting the need for ongoing optimization to ensure consistent stability. The positive feedback on user-friendliness suggests a well-designed and intuitive interface, although there is room for refinement in navigation. High ratings for visual and audio quality reflect successful technical implementations, warranting continued focus on maintaining and potentially enhancing these aspects. The data on loading times indicates an efficient content delivery system with opportunities for monitoring and optimization. Acknowledgment of occasional technical challenges underscores the importance of continuous testing and refinement. Demographic insights, particularly age distribution, provide crucial considerations for tailoring technical features and content to meet diverse user expectations. Understanding user behavior regarding duration and physical involvement offers technical insights for optimizing hardware and software to accommodate varied usage patterns.

In summary, while the VR experience demonstrates technical success, addressing occasional performance issues, refining user guidance, and maintaining optimization efforts are essential for continuous improvement and meeting the expectations of a diverse user base. Building upon the positive aspects, future developments should aim to address occasional technical issues, ensuring a consistently smooth performance for all users. Additionally, by leveraging the success of creating an immersive and realistic environment, there is an opportunity to expand content diversity to cater to a broader audience. Further research into user preferences and age-specific content could optimize engagement across different age groups. As the VR environment continues to advance, it will be crucial to concentrate on enhancing both technical capabilities and content quality to ensure the ongoing satisfaction and improvement of user experiences. Ultimately, a forward-looking strategy that integrates user feedback and technological advancements will contribute to the continued success and growth of VR experiences.

ACKNOWLEDGMENTS

The All India Council for Technical Education (AICTE) has granted funds to develop a physical AICTE IDEA Lab at K. K. Wagh Institute of Engineering Education and Research, Nashik (AICTE Permanent ID: 1-5942025).

REFERENCES

1. Di Natale, A.F., Repetto, C., Riva, G., and Villani, D., "Immersive virtual reality in K-12 and higher education: A 10-year systematic review of empirical research," *Br. J. Educ. Technol., 51*: 2006–2033, 2020. https://doi.org/10.1111/bjet.13030
2. Alatrash, S., Arnab, S., and Antlej, K., "Communicating engineering heritage through immersive technology: A VR framework for enhancing users' interpretation process in virtual immersive environments," *Comput. Educ. X Reality, 3*: 100040, 2023, ISSN 2949–6780. https://doi.org/10.1016/j.cexr.2023.100040. (https://www.sciencedirect.com/science/article/pii/S294967802300034X)
3. Oje, A. V., Hunsu, N. J., and May, D., "Virtual reality assisted engineering education: A multimedia learning perspective," *Comput. Educ. X Reality, 3*: 100033, 2023, ISSN 2949–6780. https://doi.org/10.1016/j.cexr.2023.100033. (https://www.sciencedirect.com/science/article/pii/S2949678023000272)

4. Kavanagh, S., Luxton-Reilly, A., Wuensche, B., and Plimmer, B., "A systematic review of virtual reality in education", *Themes Sci. Technol. Educ.*, *10*(2): 85–119, 2017. https://earthlab.uoi.gr/theste

5. Muhammad, A., Naz, A., Prathamesh, C., and Mohammad, T. "Virtual reality as pedagogical tool to enhance experiential learning: A systematic literature review," *Educ. Res. Inter. 2021*: 1–17, 2021. https://doi.org/10.1155/2021/7061623

6. Campos, E., Hidrogo, I., and Zavala, G. "Impact of virtual reality use on the teaching and learning of vectors," *Front. Educ. 7*: 965640, 2022. https://doi.org/10.3389/feduc.2022.965640

7. Vlahovic, S., Suznjevic, M., and Skorin-Kapov, L, "A survey of challenges and methods for quality of experience assessment of interactive VR applications," *J. Multimodal. User Interfaces 16*: 257–291, 2022. https://doi.org/10.1007/s12193-022-00388-0

8. Wienrich, C., Döllinger, N., Kock, S., Schindler, K., and Traupe, O., "assessing user experience in virtual reality: A comparison of different measurements," In: Marcus, A. and Wang, W. (eds) *Design, User Experience, and Usability: Theory and Practice. DUXU 2018*, Lecture Notes in Computer Science, vol. 10918. Springer, Cham, 2018. https://doi.org/10.1007/978-3-319-91797-9_41

7 Calibrating the IoT-Based System for Monitoring Water Quality with Known pH and Turbidity

Hasifah Namatovu and Fiona Ssozi

7.1 INTRODUCTION

Water is among the most abundant natural resources on the planet [1] and Uganda, as a country, is well endowed with extensive freshwater resources [2] like Lake Victoria, the largest freshwater lake in Africa, the world's largest tropical lake and the world's second-largest freshwater lake [3]. However, like the rest of the developing world, the country is grappling with the challenges of ensuring safe water for its people [4]. Several types of impurities such as dissolved salts, floating objects, suspended matter, pathogens, microbiological organisms, and others are usually present in water [5] either temporarily or permanently depending on various factors, which often make such water unfit for certain uses. Pollution of water may be from industrial wastes, sewer systems, and water mixing with polluted soils, atmosphere, and groundwater systems. Major pollutants include bacteria, viruses, parasites, fertilizers, pesticides, fecal matter, pharmaceutical products, and plastics [6].

Water is very important for the existence of life, and the provision of safe water for human consumption is Sustainable Development Goal 6 of the United Nations Resolutions for 2030 [7]. Financial constraint is the key challenge in providing and ensuring access to safe water for everyone [8]. The problem of access to safe water largely affects low-income households in both urban and rural communities [9]. Of the 44.2 million people in Uganda, sixty-five percent have access to improved water sources [10]. As water from these sources is used while untreated, there is a need for continuous regular monitoring to ensure its quality and safeguard the users from possible health problems arising from consuming contaminated water [11]. World Health Organization has estimated that almost 10% of the global disease burden could be prevented through water, sanitation, and hygiene intervention [8].

The Ministry of Water and Environment (MWE) recommends routine monitoring of water sources in risky areas. However, due to budget constraints, water testing is only done once per quarter for approximately 20% of water sources in each area through a labor-intensive manual process [12]. Challenges associated with manual water quality monitoring include, but are not limited to, the following:

DOI: 10.1201/9781032664828-9

it's time consuming and cost-intensive; humans make errors, which decreases the data quality; and training the workforce to take and interpret on-site readings is expensive [13].

Technological advancement has made the detection of water pollutants easy. For instance, the smartphone sensor uses DNA–magnetic particle technology to detect bacteria by taking an electrochemical measurement in the sample [14]. Like in other sectors, for instance, healthcare [15], transport [16], education [17], agriculture [18], air quality monitoring [19], and supply chain management [20], technologies like Internet of Things (IoT) have been widely used to address issues of water quality [21]. Water quality monitoring is the assortment of water at specific time intervals and testing it to ascertain its present condition [6]. In water quality, physical, chemical, and microbial elements are tested to ensure they are within the permissible ranges. The influx of sensor networks has created a need for a fusion between water quality monitoring and sensor systems. Leveraging IoT systems enables real-time tracking and recording of water quality data which is currently impossible with the existing laboratory-based methods which are labor-intensive, time consuming, and expensive.

This research aimed to test the accuracy and efficacy of the IoT system for monitoring water quality with solutions of known pH and turbidity. The rest of the paper is organized as follows: related work in Section 7.2, methods and materials in Section 7.3, and results in Section 7.4. Finally, Section 7.5 discusses the main findings, concluding with the practical implications and limitations.

7.2 RELATED WORK

IoT has been widely adopted in different sectors as a way to improve and fasten the flow of information. IoT is a technology system where objects such as cars, phones, appliances, gadgets, and sensors connect to each other and share information, and perform tasks over a network connectivity [22]. Wireless technologies have improved large-field data acquisition and reliability in harsh environmental conditions [23]. By tapping into the robustness of IoT solutions, water quality monitoring can improve at the household, community, and national levels.

Lakshmikantha et al. [6] developed an IoT system to monitor pH, turbidity, conductivity, carbon dioxide, humidity, and temperature. The sensors were integrated with the microcontroller unit (Arduino ATMEGA328) that converts the analog values into digital, whose values were displayed on the Liquid Crital Display (LCD). The Wi-Fi module offered a connection between the hardware and software that was built on the ATMEGA328 board. All water quality parameters checked were uploaded to the cloud server and displayed on the LCD. AlMetwally, Hassan, and Mourad [24] developed a real-time IoT system that measured the physical properties of water for home consumption. The parameters monitored were temperature, pH, turbidity, water flow, and water level. For each parameter, a sensor was integrated into an Arduino UNO that stores and retrieves numerical data gathered by the sensors. Their architecture further had a solenoid water valve, used to change the water flow, and water filter cartridges, used to treat the water to remove dust, mud, chlorine, and other impurities. Abbas Fadel and Ibrahim Shujaa [25] created an IoT system that monitored total dissolved solids and pH in the Tigris river and the results did not deviate from the earlier ones that had been tested in the ministry lab.

Dhruv Bhardwaj and Mayank Gathole [26] also developed an IoT water quality system that detected four parameters, that is, pH, temperature, turbidity, and water level. The sensor technology was tested with ten solutions comprising salt, baking soda, sugar, tap water, bleach, lime, soap, eggs, RO-water, and Savlon, and the system was successful in testing the prepared solutions. Yasin, Yunus, and Wahab [27] developed an IoT solution to monitor pH and turbidity and the system effectively managed to detect the parameters it was designed to test. The sensor data were processed and sent to the cloud server for storage and analysis. Ajith Jerom and Manimegalai [23] developed a floating buoy system to monitor seawater temperature, humidity, carbon dioxide emission, pH, dissolved oxygen (DO), and soil moisture. The floating buoy was surrounded by air-filled balls with a wireless module placed on top of the buoy to continuously collect data. The buoy was self-balanced over the water surface to ensure that the sensors were immersed in water. The wireless nodes were programmed with a sleep awake mechanism to conserve power while increasing network lifetime.

Sung, Fadillah, and Hsiao [28] developed an IoT system to monitor turbidity, temperature, pH, conductivity, and total dissolved solids, which sensors were embedded on the Arduino board that processed and sent data to the Wi-Fi module to further transmit it to the ThingSpeak cloud service for display and further processing. Results indicated that the water was within the acceptable ranges of consumption. Roger et al. [29] developed a water quality system for monitoring water along Portugal's Cavado River. CupCarbon integrated with OSM (OpenStreetMap) API and Google Maps allowed the development of a prototype based on real locations. They used four standalone machine learning algorithms as well as 12 hybrid data mining algorithms, and CV parameter combinations to create Iran WQI predictions.

7.3 EXPERIMENTAL DESIGN, MATERIALS, AND METHODS

In August 2022, water samples were collected from the Wakiso administrative unit in Figure 7.1, a district found in central Uganda. This district has a population of close to two million comprising 17 subcounties and is approximately 17 km on road from Kampala, the capital of Uganda. Wakiso is the most populated district in Uganda, bordered by Luwero, Nakaseke, Mityana, Mpigi, and Kalangala. With a total household population of 314,978, 11% of the households use borehole water [30].

A total of eight water samples were collected in seven villages of Kabojja A, Kabojja B, Kinaawa, Kikajjo, Kasenge, Katalemwa, and Nakawuka. The choice of this district was based on the fact that it has a mix of urban, rural, and peri-urban population. Also, there are a lot of economic activities ranging from horticulture, fish farming to livestock farming, and it is densely populated. With the help of the community members, a snowballing technique was used to identify the water sources. Water samples were drawn from boreholes, shallow wells, protected springs, hand-dug wells, and a stream. A water point was selected based on its proximity to a possible pollution source, whether or not it was functioning, and the population it served.

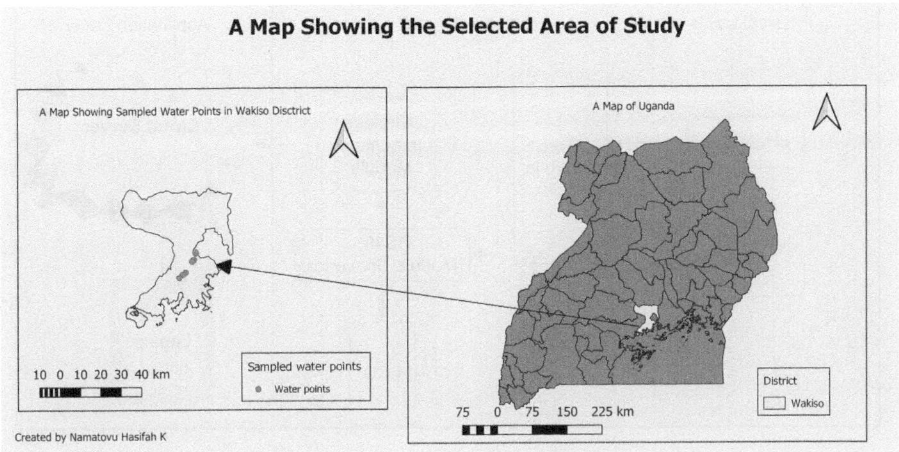

FIGURE 7.1 Map of the study area.

This study was conducted in five phases. Phase one consisted of designing and implementing an IoT system to monitor pH and turbidity in selected water points. The second phase consisted of abstracting water from the eight water points to perform a full profile testing at the National Water Quality Reference Laboratory (NWQRL) to ascertain the physico-chemical and microbial contaminants in the water, which later informed our intervention. The third phase comprised preparing nine solutions, whose pH and turbidity were tested in the College of Agriculture and Environment Sciences (CAES) lab using a water data sonde. In the fourth phase, the IoT system was used to test the nine solutions prepared in phase three in a bid to ascertain the efficacy of the system. This was to establish if the IoT results correlated with the water data sonde results, and if there are variations, establish the degree. The fifth phase saw the development team going back to the field to test the IoT system with water points selected in the second phase. For four consecutive months, phase five was repeated to test the level of precision of the IoT system, and the results are presented in Table 7.3.

7.3.1 The Adopted IoT Architecture

This high-level conceptual model demonstrated in Figure 7.2 describes the implementation of the IoT solution that was adopted to monitor water quality. Because of budgetary constraints, the team opted to monitor pH and turbidity with the possibility of future scaling to other parameters. The probes were immersed in a solution, information captured by sensors then converted into electrical signals. The microcontroller, in this case, an Arduino-integrated circuit board (R3 ATmega 328p-pu), was used to convert the analog signals to digital, with the UART TTL to RS converter, a bidirectional module that allowed the TTL interface of the microcontroller to be transferred to RS485 module. The signals were then wirelessly transmitted to the cloud (ThingSpeak) via a Global System for Mobile Communications (GSM) module. The board was powered via a USB connection from the laptop and Arduino IDE was used to write code to the microcontroller. The choice of using ThingSpeak

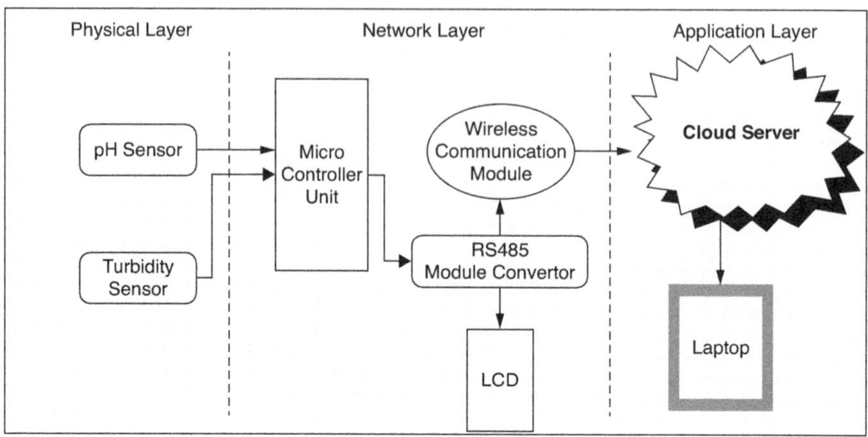

FIGURE 7.2   IoT architecture.

cloud services is the fact that it's open source and enables analytics such as aggregation and visualization to be performed on live data streams. The LCD displayed information coming from the sensors, while the laptop was used to display analyzed information from ThingSpeak.

The pH sensors were used to measure the level of alkalinity or acidity in the water while turbidity sensors were used to measure the relative clarity of water. pH ranges from 0 to 14 where 7 is considered neutral, 6.5 or less is acidic while anything above 7.5 is considered alkaline [31]. The maximum permissible standards for pH and turbidity as per the Uganda National Bureau of Standards is 5.5–9.5 pH units and 25 NTU, respectively [32]. pH is measured in pH units while turbidity is measured in nephelometric turbidity unit (NTU).

7.3.2   IMPLEMENTATION

As demonstrated in Figures 7.2 and 7.3, turbidity sensors (ZDYG-2088-01QX) and pH sensors (PH-485-PH) using Modbus RS485 communication protocol were integrated on the Arduino board (R3 ATmega 328p-pu), supported by the RS to TTL converter together with the GSM module (SIM 800C) and a laptop. After the Arduino board was programmed, sensors were calibrated to test their efficacy. Four solutions with different turbid levels and five solutions of different pH were prepared in the CAES Lab at Makerere University. The four solutions of varying turbidity were cow milk, tap water, muddy water, and Coca-Cola soda. Solutions with pH levels 2, 4, 7, 9, and 14 were prepared in Florence flasks at the lab.

Testing began with pH where probes were immersed into a solution of pH 7. Readings were taken on the laptop and for the first four immersions, pH values were not corresponding to the neutrality of water. After a few calibrations, the probes were able to read the correct pH values. This happened to all the solutions (pH and turbidity) that were being tested. Figure 7.4 shows some of the testing taking place and water abstraction during the field survey. The calibration results from the lab are demonstrated in Table 7.2.

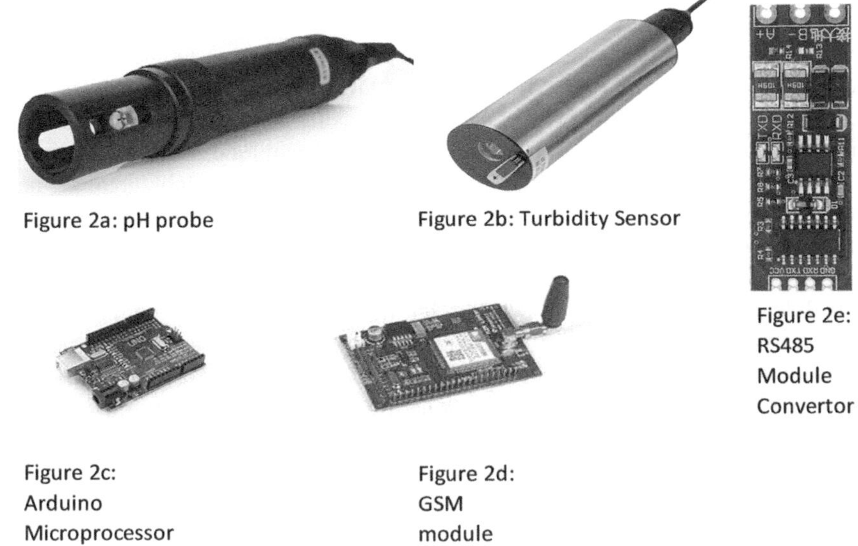

FIGURE 7.3 Equipment used in the IoT development.

FIGURE 7.4 Water abstraction and IoT system calibration.

7.3.3 WATER QUALITY ANALYSIS

In phase two of this study, water samples were collected and tested to determine the level of water pollution. Some parameters like DO, pH, temperature, electrical conductivity (EC), and turbidity were tested in situ using a water data sonde. Samples were collected in 0.5 L plastic bottles each labeled with a unique code, and transported to the NWQRL in a 45-L ice cooler box with ice packs. At the lab, metal elements, anions, and nutrients were tested. Anions (nitrites, nitrates, fluorides, chlorides, sulfates, and phosphates) and ammonium in water were determined using a Gallery plus-Thermos Fisher discreet analyzer. Sodium and potassium were measured using a BWB frame photometer while water hardness, calcium, and magnesium were determined using the standard EDTA titration method. Quality control was achieved using an independent standard measured in the same manner. Performing the water quality test in the lab helped us to understand the state of water contamination as well as ascertain the level of accuracy of the IoT technology that was developed since the results of the latter were compared with those of the former. The lab results coincided with the IoT results with very slight variations as demonstrated in Tables 7.1 and 7.3.

7.4   RESULTS

Results in Table 7.1 indicate that five out of eight sampled points had a pH lower than the permissible levels of 5.5–9.5 pH units. Nitrate was present in four out of the eight sample points, with Kasenge Primary School registering the highest level at 18 mg/L, 8 mg more than the recommended level. Total iron ranged between 0.12 and 3.67 mg/L, with an average of 1.89 mg/L. It should be noted that two out of the eight sampled water points had total iron higher than the permissible limit, registering at 1.68 and 3.67 mg/L at Kinaawa motorized hand-dug well and Nakawuka stream, respectively. All water points had a presence of manganese with the highest registered at 0.695 mg/L, a value higher than 0.1 mg/L which is the recommended value. E.coli was present in six out of the eight sampled water points, with concentrations ranging from 10CFU to 210CFU. Total coliforms were also considered high in almost all the water points apart from the Kinaawa motorized hand-dug well. The levels of dissolved oxygen, turbidity, electrical conductivity, total dissolved solids, total hardness as $CaCO_3$, calcium hardness as $CaCO_3$, magnesium hardness as $CaCO_3$, calcium, magnesium, sodium, potassium, total alkalinity, bicarbonates, fluoride, sulfates, and phosphate were below the detection limit at all sampling points.

During the calibration of the IoT system, it was noted that apart from tap water (1.3NTU) and the five pH solutions that had turbidity levels of less than 25NTU, which is the recommended standard, the rest (cow milk 30NTU, muddy water 46NTU, and Coca-Cola soda 28NTU) had turbidity way higher than the acceptable limit for potable drinking water. pH was considered to be high in cow milk, tap water, and muddy water, and the two pH solutions (pH 9 and pH 14) were alkaline, while the rest of the solutions were acidic. The readings were live feeds displayed on the laptop and generated every time the sensors were immersed in a liquid. Based on those findings, it can be inferred that the IoT system effectively identified pH levels and turbidity with precision.

TABLE 7.1

Analysis Results from the National Water Quality Reference Laboratory

		Nalongo spring	Kabojja B Spring	Kinawa Motorized hand-dug well	Kikajjo Spring	Kasenge Pri Sch Shallow well	Katalemwa Pri. School	Nakawuka Mpunge Shallow well	Nakawuka Stream	
Source Name										
Village		Kabojja A	Kabojja B	Kinawa	Kikajjo	Kasenge	Katalemwa	Katalemwa	Nakawuka	
Sub-County		Kyengera T.C	Kyengera T.C	Kyengera T.C	Kyengera T.C	Kyengera T.C	Nakawuka	Nakawuka	Nakawuka	Potable water standards
GPS		36N 0446673 32357	36N 0446407 33011	36N 0445662 29308	36N 0445662 29308	36N 0444870 28353	36N 0441632 22888	36N 0440300 21871	36N 0438489 20312	(DEAS12:2018 Maximum permissible for Natural potable Water)
Lab Identifier Code		E49103	E49104	E49105	E49106	E49107	E49108	E49109	E49110	
Field Temperature	°C	24	24	25.5	24	25	25	24	20.5	
Dissolved Oxygen	mg/L	2.85	1.1	6.2	2.8	3.01	3.7	7.5	1.5	
Turbidity	NTU	1.4	1.5	1.5	1.6	1	2.6	1.8	8.8	25
pH	Units	4.9	4.9	4.6	5.2	4.7	5.8	6.1	6.5	5.5–9.5
Electrical Conductivity	µS/cm	138	201	206	236	248	73	97	80	2,500
Total Dissolved Solids	mg/L	97	141	144	165	174	51	68	56	1,500
Total Hardness as CaCO$_3$	mg/L	31	32	46	61	45	26	36	30	600
Calcium Hardness as CaCO$_3$	mg/L	13	17	20	27	24	15	28	18	600

(Continued)

TABLE 7.1 (*Continued*)

Analysis Results from the National Water Quality Reference Laboratory

Magnesium Hardness as CaCO$_3$	mg/L	18	15	26	34	21	11	8	12	600
Calcium	mg/L	5	7	8	11	10	6	11	7	150
Magnesium	mg/L	4	4	6	8	5	3	2	3	100
Sodium	mg/L	11	20	17	14	26	1.7	4.3	3.6	200
Potassium	mg/L	3.4	5.3	4.1	5.7	6.6	1.8	1.4	1.6	50
Total Alkalinity	mg/L	9.5	7.5	3.6	13	3.5	19	36	35	–
Bicarbonates	mg/L	12	9	4	16	4	23	44	43	400
Fluoride	mg/L	0.11	0.05	0.28	0.14	0.14	0.31	0.12	0.28	1.5
Sulfates	mg/L	1.4	1	1.5	5.6	0.8	1.7	0.5	0.8	400
Chlorides	mg/L	13	21	22	28	30	2.8	0.9	3.5	250
Nitrates as N	mg/L	9.5	11	15	11.5	18	3	3.2	0.3	10
Nitrites as N	mg/L	<0.001	<0.001	0.01	<0.001	<0.001	0.01	0.001	0.001	0.9
Ammonium as N	mg/L	0.35	0.47	0.231	0.066	0.309	0.112	0.262	0.072	0.5
Phosphates as P	mg/L	0.051	0.3	0.021	0.4	0.008	0.02	0.018	0.13	0.7
Total Iron	mg/L	0.12	0.28	1.68	0.26	0.12	0.39	0.22	3.67	0.5
Manganese	mg/L	0.25	0.64	0.278	0.258	0.489	0.182	0.342	0.695	0.1
Copper	mg/L	<0.002	<0.002	<0.002	<0.002	<0.002	<0.002	<0.002	<0.002	1
Zinc	mg/L	<0.001	0.1524	<0.001	0.0009	<0.001	2.3155	0.1962	<0.001	5
Mercury	mg/L	<0.001	<0.001	<0.001	<0.001	<0.001	<0.001	<0.001	<0.001	0.001
E. coli	CFU/100 mL	10	78	<1	88	<1	11	34	210	<1
Total coliforms	CFU/100 mL	165	1,553	<1	1,046	165	165	165	>2,420	<1

TABLE 7.2

Test Results from the Calibration of the IoT System

Solution	Water Data Sonde Results		IoT Results	
	Turbidity (NTU)	pH (pH Units)	Turbidity (NTU)	pH (pH Units)
Cow milk	32	6.6	30	6.4
Tap water	1.1	7.2	1.3	7.0
Muddy water	41	9.0	46	9.0
Coca-Cola soda	29	2.5	28	2.3
pH 2 solution	1.0	2.0	1.0	2.2
pH 4 solution	1.3	3.9	1.1	3.8
pH 7 solution	1.1	7.0	1.1	6.9
pH 9 solution	1.4	9.0	1.3	9.0
pH 14 solution	1.2	13.9	1.3	13.9

Figure 7.5 demonstrates the calibration that took place in the field and the lab. Figure 7.5a and b shows the turbidity results from Katelemwa and Kinaawa water points obtained from the ThingSpeak cloud using the IoT system at time T2 (Jan) and T4 (Mar), respectively. Figure 7.5c shows the pH and turbidity validation results using the water data sonde. Validating the IoT system was imperative by first getting the actual values of pH and turbidity in situ. Figure 7.5d shows sensor pH readings displayed on the laptop for the pH solution 4 that was prepared in the lab.

The results shown in Table 7.3 have been clustered into four timeframes (T1, T2, T3, and T4) indicating when tests were conducted for each timeframe, which was exactly a month apart. In time T1 and T2, seven out of the eight sampled points had pH outside the recommended level of 5.5–9.5 pH units. In T3, all the samples were outside the recommended levels while T4 had two water samples that were within the permissible ranges. Figures show that the water is largely acidic and not permissible for consumption. The average temperature of T1, T2, T3, and T4 was 24.75°°C, 24.05°C, 24.01°C, and 24.07°C, respectively. The turbidity levels were all good, falling below 25NTU, which is the recommended standard for potable drinking water. The small variations between the IoT readings and the lab results could be attributed to the fact that IoT results were obtained from the field where temperatures oscillated between 20°C and 27°C, yet temperature is known to affect pH [33] and turbidity [34].

7.5 DISCUSSION

The results in Table 7.1 show the existence of nitrate, fluctuating pH levels, *Escherichia coli*, total coliforms, and manganese in certain water points. Notably, the Nakawuka stream exhibited elevated levels of all impurities, making this water unsuitable for both human and animal consumption. For instance, *E. coli* was 210 times more than the recommended consumption levels. Channah et al. [35] have noted that the presence of *E. coli* in water is an indication of recent human sewage

TABLE 7.3

Results from the IoT System during Field Visits

Source Name	Nalongo spring	Kabojja B Spring	Kinawa Motorized hand-dug well	Kikajjo Spring	Kasenge Pri Sch Shallow well	Katalemwa Pri. School	Nakawuka Mpunge Shallow well	Nakawuka Stream
Village	Kabojja A	Kabojja B	Kinawa	Kikajjo	Kasenge	Katalemwa	Katalemwa	Nakawuka
Sub-county	Kyengera T.C	Kyengera T.C	Kyengera T.C	Kyengera T.C	Kyengera T.C	Nakawuka	Nakawuka	Nakawuka
GPS	36N 0446673 32357	36N 0446407 33011	36N 0445662 29308	36N 0445662 29308	36N 0444870 28353	36N 0441632 22888	36N 0440300 21871	36N 0438489 20312
T1 (Dec.)								
Temp.	23	23.5	24	23	26	25.5	27	26
Turbidity (NTU)	1.5	1.5	1.5	1.6	1.1	2.4	1.8	9.0
pH (Units)	4.9	4.9	4.6	4.2	4.7	5.1	5.8	5.2
T2 (Jan.)								
Temp.	22	20.4	23.9	26.6	26	24.9	25.6	23
Turbidity (NTU)	1.4	1.3	1.5	1.7	1.0	1.9	1.4	10.0
pH (Units)	4.7	4.7	4.6	4.0	5.0	5.4	5.8	5.0
T3 (Feb.)								
Temp.	20.5	25	24.1	23.8	24	27	22.7	25
Turbidity (NTU)	1.6	1.7	1.4	1.1	1.4	1.4	1.3	9.3
pH (Units)	4.9	4.6	4.5	4.1	4.8	5.3	5.4	5.3
T4 (Mar.)								
Temp.	27	24.6	24.4	23	22	26	25.6	20
Turbidity (NTU)	1.2	1.3	1.0	1.4	1.2	2.4	1.7	8.9
pH (Units)	4.8	4.7	4.4	4.0	4.8	5.5	5.5	5.4

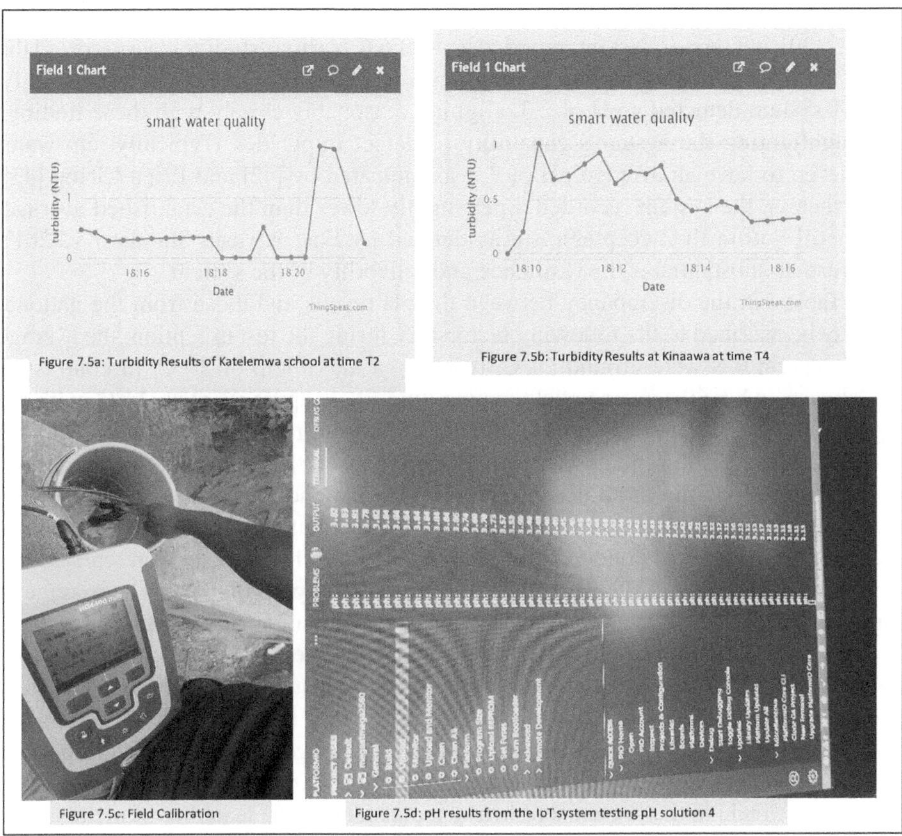

Figure 7.5a: Turbidity Results of Katelemwa school at time T2

Figure 7.5b: Turbidity Results at Kinaawa at time T4

Figure 7.5c: Field Calibration

Figure 7.5d: pH results from the IoT system testing pH solution 4

FIGURE 7.5 Field and lab calibration results. (a) Turbidity of Katelemwa school at time T2. (b) Turbidity result at Kinaawa at time T4. (c) Field calibration. (d) pH results from the IoT system testing pH solution 4.

or animal waste contamination and health hazards such as urinary tract infections in humans and animals are probable [36]. The existence of nitrate, manganese, and *E. coli* may be linked to the extensive khat plantations observed during the water abstraction process. The presence of such agricultural farms implies the use of pesticides and manure (animal and artificial) for fertilization, which are known to be present in groundwater from surface runoff [37,38]. Moreover, this area includes highly urbanized zones, and certain areas exhibited broken sewer lines, resulting in sewer overflows from manholes. These overflows, when combined with rainfall-runoff, flow into streams and infiltrate the ground. This could provide additional clarification for the detection of *E. coli* in the water. The deteriorated condition of certain boreholes may account for the occurrence of total iron in specific sampling points. In a study conducted by Usoro et al. [37], nitrate was found to be present in borehole water than in packaged water, implying a strong correlation between boreholes and nitrate presence.

The results in Table 7.2 suggest that the created IoT solution possesses the ability to assess water quality with minimal variations compared to traditional lab-based testing. As an illustration, cow milk is generally recognized to have a pH range of 6.7–6.9 [39], but the IoT system recorded a pH of 6.6. Various studies have reported the pH of Coca-Cola soda, with values ranging from 2.37 [40] to 2.47 [41]. Interestingly, the IoT system detected a pH of 2.3, aligning reasonably closely with these findings and highlighting the system's capability to detect impurities. Typically, tap water is believed to have an average pH of 7.5, as indicated by [42] and Erica Cirino [43]. Nevertheless, the system recorded a pH slightly lower than the established average, albeit still within the acceptable limits defined by East African Standard 12:2017. This further substantiates the resilience and reliability of the system.

In Table 7.3, the discrepancy between the pH results and those from the national lab may be ascribed to the following factors: (i) during the test execution, the average field temperatures were around 24°C, reaching a maximum of 27°C, in contrast to the lab tests, where the average field temperatures were approximately 24°C, with the highest at 25°C. Temperature is recognized to have an inverse relationship with pH [44], indicating that an increase in temperature leads to a reduction in pH, resulting in the water becoming more acidic due to ionization and the formation of H+ ions. Additionally, the disparity in pH results could be attributed to the time gap between the tests; lab tests were conducted in August 2022, while IoT tests were carried out from December 2022 to March 2023. This essentially implies that, with the passage of time, various factors such as chemicals, bedrock composition, and minerals may have become incorporated into the water, consequently influencing its pH. In the second phase, there was a minor difference in the turbidity results between the IoT system and the NWQRL lab. This underscores the precision of the IoT system and suggests its potential as a viable solution to water quality challenges in Uganda. It's important to highlight that a very low pH indicates water acidity, and the repercussions of such acidic water include the presence of heavy metals due to the leaching of metals from the surrounding environment [45]. Also, acidic water is likely to prevent the fish from hatching, and irritate fish and aquatic insect gills [46]. Subsequently, high pH means that water is alkaline and studies have shown that high levels of alkalinity pose a lot of threat to the aquatic biodiversity registering death in severe instances [46].

When comparing the laboratory results in Table 7.1 with the IoT outcomes in Table 7.3, a minute and inconsequential difference was observed. It's essential to emphasize that the results in the latter table corroborated those in the former dataset after undergoing a meticulous calibration, as illustrated in Table 7.2, which constituted the primary goal of this research. With the findings presented in Table 7.2, it can be affirmed that our IoT technology is efficient and precise in assessing both pH and turbidity.

7.6 CONCLUSION

This study aimed to calibrate and validate the authenticity and precision of IoT solutions in the monitoring of water quality. The initial step involved the creation of an IoT system specifically designed to assess pH and turbidity. Subsequently, water samples were gathered and analyzed at the NWQRL in Entebbe. In the CAES lab, nine solutions were meticulously prepared, undergoing pH and turbidity testing. These solutions were then employed to assess the performance of the IoT system before its deployment in the field. The goal was to ensure that the developed system effectively examines the targeted parameters. After a few calibrations and confirming the system's functionality as anticipated, samples were extracted from the eight water points and subjected to testing using the IoT system. The findings demonstrated a strong correlation among the IoT results, the nine solutions crafted in the CAES lab, and the results obtained from the NWQRL. Based on these experiments, we can confidently affirm that IoT serves as a superior and cost-effective solution for water quality monitoring.

7.6.1 PRACTICAL IMPLICATIONS

Policymakers ought to allocate resources to affordable water quality systems capable of mitigating the existing gap in water quality monitoring, which is currently compromising overall quality and exposing many to health risks. The existing gap caused by the inability of relevant authorities to conduct water quality checks can be filled by embracing real-time sensing technologies. As IoT systems are rich in information, analyzing the data they generate can offer valuable insights into water quality trends and patterns. This information can then be utilized by policymakers to make well-informed decisions regarding land use management, pollution control, and settlement management. Policymakers have the opportunity to utilize real-time data from IoT systems for monitoring and inspecting facilities to ensure compliance with water quality standards and environmental pollution regulations. Providing real-time updates on water quality status informs consumers about the safety of the water and empowers them to take appropriate actions.

7.6.2 LIMITATION

Due to budgetary constraints, the monitoring parameters were restricted. Nitrate and iron, despite their broader impact compared to pH and turbidity, couldn't be included in the system due to the high cost of the required probes.

7.6.3 FUTURE WORK

Integrating a mobile phone for widespread access to information and implementing an alert system for exceeding permissible thresholds in drinking water parameters are essential. Scaling the system is necessary as NWQRL water quality checks revealed the presence of iron, nitrate, manganese, and *E. coli* in numerous water points.

ACKNOWLEDGMENTS

I extend sincere appreciation to the NWQRL management, particularly the chemists who aided in water sample collection and maintained quality from the field to the lab. Additionally, gratitude is expressed to the management of the College of Computing and Information Sciences for the RISE seed grant, which provided partial funding for this initiative.

REFERENCES

1. T. Oki and S. Kanae, "Global hydrological cycles and world water resources," *Science (1979)*, vol. 313, no. 5790, pp. 1068–1072, Aug. 2006, doi: 10.1126/science.1128845

2. F. N. W. Nsubuga, E. N. Namutebi, and M. Nsubuga-Ssenfuma, "Water resources of Uganda: An assessment and review," *Journal of Water Resource and Protection*, vol. 06, no. 14, pp. 1297–1315, 2014, doi: 10.4236/jwarp.2014.614120

3. S. Hamilton, A. T. Munyaho, N. Krach, and S. Glaser, "Bathymetry TIFF, Lake Victoria Bathymetry, raster," *Harvard Dataverse*, vol. 07, pp. 10, 2017, doi: 10.7910/DVN/SOEKNR

4. World Health Organization and UNICEF, "Progress on Drinking Water, Sanitation and Hygiene," WHO, Geneva, Jul. 2017. Accessed: Sep. 16, 2023, Online. Available: https://data.unicef.org/wp-content/uploads/2017/07/JMP-2017-report-launch-version_0.pdf

5. R. Haruna, F. Ejobi, and E. K. Kabagambe, "The quality of water from protected springs in Katwe and Kisenyi parishes, Kampala city, Uganda.," *African Health Sciences*, vol. 5, no. 1, pp. 14–20, 2005.

6. V. Lakshmikantha, A. Hiriyannagowda, A. Manjunath, A. Patted, J. Basavaiah, and A. A. Anthony, "IoT based smart water quality monitoring system," *Global Transitions Proceedings*, vol. 2, no. 2, pp. 181–186, Nov. 2021, doi: 10.1016/j.gltp.2021.08.062

7. United Nations, "The Sustainable Development Goals Report 2022," New York, 2022. Accessed: Sep. 16, 2023, Online. Available: https://reliefweb.int/report/world/sustainable-development-goals-report-2022?gclid=CjwKCAjwpJWoBhA8EiwAHZFzfti3sire7qJLntElqAYLEFLSq9Fouozre_bGLzH611SafwTB8ArG2RoChskQAvD_BwE

8. B. Peter, K. Celine, and G. Valerie, *"Meeting the Challenge of Financing Water and Sanitation,"* OECD, Paris, Oct. 2011.

9. UNICEF, "The Health Sector Is Critical in the Attainment of Uganda's Vision 2040 by Producing a Healthy and Productive Contributes to Socio Economic Growth," vol. 3, pp. 25–28, 2019.

10. Uganda Bureau of Statistics, "2022 Statistical Abstract," Kampala, 2022. Accessed: Sep. 16, 2023. Online. Available: https://www.ubos.org/2022-statistical-abstract/

11. D. Eurien et al., "Cholera outbreak caused by drinking unprotected well water contaminated with faeces from an open storm water drainage: Kampala City, Uganda, January 2019," *BMC Infectious Diseases*, vol. 21, no. 1, p. 1281, Dec. 2021, doi: 10.1186/s12879-021-07011-9

12. Ministry of Water and Environment, "Water and Environment Sector," Kampala, Mar. 2020.

13. BizIntellia, "A Concise Guide for IoT based Water Quality Monitoring." Accessed: Apr. 27, 2022. Online. Available: https://www.biz4intellia.com/blog/undeniable-benefits-of-water-quality-monitoring-solutions/

14. V. Hoang-Nam, "New Technologies Hope to Make Detection of Biological Water Contamination Easier," Save the Water. Accessed: Aug. 24, 2023. Online. Available: https://savethewater.org/water-contamination-detection/

15. A. Rejeb et al., "The Internet of Things (IoT) in healthcare: Taking stock and moving forward," *Internet of Things*, vol. 22, p. 100721, Jul. 2023, doi: 10.1016/j.iot.2023.100721

16. D. Ushakov, E. Dudukalov, E. Kozlova, and K. Shatila, "The Internet of Things impact on smart public transportation," *Transportation Research Procedia*, vol. 63, pp. 2392–2400, 2022, doi: 10.1016/j.trpro.2022.06.275

17. V. Terzieva, S. Ilchev, and K. Todorova, "The role of internet of things in smart education," *IFAC-PapersOnLine*, vol. 55, no. 11, pp. 108–113, 2022, doi: 10.1016/j.ifacol.2022.08.057

18. J. Xu, B. Gu, and G. Tian, "Review of agricultural IoT technology," *Artificial Intelligence in Agriculture*, vol. 6, pp. 10–22, 2022, doi: 10.1016/j.aiia.2022.01.001

19. S. M. S. D. Malleswari and T. K. Mohana, "Air pollution monitoring system using IoT devices: Review," *Materials Today: Proceedings*, vol. 51, pp. 1147–1150, 2022, doi: 10.1016/j.matpr.2021.07.114

20. A. Rejeb, S. Simske, K. Rejeb, H. Treiblmaier, and S. Zailani, "Internet of Things research in supply chain management and logistics: A bibliometric analysis," *Internet of Things*, vol. 12, p. 100318, Dec. 2020, doi: 10.1016/j.iot.2020.100318

21. S. Geetha and S. Gouthami, "Internet of things enabled real time water quality monitoring system," *Smart Water*, vol. 2, no. 1, p. 1, Dec. 2016, doi: 10.1186/s40713-017-0005-y

22. P. Raja, S. Kumar, D. S. Yadav, and T. Singh, "The Internet of Things (IOT): A review of concepts, technologies, and applications," *International Journal of Information technology and Computer Engineering*, vol. 32, pp. 21–32, Mar. 2023, doi: 10.55529/ijitc.32.21.32

23. B. Ajith Jerom and R. Manimegalai, "An IoT Based Smart Water Quality Monitoring System using Cloud," In *International Conference on Emerging Trends in Information Technology and Engineering, IC-ETITE 2020,* Institute of Electrical and Electronics Engineers Inc., Feb. 2020. doi: 10.1109/ic-ETITE47903.2020.450

24. S. A. H. AlMetwally, M. K. Hassan, and M. H. Mourad, "Real time Internet of Things (IoT) based water quality management system," *Procedia CIRP*, vol. 91, pp. 478–485, 2020, doi: 10.1016/j.procir.2020.03.107

25. A. Abbas Fadel and M. Ibrahim Shujaa, "Water quality monitoring system based on IOT platform," *IOP Conf Ser Mater Sci Eng*, vol. 928, no. 3, p. 032054, Nov. 2020, doi: 10.1088/1757-899X/928/3/032054

26. D. Bhardwaj and M. Gathole, "Water quality detection system using IOT," *SRM Institute of Science and Technology*, vol. 2, no. 4, p. 168, 2021.

27. S. N. T. M. Yasin, M. F. M. Yunus, and N. B. A. Wahab, "The development of water quality monitoring system using internet of things," *Journal of Educational and Learning Studies*, vol. 3, no. 1, p. 14, Mar. 2020, doi: 10.32698/0852.

28. W. T. Sung, F. N. Fadillah, and S. J. Hsiao, "IoT-based water quality monitoring," *Sensors and Materials*, vol. 33, no. 8, pp. 2971–2983, 2021, doi: 10.18494/SAM.2021.3342.

29. M. A. P. Roger, R. Ap, R. Vijay, R. Surya, and V. Sowmethran, "Review of water quality monitoring using internet of things (IoT)," *International Journal of Research and Analytical Reviews*, vol. 4, no. 1, pp. 2971, 2022, Online. Available: www.ijrar.org

30. UBOS, "Statistical Abstract for Kampala City," pp. 2–25, 2019, Online. Available: https://www.kcca.go.ug/media/docs/Statistical-Abstract-2019.pdf

31. K. McGrane, "Acidic Water: Risks, Benefits, and More." 2020.

32. Uganda National Bureau of Standards, *"Draft Uganda Standard - Potable Water Specifications,"* Uganda National Bureau of Standards, Kampala, 2017.

33. C. Gillespie, "The Effects of Temperature on the pH of Water." 2019.

34. M. Shi, J. Ma, and K. Zhang, "The impact of water temperature on in-line turbidity detection," *Water (Basel)*, vol. 14, no. 22, p. 3720, Nov. 2022, doi: 10.3390/w14223720

35. R. Channah, M. Jean, R. Paula, B. Natalie, and D. Jessica, *"E. coli, Water Quality, Food Safety, and Human Health,"* The University of Arizona, Tucson, Arizona, Apr. 2018.

36. A. M. Hammerum and O. E. Heuer, "Human health hazards from antimicrobial-resistant *Escherichia coli* of Animal Origin," *Clinical Infectious Diseases*, vol. 48, no. 7, pp. 916–921, Apr. 2009, doi: 10.1086/597292

37. C. Usoro, L. Ibiang, I. Usoro, A. Nsonwu, and M. Etukudo, "Nitrates and nitrites content of water boreholes and packaged water in Calabar metropolis," *Mary Slessor Journal of Medicine*, vol. 5, no. 1, pp. 11036, Aug. 2005, doi: 10.4314/msjm.v5i1.11036

38. World Health Organization, *"Nitrate and Nitrite in Drinking-Water,"* WHO, Geneva, 2003.

39. I. Noreen, "What's the pH of Milk, and Does It Matter for Your Body?" Accessed: Aug. 26, 2023, Online. Available: https://www.healthline.com/health/ph-of-milk#bottom-line

40. A. Reddy, D. F. Norris, S. S. Momeni, B. Waldo, and J. D. Ruby, "The pH of beverages in the United States," *The Journal of the American Dental Association*, vol. 147, no. 4, pp. 255–263, Apr. 2016, doi: 10.1016/j.adaj.2015.10.019

41. D. H. J. Jager, A. M. Vieira, J. L. Ruben, and M. C. D. N. J. M. Huysmans, "Estimated erosive potential depends on exposure time," *Journal of Dentistry*, vol. 40, no. 12, pp. 1103–1108, Dec. 2012, doi: 10.1016/j.jdent.2012.09.004

42. K. Kulthanan, P. Nuchkull, and S. Varothai, "The pH of water from various sources: an overview for recommendation for patients with atopic dermatitis," *Asia Pacific Allergy*, vol. 3, no. 3, pp. 155–60, Jul. 2013, doi: 10.5415/apallergy.2013.3.3.155

43. Erica Cirino, "What pH Should My Drinking Water Be?" Accessed: Aug. 26, 2023, Online. Available: https://www.healthline.com/health/ph-of-drinking-water

44. Savitri, "Does Temperature Affect pH?" Accessed: Aug. 25, 2023, Online. Available: https://techiescientist.com/does-temperature-affect-ph/

45. Water Science School, (2019) "pH and Water." Accessed: Aug. 26, 2023, Online. Available: https://www.usgs.gov/special-topics/water-science-school/science/ph-and-water?qt-science_center_objects=0#qt-science_center_objects

46. M. Nancy and G. John, "What is pH?" Utah, Dec. 2010.

8 Prototype Design for Smart Birdnest and Related Android App

Ayush Kurapati, Ganesh Bhutkar, Shreyas Kapse, Tanishk Babar, and Chittaranjan Mahajan

8.1 INTRODUCTION

Nature consists of a wide range of living as well as non-living aspects, each of them having their own significance. In the living world, nature manages to encompass living things right from the tiniest of microbes up to huge trees, covering miles of landscape. Among these living things, birds have fascinated humans for ages due to their diverse ranges and appealing habitats. Birds have always been an integral part of the ecosystem for aeons. The beauty of birdnest is found right in the centre of nature's one of the greatest treasures. Nest-building is known to ornithologists and wildlife enthusiasts as a key aspect of the lives of birds. A birdnest plays a major role in the life of birds, as it provides shelter, comfort and a place to take care of their young ones. Birds build these birdnests in various natural as well as human-related habitats such as forests, wetlands, grasslands, farmland, building areas, parks and others [1].

Nevertheless, the typical habitats of these birds have been negatively affected due to changes in various environmental conditions as a result of human activities and natural shifts. In these habitats, rampant cutting of trees has led to deforestation at a large scale. Due to deforestation, birds are facing problems in finding suitable trees for nesting. Apart from this, modern industrialisation has deteriorated the quality of natural resources, required for building nests [2]. Thus, the number of birds in different habitats has been greatly affected in today's times.

Unnerving numbers from the report—State of the World's Birds 2022—shows that about 49% of bird species worldwide are experiencing population declines. The situation is far more serious in India, where between 2000 and 2018, there was a 62% reduction in forest-dwelling birds, a 59% decline in birds inhabiting grassland shrubs and a 47% decline in wetland-dwelling birds [3]. In the state of Maharashtra, Mumbai city saw a 44% downfall in the number of migratory birds. This drop in the number of birds was brought on by the establishment of multiple factories there [4]. Apart from this, in Navegaon National Park, various bird species such as Black-tailed Godwit (Limosa limosa) and Sarus Crane have declined by 25% from 2000 to 2003 [5].

Considering the numbers about birds and their habitats, it appears that there is a necessity to preserve and raise the population of birds in different habitats. To do so, there is a need to create awareness among people regarding bird conservation.

DOI: 10.1201/9781032664828-10

FIGURE 8.1	Nestbox for birds in the home balcony.

It is harder for people to improve the nesting conditions of birds in nature-related habitats. So, they can be encouraged to actively create nesting spaces for bird conservation around their homes and workplaces. These nest spaces can be left empty for birds to build their nests, or a birdnest in the form of a nestbox can be provided to the birds [6]. Figure 8.1 depicts an arrangement of a nestbox for birds; placed in the home balcony. The nestbox is used by local birds for living and for laying their eggs.

In today's technologically advanced era, the idea of birdnest can be further enhanced by incorporating Internet of Things (IoT) technology into it. These functionalities will make the birdnest more comfortable and efficient to use. Various forms of birdnests have been designed using Information and Communication Technology (ICT) in recent years. This paper discusses in detail about the literature survey related to the hardware aspects of different smart birdnest models as well as the software aspects of applications related to birds. It also presents a review of different Android apps, which assist the users in handling these smart birdnests. The paper also describes the IoT components and design of the proposed smart birdnest along with its vital features. Finally, the paper presents the prototype design of an Android app with an HCI approach and various functionalities.

## 8.2	LITERATURE SURVEY

Researchers have designed various systems related to smart birdnest over the years along with their associated software applications. This section describes the comparative aspects of such systems related to birdnests in two subsections on birdnest hardware design and related software applications.

### 8.2.1	BIRDNEST HARDWARE DESIGN

The first paper presented the idea of creating an artificial nestbox. Birds within this nestbox can be monitored as they arrive and leave using an infrared (IR) sensor. Apart from this, temperature and humidity inside the nest can also be monitored. Solar energy has been used to provide power supply to this system. Image/Video footage capturing is provided using a camera module—ESP32-CAM [7]. The system has been tested by deploying it in a natural environment. It was found to have 86% accuracy in detecting and taking the images of the birds.

In the second paper, Radio-Frequency IDentification (RFID) technology has been incorporated into the smart birdnest for the detection of specific birds as they arrive or leave. Temperature monitoring and video capturing using monochromatic industrial cameras are implemented in this birdnest [8]. The birdnest is powered using a 60 Ah 12 V nickel–cadmium battery. The entire system is tested with an accuracy of 74% in bird entry detection.

The third paper describes in great detail about the RFID technology for detecting specific birds and solar energy to power the birdnest. Sensor monitoring and video capturing features are absent in this birdnest presented in this paper [9]. The paper explains about the pilot and field-testing methods utilised and the accuracy of 95% in bird entry detection.

Another paper presents the idea of birdnest, which monitors environmental parameters of temperature and humidity within the birdnest. Bird entry detection is done using an IR sensor. For image/video capturing, they have used different sets of lenses and a camera module - Arducam IMX477 [10]. The paper even talks about testing the designed model of birdnest. Solar energy is used for powering the system.

The next paper exhibits a bird cage with several IoT-related functionalities meant for the proper growth and care of turtle doves. It makes use of solar energy and maintains the proper temperature required for turtle doves [11]. It lacks entry detection of birds and video capturing, but the overall system is well-tested.

The last paper delves around a dummy model of a birdnest, which detects the entry of specific birds using RFID and is powered by solar energy. The image/video capturing feature is absent, but they have incorporated a Proximity IR (PIR) sensor for monitoring the movements of birds within the nest [12].

Table 8.1 depicts the distinguishable and comparative aspects among the different research papers related to IoT-based birdnests and their hardware design. The major observations based on data in Table 8.1 are elaborated as follows:

- Most research papers (five out of six papers) have implemented some way of **monitoring environmental parameters** within the birdnest. Three papers have measured only **temperature**, while two papers have also measured **humidity**.
- Most research papers (five out of six papers) have made use of **IoT-based technology for bird entry detection**. Out of these papers, three papers have made use of **RFID technology**, while two papers have made use of **IR Sensors** for bird entry detection.
- Most research papers (five out of six papers) have incorporated **solar energy** to supply required power within the system.
- Most research papers (five out of six papers) have described the **testing** of their birdnest model in different ways. The testing methods generally used have two phases. The first phase is **pilot testing** and the second phase is **field testing**.

TABLE 8.1

Comparative Aspects in Literature Review Related to Birdnest Hardware Design

Aspects	Abdelouahid et al.	Arybnicka et al.	Del-Rio-Ruiz et al.	Bianchi et al.	Oktafian Sultan Hakim et al.	Garcia-Rodriguez et al.
Bird entry detection technology	Yes (IR sensor)	Yes (RFID)	Yes (RFID)	Yes (IR sensor)	No	Yes (RFID)
Environmental parameter monitoring	Yes (temperature and humidity)	Yes (temperature)	No	Yes (temperature and humidity)	Yes (temperature)	Yes (temperature)
Image/video footage capturing	Yes (ESP32-CAM)	Yes (monochromatic industrial cameras)	No	Yes (Arducam IMX477)	No	No
Tracking of specific birds	No	Yes	Yes	No	No	Yes
Solar power usage	Yes (5W solar panel)	No	Yes	Yes (9W solar panel)	Yes	Yes
System testing	Yes (86% accuracy in bird entry detection)	Yes (74% accuracy in bird entry detection)	Yes (95% accuracy in bird entry detection)	Yes	Yes	No

- Some research papers (three out of six papers) even mention their respective accuracies for bird entry detection with respect to a specific technology. The **range of these accuracies is from 74% to 95%**.
- Some research papers (three out of six papers) have included **image/video capturing** in their design and made use of modules such as **ESP32-CAM, monochromatic industrial cameras** and **Arducam IMX477**.
- Some research papers (three out of six papers) have implemented **tracking of specific birds using RFID technology** during their entry into the birdnests.

8.2.2 SOFTWARE APPLICATIONS FOR BIRDS

In the first paper, a smartphone-based bird monitoring system is proposed, integrating customised bird feeders to attract birds to observe and collect data. Using video analysis software, the system uses motion detection to trigger autonomous video recording [13]. An advanced object detection algorithm is used for capturing bird movement. These recorded videos can be easily stored in the cloud or with local storage, ensuring efficient data management.

The second paper discusses swiftlet farming by providing a swiftlet birdhouse. The camera placed near the swiftlet house captures the picture semi-autonomously. The system also monitors the environmental parameters such as inner temperature and humidity. Related data is stored in the cloud [14]. The intrusion algorithm is based on artificial intelligence (AI) and is used to keep track of the entry and exit of birds.

The third paper mainly focuses on the swallow nest farming system. The proposed system is used for monitoring the temperature and humidity of nests. The data from temperature and humidity sensors is stored in the Firebase, which acts as a centralized server for this system [15]. The related Android app is developed in Flutter and is used for displaying data from the Firebase.

The fourth paper discusses the classification of the bird by making use of machine learning (ML) and AI aiming to simplify bird identification for new bird enthusiasts. Bird identification is accomplished through Flutter-based mobile apps. A trained AI engine classifies birds using internet-sourced images and a convolutional neural network (CNN) [16]. Bird identification is done through Flutter-based mobile applications. A trained AI engine classifies birds using images from the internet. The testing of this AI engine is done using five random images of flying birds. The AI engine was successful in classifying these images, which resulted in an average accuracy of 79%.

The next paper proposes a deep-learning approach for the individual identification of birds. The study utilises semi-autonomous video monitoring with an interval of 2 s. Video analysis is performed on the video files to extract images. As an ML approach, the CNN classification algorithm is then employed to classify the captured images of birds [17].

The last paper focussed primarily on the identification of Indian bird species using ML and deep-learning approaches. The study employed the CNN algorithm for the classification of tasks [18]. Bird identification was conducted using input images and video files. The model is tested on various parameters like MobileNetV1, NASNetMobile and others.

Table 8.2 shows a distinguishable and comparative analysis of research papers **related** to software applications used along with birds. The major observations based on data in Table 8.2 are as follows:

- Most of the research papers (five out of six papers) have utilised algorithms for a variety of purposes. Among these research papers, three papers have used the **CNN algorithm** for species recognition and video analysis, while the other two papers have used **Advanced Object Detection and/or Intrusion Algorithm** for bird entry detection.
- Most of the research papers (five out of six papers) perform **system testing** to check the robustness of their system. Two of the research papers (two out of six papers) have briefly mentioned **testing their mobile apps**, ensuring the functionality and reliability of their systems.
- The utilisation of AI and ML techniques is evident in most research papers (four out of six papers) for facilitating tasks like **bird/species identification and classification.**
- Most research papers (four out of six papers) have presented data **storage** functionality. They have used **cloud storage** (Firebase or Dropbox) and there is a case utilising **local storage** as well.
- Some research papers (three out of six papers) have used **image and video capturing** in either **autonomous or semi-autonomous ways**.
- Video analysis is a common feature of some research papers (three out of six papers) for several things such as **bird entry detection, image extraction and bird/species identification**.
- Few research papers (two out of six papers) have discussed **environmental monitoring facilities** in mobile apps to measure the **temperature and humidity of the birdnest**.

8.3 ANDROID APP REVIEW

Several app developers have designed and deployed various Android apps associated with birds and bird nests on Google Play Store. This section describes the comparative aspects of such Android apps related to birds and birdnests.

The first app—Nestwatch—provides a walkthrough to the new users as guidance. The app even provides information regarding different types of birds and birdnests. Apart from this, the app provides facility to upload pictures and sync user data over a number of devices. The app even provides multi-language support in 16 languages. The app even provides the locations of various birdnests in the vicinity of the user. It has a user rating of 2.6 on the Play Store. Help and support are provided to the user in the form of FAQs (Frequently Asked Questions) on a website. The website even provides the facility of submitting a request to resolve an issue by providing an email.

The second app—Bird Buddy—provides the location of different birdnests and allows users to enter information regarding the same. The app enables users to upload pictures as well as videos of birds and birdnests. The app provides direction of use to first-time users and has community support. The app provides language support in up to eight languages. It even provides data synchronisation features to the users.

TABLE 8.2

Comparative Aspects in Literature Review Related to Software Applications Used Along with Birds

Aspects	Steen et al.	Ibrahim et al.	Noverta et al.	Tian et al.	Ferreira et al.	Chaudhary et al.
Environmental parameter monitoring	No	Yes (temperature, humidity)	Yes (temperature, humidity)	No	No	No
Image/video footage capturing	Yes (autonomous)	Yes (semi-autonomous)	No	No	Yes (semi-autonomous)	No
Algorithm used	Advanced Object Detection (tracking of the entry and exit of birds)	Intrusion algorithm (tracking of the entry and exit of birds)	NA	CNN (classification of images)	CNN (classification of images)	CNN (classification of images)
Use of AI/ML	No	Yes (identification and classification)	No	Yes (identification and classification)	Yes (identification and classification)	Yes (identification and classification)
App development technology	NA	NA	Flutter	Flutter	No	NA
Data storage technology	Local and cloud (DropBox)	Cloud	Cloud (firebase)	Cloud (firebase)	NA	NA
System testing	Yes	NA	Yes	Yes (79% accuracy for image identification)	Yes	Yes

It has a user rating of three on the Play Store. The app has a separate section for help and support which redirects the user to a website. This website provides access to different FAQs and request submission textbox for the user.

The third app—eBird—provides information about different birds located in specific bird habitats. The app provides the facility of data synchronisation to the user. Help and support are provided in the form of FAQs on a website The user rating of this app is 4.5 on the Play Store.

The fourth app—GoBird—provides information on different species of birds to the user. The app even provides the location of different bird habitats for bird enthusiasts. The app supports up to 23 languages and provides data synchronisation to the users. It provides help and support to the users by means of email (tony@thenerd-birder.com) and has a user rating of 4.7 on the Play Store.

Another app—BirdNet—provides information about the location of different bird habitats to the users. It even provides an app walkthrough to the app users to assist them. It supports up to 25 languages and has a user rating of 4.5. It even provides the facility of data synchronisation. Help and support is provided to the user through the email (ccb-birdnet@cornell.edu).

The last app—BirdTrack—provides a walkthrough of the app's interface to guide the user in understanding the different functionalities of the app. It even provides the location of different bird habitats to the users. The app provides data synchronisation to users on multiple devices. In case of any query, the user can contact their email (birdtrack@bto.org). The app is relatively new on Play Store so there is no user rating for the app.

Table 8.3 shows the distinguishable and comparative analysis of Android apps associated with birds and birdnest. The major observations based on data in Table 8.3 are as follows:

- All apps (six out of six apps) have used **location-based services** to provide precise location of bird habitat. Four of the apps (four out of six apps) provide the location of different **bird habitats** for bird sightings, and the remaining two apps (two out of six apps) provide the **location of birdnest**.
- All apps (six out of six apps) have used **data synchronisation** to restore the user data on different devices by making a user account.
- Most of the apps (five out of six apps) have provided help and support. Three of the apps (three out of six apps) have provided only **email support** to resolve user queries. Two of the apps (two out of six apps) have provided **websites to present FAQs and submit the query request**.
- Most of the apps (four out of six apps) have provided directions for Android app usage.
- Most of the apps (four out of six apps) have provided **multi-language support** with at least 8 foreign languages.
- Few apps (two out of six apps) have provided **information regarding birds** and/or birdnest. One app provides information only regarding bird species; while the second app provides **information regarding birdnest** as well.
- Few apps (two out of six apps) have provided support for **uploading image and/or video** files. One paper supports uploading only pictures while another allows **uploading** of both pictures and videos.

TABLE 8.3

Comparative Aspects Related to Software Applications Used Along with Birds

Aspects	Nest Watch	Bird Buddy	eBird	GoBird	BirdNet	Bird Track
Directions for app usage	Yes (walk through)	Yes (walk through)	No	No	Yes (walk through)	Yes (walk through)
Information regarding bird/birdnest	Yes (bird & birdnest)	No	No	Yes (bird)	No	No
Location of bird habitats/ birdnests	Yes (birdnest)	Yes (birdnest)	Yes (bird habitats)	Yes (bird habitats)	Yes (bird habitats)	Yes (bird habitats)
Image/video uploading	Yes (image)	Yes (image & video)	No	No	No	No
Help and support	Yes (website for FAQs and query requests)	Yes (website for FAQs and query requests)	Yes (website for FAQs)	Yes (email support)	Yes (email support)	Yes (email support)
Birdnest peer review	Yes	No	No	No	No	No
Multi-language support	Yes (16 languages)	Yes (8 language)	No	Yes (23 language)	Yes (25 language)	No
Data synchronisation	Yes	Yes	Yes	Yes	Yes	Yes
Community interaction	No	Yes	No	No	No	No
User rating (out of five)	2.6	3.0	4.5	4.7	4.5	NA

- Only one app (one out of six apps) provides **peer review** from users.
- The average **user rating** for the reviewed apps is 3.86.

8.4 IOT-BASED DESIGN OF SMART BIRDNEST

Smart birdnest is an IoT-integrated birdnest, which enables users to constantly monitor the birds residing within the nest. The smart birdnest manages to monitor the environmental condition inside the nest while keeping track of the entry and exit of birds from the birdnest. It even provides live footage of activities inside the nest periodically. The users of these birdnests can undertake measures to improve the quality of life of birds within the nest by providing food or any other assistance required to the birds. This section elaborates on the design of smart birdnest, along with its important features and working of the system.

8.4.1 IoT Components Used in Smart Birdnest Design

The proposed smart birdnest design makes use of different IoT components to provide various features to the birdnest users. These IoT components help in collecting, transmitting and receiving different forms of bird/birdnest-related data. The major IoT components shown in Figure 8.2 are as follows:

Humidity and Pressure Sensor—BME 280: It is a high-accuracy sensor, which determines the temperature, humidity and air pressure inside the birdnest. The sensor is capable of measuring the temperature from $-40°C$ up to $+85°C$. It measures the humidity in the atmosphere relatively in terms of percentage; from 0 to 100%. It can measure air pressure in the range of 300–1,100 hPa.

ESP32—CAM: This is a **camera module** which is based on the ESP32 chip. This camera module is programmed using a Future Technology Devices International (FTDI) programmer. It has a resolution of up to $1,600 \times 1,200$ pixels, which can be scaled down as per user requirements for monitoring bird activities. This camera module has a 4 GB SD card which can be used to store images and videos of birds and birdnest locally. It can be assigned a fixed IP address for sending the captured data over the internet.

LM393 Chip IR Sensor: This sensor is an **Obstacle Avoidance IR Sensor Module**, which detects objects within a specified distance, and it is used in birdnest to detect a bird entry. The range of this IR sensor is from 2 to 30 cm. A microcontroller can be used to program this sensor for detecting objects at a specified distance within 30 cm. The IR sensor consists of an emitter and a receiver which emits and receives the infrared waves, respectively. The emitted infrared waves collide with the bird in its path and get reflected. These reflected waves are then detected by the receiver. The time taken for this collision and reflection is used to determine the distance of the bird in front of the IR sensor.

Neo-6M GPS: It is a **Global Positioning System (GPS) Module,** which makes use of satellites to determine its location. It connects to a nearby satellite and constantly collects the latitude and longitude values of the place where it is located. These values are used to get the location of the birdnest.

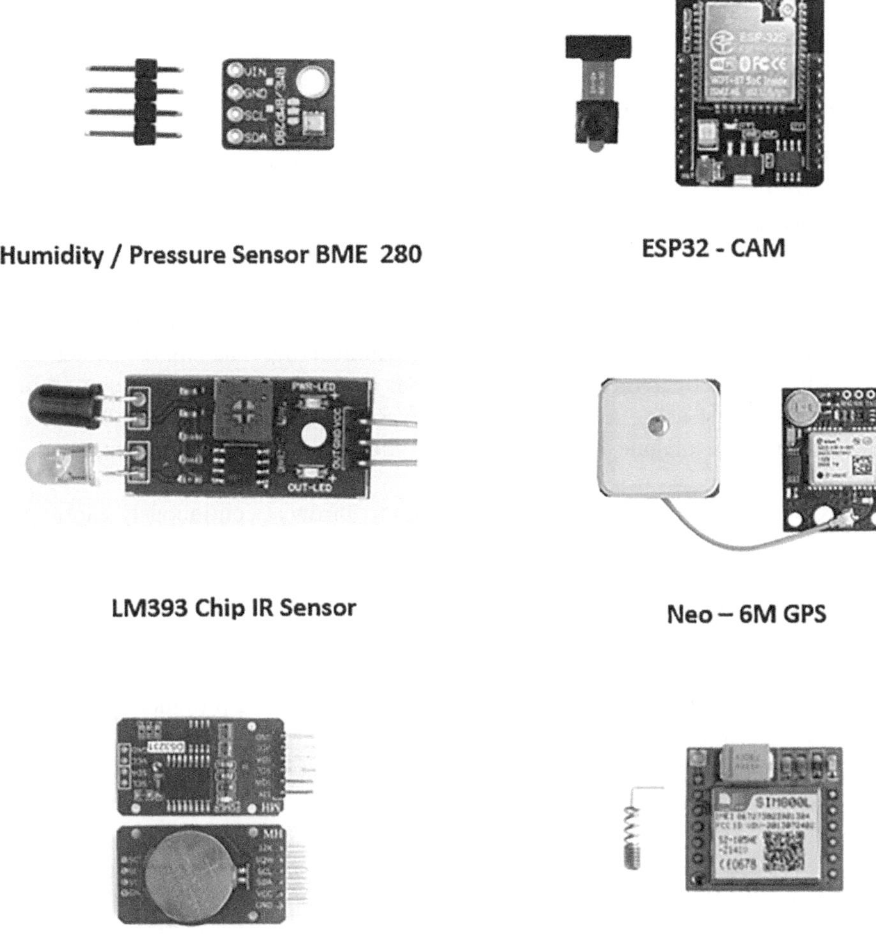

FIGURE 8.2   IoT components used for building smart birdnest.

DS3231 RTC: It is a **Real-Time Clock (RTC)**, which is an IoT device which keeps track of time in various IoT applications. It has an in-built slot for a button cell, to make sure it stays powered on at all times. Even when the power supply to the entire system is cut off the RTC makes sure to keep track of time.

SIM800L GSM: It is a Global System for Mobile Communications (GSM) module, which is capable of carrying out different basic functionalities of a mobile phone. These functionalities include SMS, internet connectivity, calling and others. A SIM card of any telecom company, which works on the GSM spectrum is placed into this module. When this GSM module gets powered on, it uses the SIM card to establish a connection with the mobile network. This connection provides internet connectivity for the birdnest, required for sending and receiving sensor data over the internet.

8.4.2 IMPORTANT FEATURES OF SMART BIRDNEST

The vital features of smart birdnest are provided as follows:

- **Precise Bird Entry Detection Technology:** The smart birdnest is capable of detecting the presence of birds at the entry of birdnest using an **IR sensor**. The image of the bird during its entry and exit from the birdnest is also captured with a precise timestamp using **ESP32-CAM and RTC**.
- **Effective Environmental Parameter Monitoring:** The smart birdnest is capable of effectively monitoring various environmental parameters within the birdnest using a **humidity and pressure sensor—BME 280**. These parameters include temperature, humidity and air pressure within the birdnest.
- **Live Image/Video Capturing:** The smart birdnest continuously captures the video footage of activities occurring within the birdnest using a camera module—**ESP32-CAM**. This video footage is consistently uploaded to the Firebase for user access.
- **Periodic Location Tracking of Smart Birdnest:** The location of smart birdnest is periodically tracked using **Neo-6M GPS** in the form of latitudinal and longitudinal coordinates. These coordinates help in identifying the exact location of the smart birdnest.
- **Renewable Power Supply Using Solar Energy:** The entire smart birdnest is powered using a **solar panel** (photovoltaic cells) that stores solar energy into the **rechargeable lithium-ion batteries**.

8.4.3 BLOCK DIAGRAM AND WORKING OF SMART BIRDNEST

Figure 8.3 depicts the block diagram of smart birdnest. Related working of the system is discussed in this subsection.

Firstly, the solar panels collect the solar energy and store it in lithium-ion batteries. These batteries supply power to the different functional components of smart birdnest. Then, the SIM800L GSM module gets turned on and provides a stable internet connection for data transfer from all the sensors to the Firebase. ESP32-CAM gets powered up after the GSM module and starts capturing the footage of activities occurring within the smart birdnest. This captured footage is sent to the Firebase from where it can be accessed in real-time. After ESP32-CAM, other independently working sensors get powered up. These sensors include the Neo-6M GPS module, humidity and pressure sensor BME 280, RTC and the IR sensor. The data from these sensors can be collected and processed using an Arduino Microcontroller. The Neo-6M GPS module starts detecting the location of smart birdnest by connecting to a nearby satellite and constantly sends the location coordinates to the Firebase. Alongside this, the humidity and pressure sensor BME 280 periodically measures

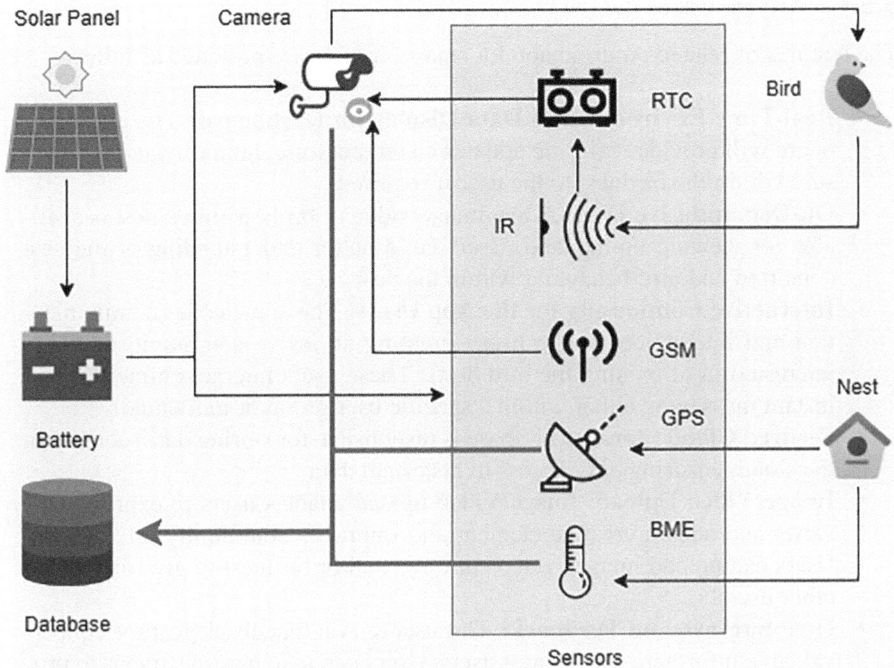

FIGURE 8.3 Block diagram of smart birdnest.

the environmental factors within the nest and sends all this data to the Firebase. The IR sensor emits infrared waves, which reflect when they collide with an obstacle in the provided range of distance. An IR sensor is fitted at the entry point of the smart birdnest to detect when the bird enters or leaves while passing through that range. As soon as the bird is detected, the Real-Time Clock and ESP32-CAM work in unison to capture an image of the entry point with a timestamp. This image along with its timestamp is then sent to the Firebase along with the rest of the data.

8.5 PROTOTYPE OF PROPOSED ANDROID APP FOR SMART BIRDNEST

The IoT-integrated smart birdnest can be effectively used along with the proposed Android app. The app enables users to monitor real-time environmental data as well as an on-demand live feed of data. The app even provides an interactive community and secured cloud storage to the users. The prototype works on mobile phones having Android OS version 8.0 (Oreo) and above. This app prototype is developed in the Kotlin language. The size of apps developed in Kotlin is comparatively smaller as compared to Flutter. The Android app uses Firebase as a cloud storage provider and Firebase even provides a number of features for testing purposes. This data consists of user information, user reviews, images and videos related to the birdnest, surveys and other information related to birds. This section elaborates on the features and design of the proposed prototype of the Android app.

8.5.1 App Features

The features of related Android app for smart birdnest are provided as follows:

- **Real-Time Environmental Data Display on Dashboard:** The app dashboard will provide real-time updates on temperature, humidity and air pressure within the birdnest to the user if required.
- **On-Demand Live Feed:** A live image/video of birds within a nest is available for viewing on-demand. Users get a better understanding of the nest condition and bird behaviour within the nest.
- **Interactive Community for the App Users:** The app enables communication and interaction among users, creating an active community of users interested in or owning the bird nests. These users interact online through instant messaging either within a specific user group or individually.
- **Secured Cloud Storage:** Firebase is responsible for storing data securely in the cloud, ensuring easy access to historical data.
- **Image/Video Upload:** Image/Video upload enables users to express their views and adds a creative element and improves community interactions. Users can upload images/videos of birds and/or birdnest to give updates to other users.
- **User Surveys and Feedback:** The user survey/Feedback feature collects valuable information from nest users. Users can take part in surveys to provide their perspectives and insights related to birdnest.

8.5.2 Design of Android App—Haven

This section presents the design of an Android app for smart birdnest. This Android app is named Haven. Haven refers to 'any safe place where birds can rest'. It focuses on building a community, which promotes bird conservation. Thus, it acts as a Haven for the birds. The Haven app prototype design is based on the prototype of the 'Mothers of Preterm Infants' app [19]. The app consists of various screens that provide a number of features to the nest users. These screens include 'Login Screen', 'Home Screen', 'Birdnest Preview Screen', 'Explorer Screen', 'Nest Information Screen', 'User Profile Screen', 'Birdnest Community Screen' and 'Survey Screen'.

Haven has been designed with a Human-Computer Interaction (HCI) approach, prioritising user-friendliness and user engagement. The HCI principles, guidelines and/or concepts include Miller's Principle, Hamburger Menu, Sequential Colour Palette, Gestalt's Principles, Visibility of System Status, Clean Design, Responsive Design, Typography Principles, Consistency, Social Presence Theory, Thumb's Rule, and Hierarchy and Information Architecture Principle. The design of selected screens of the Android app for birdnest is discussed in the following subsections.

8.5.2.1 'Login Screen' and Menu

The Haven app starts with a user 'Login Screen' as shown in Figure 8.4a. If the user is not registered, then the navigation for the 'Signup Screen' is provided at the bottom of the 'Login Screen'. The user authentication is done through the Firebase authenticator. After successful login, the 'hamburger menu' is visible to the user as shown

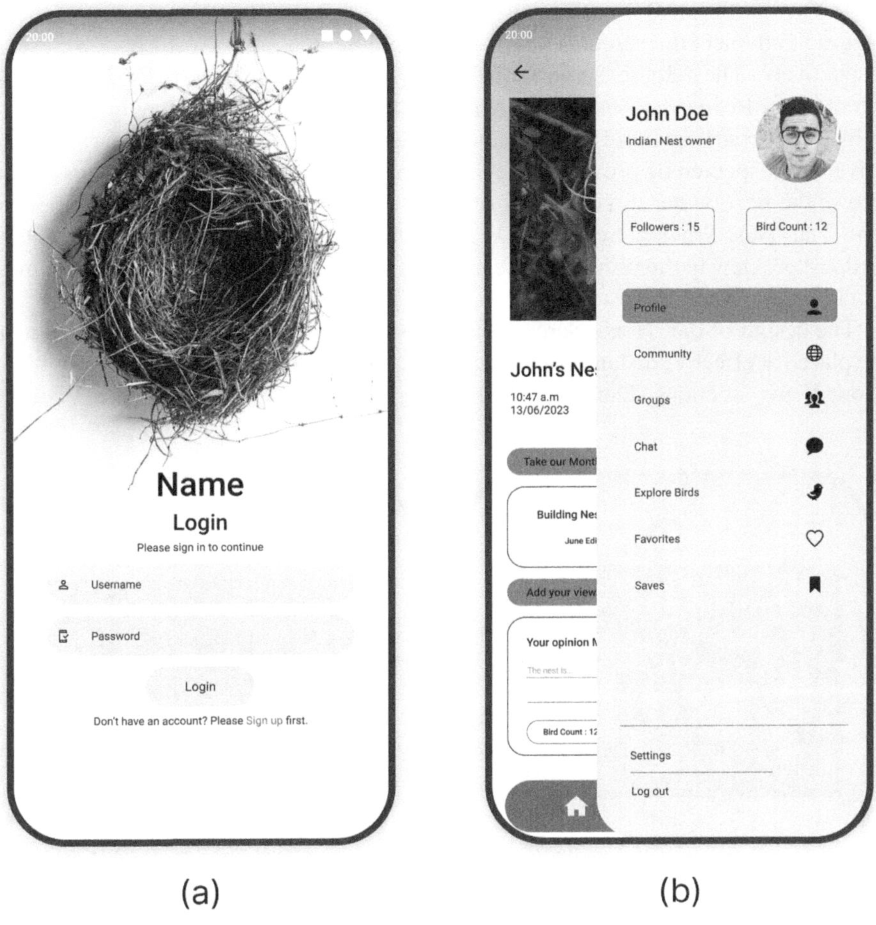

(a)   (b)

FIGURE 8.4 Screenshots of (a) login screen and (b) Hamburger menu.

in Figure 8.4b. This menu serves as a quick access route to all vital app features and functionalities. This menu showcases the number of followers of the nest user and the bird count in one's bird nest. The hamburger menu at the top of the screen provides navigation to profile, community, chat, explorer, favourites and saves. At the bottom of the menu, 'Settings' and 'Logout' options are provided.

The cup-shaped birdnest picture made of grass and shoots, is used as a theme in the login screen. This theme highlights the connection between the app's purpose and the initial interaction of the user as shown in Figure 8.4a. A **hamburger menu** is used to help in reducing the cognitive load of the user. It stacks seven relevant options one below the other in a sidebar [20]. The listing of these seven options in the **hamburger menu** is designed based on **Miller's 7 ± 2 rule** [21]. The **sequential colour palette** of different combinations of white and brown colours signifies clean consistency [22]. It even provides a nest-like appearance to the app Haven, and it fits with the natural colour theme for Haven.

8.5.2.2 'Home Screen' and 'Birdnest Preview Screen'

Figure 8.5 depicts the screenshots of the 'Home Screen' and 'Nest Preview Screen', respectively. The 'Home Screen' of the app serves as a platform for various user interactions to increase user engagement. Through the 'Home Screen', users can upload files, participate in surveys and get reviews from other nest users. The user privacy is respected by providing the feature of 'Who Viewed My Posts'. This feature allows the nest owner to keep track of other users, who viewed his/her profile content. The 'Nest Preview Screen' in Figure 8.5b is responsible for providing a smart birdnest preview to the users. Users can have a closer look at the interior of a birdnest, by clicking on the birdnest picture.

The design of the 'Home Screen' makes use of **Gestalt principles** [23]. The icons are placed at effective distances and a sense of symmetry is applied to the 'Surveys' and 'Your Views' sections. The 'Nest Preview Screen' and 'Home Screen' use **Visibility**

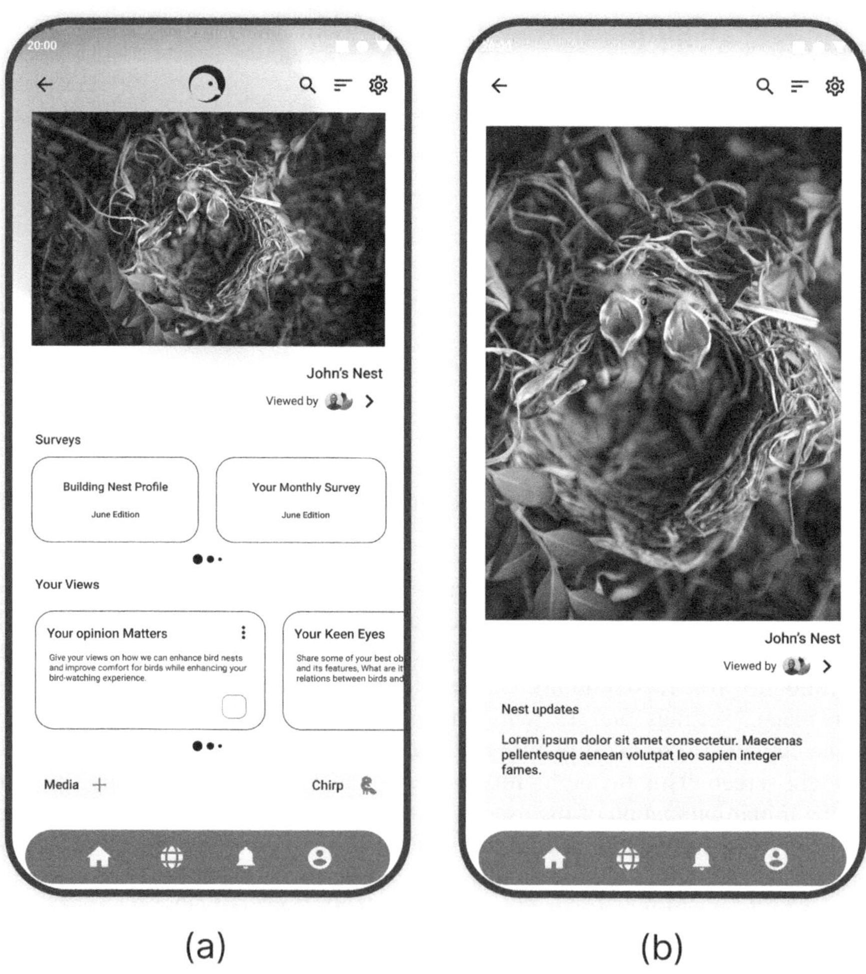

(a)   (b)

FIGURE 8.5   Screenshots of (a) Home Screen and (b) nest preview screen.

of System Status for the 'Who Viewed My Posts' feature which provides the status of users who visited the user profile [24].

8.5.2.3 'Explorer Screen' and 'Nest Information Screen'

Figure 8.6 depicts the screenshots of the 'Explorer Screen' and 'Nest Information Screen', respectively. The 'Explorer Screen' as shown in Figure 8.6a offers different sections of 'Birds', 'People & Nest', 'Bird Groups' and 'Explore'. Each section contains related pictures which provide respective information to the user. The 'Birds' section provides information regarding different bird species. The 'People & Nests' section provides information about different nests and their owners as shown in Figure 8.6b. The pictures in the 'Bird Groups' section redirect users to respective bird enthusiast communities. The 'Explore' section provides random posts from different app users. In the 'Nest Information Screen', users can see peer birdnests and

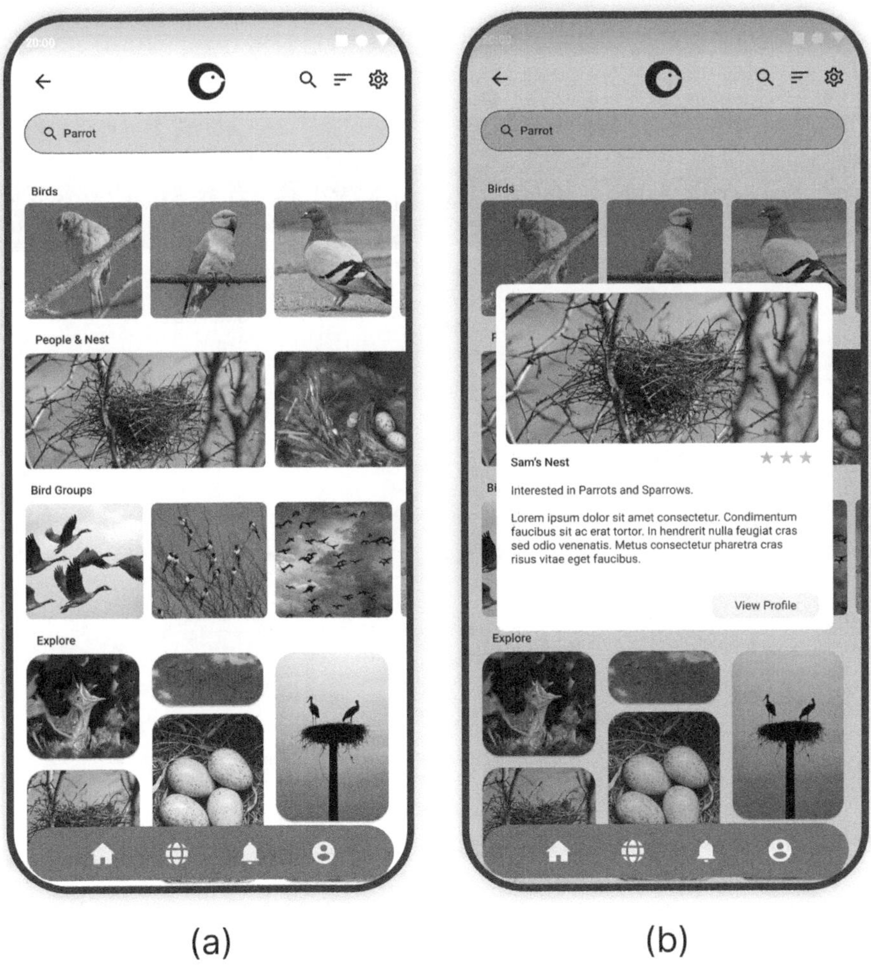

FIGURE 8.6 Screenshots of (a) explorer screen and (b) nest information screen.

information related to nests and their owners. Users can also rate the birdnests from other owners, based on factors like structure, bird activity and surroundings.

The 'Explorer Screen' promotes **Clean Design** by using a white colour scheme [25] and sufficient spacing between images following **Gestalt's Principles** [23]. The dialogue boxes pop up when a user clicks on a specific image. **Responsive Design** allows the dialogue boxes to adjust themselves on different devices like mobile phones and tablets, as per their display size [26].

8.5.2.4 'User Profile Screen'

Figure 8.7a depicts the screenshot of the 'User Profile'. The 'User Profile' allows the nest user to customise his/her profile created in the Haven app. Users can provide additional personal information and bird count of their nest to showcase their interest to other bird enthusiasts. Users can also post a 'text only' status as a 'Chirp' by clicking on the 'Chirp' button. This allows users to show his/her engagement with other

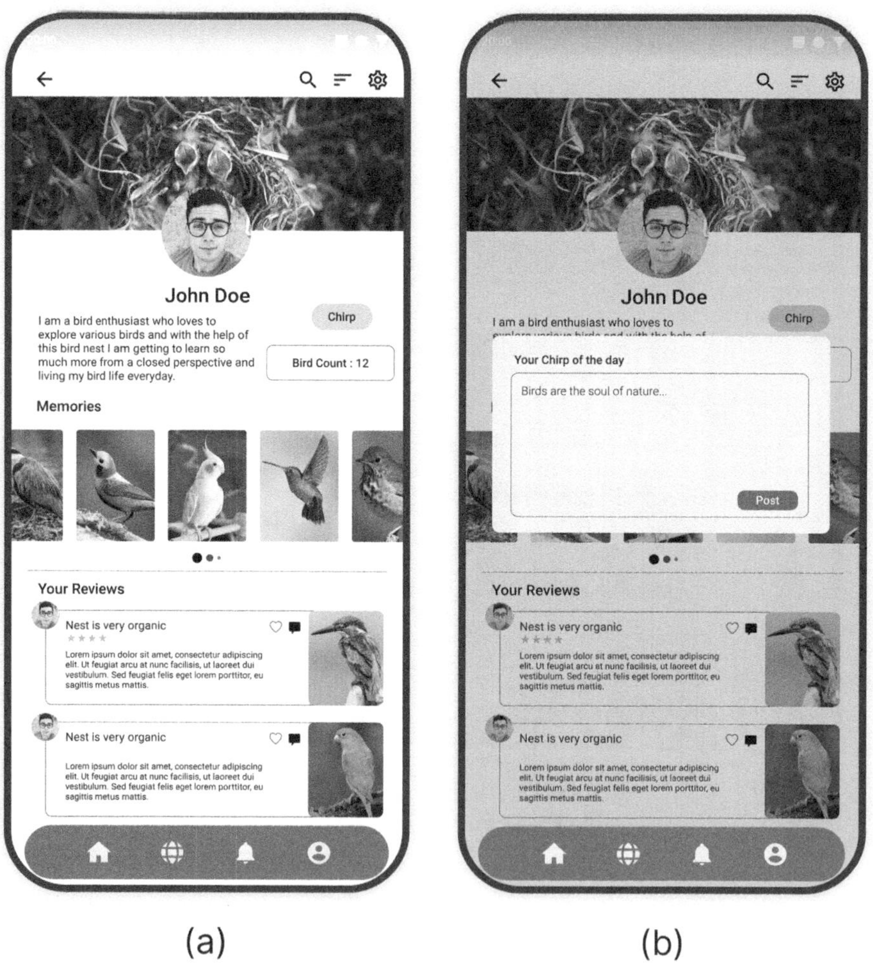

(a)   (b)

FIGURE 8.7 Screenshots of user profile screens.

users by sharing 'Chirp of the day' as shown in the pop-up of Figure 8.7b. Users also have an option to visit their previous reviews of other posts. This option allows them to go back and reconsider their input and observations. Additionally, users can browse through their past contributions to the birdnest community in the form of pictures, posts or observations under the section 'Memories'.

The text style, font size and spacing are carefully chosen to improve readability and visual hierarchy in the 'User Profile Screen' of birdnest by following **Typography Principles** [27]. The **Horizontal Scroll** with the grid structure in the memories section provides another distinguishable aspect of the user interface [28]. This aspect provides a mechanism to align multiple elements in a limited horizontal space.

8.5.2.5 'Birdnest Community' and 'User Interaction'

The 'Explorer Screen' serves as a way to connect with bird enthusiasts. By clicking on respective communities, the user can visit the 'Birdnest Community Screen' as depicted in Figure 8.8a. Users can browse through various pictures/videos uploaded

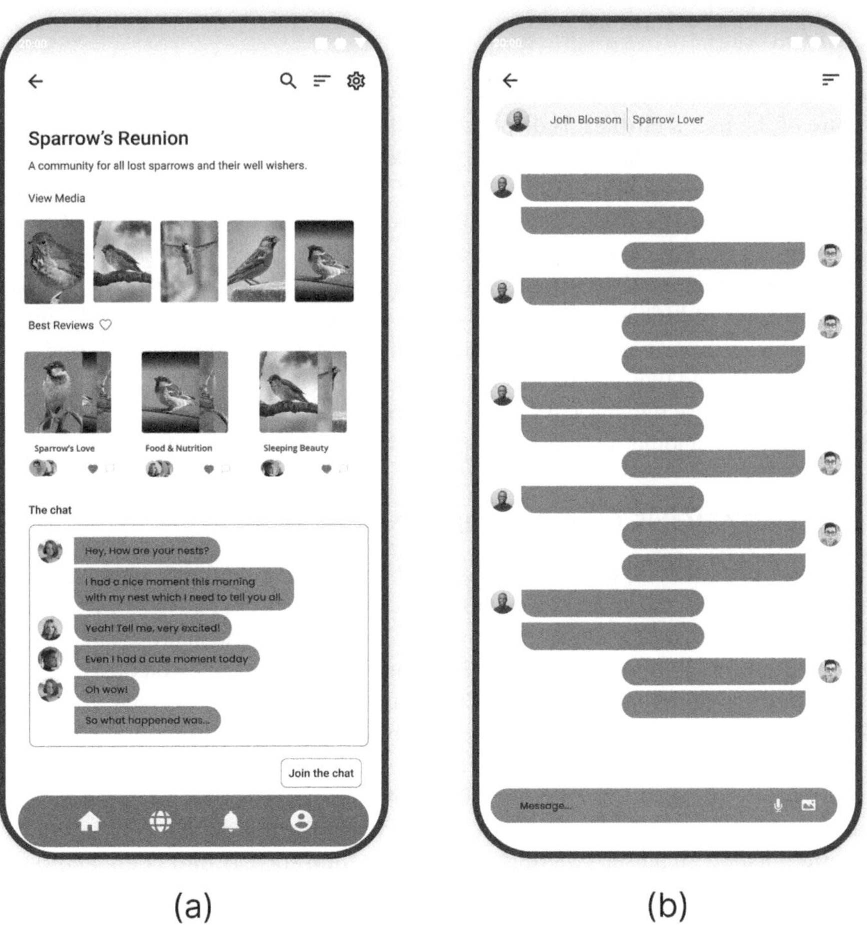

(a)   (b)

FIGURE 8.8 Screenshot of (a) birdnest community screen and (b) instant message.

by community members, under the 'View Media' section. Users can even browse through the best reviews from the 'Best Reviews' section inside every community. A separate 'The chat' section is given to encourage engagement and discussion among community members. The chat window is provided for the community members to interact with each other. Active participating users can take part in community discussion by clicking on the 'Join the Chat' button. Figure 8.8b shows that users can engage in real-time discussion and share information and conversation among birdnest users.

The messages of the sender and receiver are distinguished using the user's profile picture. The profile picture of the sender is displayed on the left side and the profile picture of the receiver on the right side of the screen. The use of user profile pictures for text message flow even provides a sense of being in close proximity with another user by following the **Social Presence Theory** in HCI [29]. **The Hierarchy and Information Architecture principle** is used to align the 'View Media', 'Best Reviews' and 'The chat' sections within the user interface as per users' priority to sections [30].

8.5.2.6 'Survey Screens'

Users can click on a survey from the 'Surveys' section provided on the 'Home Screen'. This action redirects the users to the 'Survey Screen' depicted in Figure 8.9a. The picture of the user's birdnest is displayed on the top of the screen. The 'View' button on Figure 8.9a takes the user to the complete survey form depicted in Figure 8.9b. 'Survey Screen' is used to fill the Multiple-Choice Questions (MCQs) asked in the survey regarding their nest. These MCQs help in collecting information from the nest users for improving the app in future.

The visual element of the birdnest picture is kept on the topside and textual elements are in the form of MCQs in the middle of the screen. These questions are the interactive element of the screen. The positioning of these interactive elements is based on the **Thumb's rule**, and they can be easily accessed by the user's thumb [31]. MCQs on the 'Survey Screen' use light brown and white colour combinations to distinguish different MCQs from one another following the principle of **Consistency** [32].

8.6 LIMITATIONS OF SMART BIRDNEST

The developed prototypes of smart birdnest and related Android apps have a few limitations, which may affect its performance occasionally. The Neo-6M GPS module on activation takes around 15–20 min to connect with the nearest satellite. This may create a 15–20 minutes delay for location data initially. This inconsistency gets removed after the delay time of 15–20 minutes and all data received after that is always consistent. The images and videos of birds within the birdnest are adjusted to a specific lighting while capturing. The increase or decrease in light exposure during the afternoon or night-time affects the quality and visibility of images and videos captured. The related Android app does not provide the real-time locations of different bird hotspots in a user's vicinity. The information such as bird count and type of bird in the nest needs to be manually updated by the user periodically.

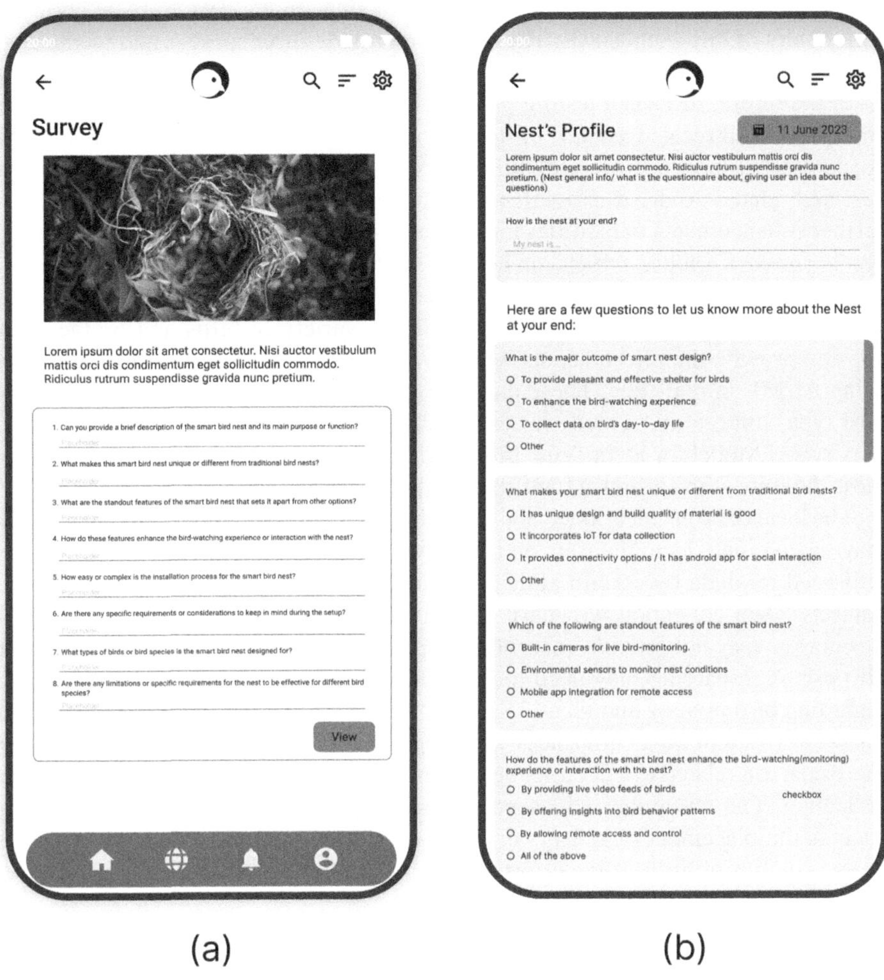

(a) (b)

FIGURE 8.9 Screenshots of survey screens.

8.7 CONCLUSION AND FUTURE WORK

The idea of improving the sustainability and effectiveness of birdnest by providing a smart birdnest and related Android app is presented in this paper. Sustainability and effectiveness are achieved by providing vital features and facilities to the smart birdnest users. They include sensing environmental parameters, capturing live footage, location tracking and record-keeping of the entry and exit of birds. Alongside this, a prototype of an Android app is developed, which works with this smart birdnest. The app is designed based on different HCI aspects for prioritising user-friendliness and user engagement functionalities. Some of these HCI aspects include Miller's Principle, Responsive Design, Consistency, Typography Principles and many others.

This related Android app provides social interaction among nest users by building a community of bird enthusiasts. The app even provides user surveys and feedback for the purpose of improving the app design.

In the future, there is a testing plan related to this smart birdnest. About 25 bird enthusiasts will receive a smart birdnest along with the associated Android app. The purpose of birdnest testing is to verify the working of the IoT-integrated smart birdnest and related Android apps. Researchers can develop Android apps that deliver further enhancements in facilities for the nest users. AI-based bird species identification is one such feature, which can help the user in identifying the species of the bird residing in the smart birdnest. An AI-based chatbot can be integrated into the app, which will respond to user queries regarding a variety of birds and birdnests. An app can provide the bird hotspots within a city or locality based on a local dataset, using AI/ML in real-time. These datasets will contain fields such as time, location, bird type, images and videos related to different birds in a particular locality. A Supervised Model for identifying birds and bird hotspots can be developed by using these datasets with several AI/ML-based algorithms.

The local datasets mentioned above can be rare to find and utilise. The researchers may try to create local datasets on their own to tackle this issue. These researchers will need to abide by certain ethical considerations while collecting data for these datasets. Data collection will have ethical considerations because different localities may or may not provide consent for capturing images or videos required for the purpose. A researcher may need to enter into a dialogue with the local community regarding bird privacy and identification. One can even collect data by communicating with different bird enthusiasts who may be willing to participate. There is one more concern related to affecting the regular life of birds while proceeding with data collection. The researcher will need to take proper precautions to make sure stuff such as the placement of sensors or video capturing equipment will not hinder the daily activities of birds or the efforts may prove counter-productive.

ACKNOWLEDGEMENT

The hardware and technical support for the smart birdnest project is provided by Dolphin Labs, Pune, India.

REFERENCES

1. Tu, H., Fan, M., Ko, J. C.: Different habitat types affect bird richness and evenness. *Scientific Reports*, 10, 1221 (2020). https://doi.org/10.1038/s41598-020-58202-4
2. Sánchez-Bayo, F., Wyckhuys, K.: Worldwide decline of the Entomofauna: A review of its drivers. *Biological Conservation*, 232, 8–27 (2019). https://doi.org/10.1016/j.biocon.2019.01.020
3. Haskell, L.: State of world's birds 2022. Report by Birdlife International. BirdLife International, pp. 1–45 (September, 2022). https://datazone.birdlife.org/userfiles/images/SOWB2022_EN_compressed.pdf.
4. Mumbai live team: The lost migrants: 44 percent decline in the number of migratory birds visiting Mumbai. Mumbai Live (2017).

5. Paliwal, G., Bhandarkar, S.: Ecology and conservation of threatened birds in and around Navegaon National Park, Maharashtra. *ESSENCE - International Journal for Environmental Rehabilitation and Conservation*, VIII(1), 120–135 (June, 2017).

6. Reynolds, S., Ibáñez-Alamo, J., Sumasgutner, P., Mainwaring, M.: Urbanisation and nest building in birds: A review of threats and opportunities. *Journal of Ornithology*, 160(3), 841–860 (2019). https://doi.org/10.1007/s10336-019-01657-8

7. Abdelouahid, R., Debauche, O., Mahmoudi, S., Marzak, A., Manneback, P., Lebeau, F.: Smart nest box: IoT based nest monitoring in artificial cavities. *2020 3rd International Conference on Advanced Communication Technologies and Networking (CommNet)*. IEEE Xplore, New York, pp. 1–7 (September, 2020). https://doi.org/10.1109/CommNet 49926.2020.9199624

8. Zárybnická, M., Kubizňák, P., Šindelář, J., Hlaváč, V.: Smart nest box: A tool and methodology for monitoring of cavity-dwelling animals. *Methods in Ecology and Evolution*, 7(4), 483–492 (2016). https://doi.org/10.1111/2041-210x.12509

9. Del-Rio-Ruiz, R., Lopez-Garde, J., Ruiz-De-Garibay, J., Macon, J.: Smart Nests: IoT for Ornithology. *2018 Global Internet of Things Summit (GIoTS)*, Bilbao, Spain, pp. 1–6 (2018). https://doi.org/10.1109/giots.2018.8534436

10. Bianchi, T.: Designing a smart nest box for automated biodiversity monitoring. University of Twente Student Theses (2022). https://essay.utwente.nl/92674/

11. Hakim, O., Rifadil, M., Putra, P.: Smart bird cage based on STM32 for turtle dove bird using solar panel. *Interdisciplinary Social Studies,* 1(3), 280–290 (2021). https://doi.org/10.55324/iss.v1i3.43

12. Garcia-Rodriguez, A., Rodriguez-Sakamoto, R., Fernandez-Berni, J., del Rio, R., Marin, J., Baena, M., Bustamante, J., Carmona-Galán, R., Rodriguez-Vazquez, A.: Live demonstration: Low-power low-cost cyber-physical system for bird monitoring. *2018 IEEE International Symposium on Circuits and Systems (ISCAS)*, Florence, Italy, pp. 1–1 (2018). https://doi.org/10.1109/ISCAS.2018.8351434

13. Steen, R.: Bird monitoring using the smartphone (iOS) application videography for motion detection. *Bird Study*, 64(1), 62–69 (2017). https://doi.org/10.1080/00063657.20 16.1271772

14. Ibrahim, A., Ibrahim, N., Harun, A., Kassim, M., Kamaruddin, S., Witjaksono, G.: Bird counting and climate monitoring using lorawan in swiftlet farming for IR4.0 applications. *2018 2nd International Conference on Smart Sensors and Application (ICSSA)*, Kuching, Malaysia, pp. 33–37 (2018). https://doi.org/10.1109/icssa.2018.8535955

15. Noverta, R., Wahab, N., Ahsan, M.: Hybrid system on temperature and humidity for swallow nest farm via mobile application in internet of things. *9th International Graduate Conference on Engineering, Science and Humanities,* Johar Bahru, Malaysia, pp. 93–97 (2022).

16. Tian, C., Sun, Y.: AI birder: An intelligent mobile application to automate bird classification using artificial intelligence and deep learning. *8th International Conference on Artificial Intelligence (ARIN 2022)*, Sydney, Australia, Vol. 12, Number 10 (2022). https://doi.org/10.5121/csit.2022.121005

17. Ferreira, A., Silva, L., Renna, F., Brandl, H., Renoult, J., Farine, D., Covas, R., Doutrelant, C.: Deep learning-based methods for individual recognition in small birds. *Methods in Ecology and Evolution*, 11(9), 1072–1085 (2020). https://doi.org/10.1111/2041-210x.13436

18. Chaudhary, S., Saran, S.: Evaluation of pre-trained convolutional neural networks for the development of web/mobile based bird species identification applications under Indian Bioresource Information Network (IBIN) framework. *Environmental Sustainability*, 6(3), 415–421 (2023). https://doi.org/10.1007/s42398-023-00281-w

19. Patil, A., Bhutkar, G., Pendse, M., Tawade, A., Bodkhe, A., Shaha, S., Deshmukh, S.: Prototype design of android app for mothers of preterm infants. *International Federation for Information Processing*, 3, 01–14 (2019). https://doi.org/10.1007/978-3-030-05297-3

20. Tsiodoulos, D.: Comparison of hamburger and bottom bar menu on mobile devices for three level navigation. Stockholm, Sweden (2016). https://www.semanticscholar.org/paper/Comparison-of-hamburger-and-bottom-bar-menu-on-for-Tsiodoulos/9ac58840d288333a942d2f7edd875c3f085225e5

21. Cowan, N.: George Miller's magical number of immediate memory in retrospect: Observations on the faltering progression of science. *Psychological Review* 122(3), 536–541 (2015). https://doi.org/10.1037/a0039035

22. Patil, S., Bhutkar, G., Vaidya, P.: Psychological survey of color perceptions for Indian users. *18th International Conference on Humanizing Work and Work Environment HWWE-2020*, pp. 433–445 (2020). https://doi.org/10.1007/978-981-16-6982-8_39

23. Wagemans, J., Elder, J., Kubovy, M., Palmer, S., Peterson, M., Singh, M., Heydt, R.: A century of Gestalt psychology in visual perception: I. Perceptual grouping and figure–ground organization. *Psychological Bulletin*, 138(6), 1172–1217 (2012). https://doi.org/10.1037/a0029333

24. Nielsen, J.: Ten usability heuristics (2005).

25. Wich, M., Kramer, T.: Enhanced Human-Computer Interaction for business applications on mobile devices: A design-oriented development of a usability evaluation questionnaire. *2015 48th Hawaii International Conference on System Sciences*, Kauai, HI, pp. 472–481 (2015). https://doi.org/10.1109/HICSS.2015.63

26. Calderon, C.: A framework towards designing responsive public information systems. *Embodying Virtual Architecture: The Third International Conference of the Arab Society for Computer Aided Architectural Design (ASCAAD 2007)*, 28–30 November 2007, Alexandria, Egypt, pp. 767–782 (2007).

27. Kelly, S., Gregory, T.: Typography in human-computer interaction cognitive and aesthetic implications of typeface selection and presentation. Iowa State University, Computer Science, Psychology (2012). https://www.semanticscholar.org/paper/Typography-in-Human-Computer-Interaction-Cognitive-Kelly-Gregory/6c34049850817b580c502e21860e23069c7405f4

28. Braganza, C., Marriott, K., Moulder, P., Wybrow, M., Dwyer, T.: Scrolling behaviour with single-and multi-column layout. *Proceedings of the 18th international conference on World Wide Web,* Madrid, Spain, pp. 831–840 (2009, April).

29. Lee, K., Nass, C.: Designing social presence of social actors in human computer interaction. *Proceedings of the SIGCHI Conference on Human Factors in Computing Systems*, 6(1), 289–296 (2003, April).

30. Karatsolis, A., Karatsoli-Chanikian, L.: Using principles from architecture to inform HCI design. *Proceedings of the International Conference on Information Systems and Design of Communication*, pp. 8–14 (2014).

31. Qian, F.: Designing for a thumb: An ideal mobile touchscreen interface for Chinese users. *Lecture Notes in Computer Science*, 2, 44–53 (2013). https://doi.org/10.1007/978-3-642-39241-2

32. Tanaka, T., Eberts, R., Salvendy, G.: Consistency of human-computer interface design: Quantification and validation. *Human Factors*, 33(6), 653–676 (1991). https://doi.org/10.1177/001872089103300604

9 AI on Wheels
AI-Powered Predictive Models for Smart Vehicles

Sachin Vanjire, Rashmi Naveen, and Sachin Vanjire

9.1 INTRODUCTION

Interdisciplinary discoveries that bring together previously independent domains, such as artificial intelligence (AI), robotics, and networking, have emerged over the past few decades. The integration of connectivity and vehicle autonomy is set to pave the way for the development of Connected Autonomous Vehicles (CAVs). Traditional techniques for defect detection and control are becoming less effective with the rapid spread of new and advanced automobile features, necessitating the development of smart solutions. Modern vehicles are equipped with sensors, instruments, and cameras that record information about the vehicle and its components. The ability to collect and analyze asset data is what enables businesses to transition from reactive to preventive maintenance. In the transportation and logistics industries, where downtime can incur significant costs, predictive maintenance is a top priority for fleet managers. Predictive maintenance is expected to gain increasing acceptance in the automotive industry, offering potential benefits to individual vehicle owners, fleet operators, and manufacturers. The integration of AI and automotive technology has resulted in novel approaches that enhance comfort, security, and progress toward fully autonomous vehicles (AVs). Cars now feature AI-powered systems and algorithms capable of processing massive quantities of data and making intelligent decisions on the fly. Prediction models based on AI are experiencing significant uptake in practical applications, such as driverless cars. Ensuring the effective operation and security of AVs will be crucial to estimating the prediction uncertainty of these AI models. Through experiments and a paper, we demonstrate how the Drive Horizon AI-based approach can be applied in practical vehicles to estimate future vehicle models. Automotive manufacturers are increasingly adopting advanced AI-based software development solutions to realize the vision of automated vehicles. This has resulted in an industry heavily relying on AI in the design and manufacturing of vehicles, indicating that hybrid cars, electric cars, and autonomous cars are the future of the automotive industry. In this chapter, major contributions include predictive maintenance to reduce downtime, AI-powered safety features, customized driving experiences, AI-powered predictive maintenance, connected vehicles, and addressing challenges associated with using AI in the automotive industry.

DOI: 10.1201/9781032664828-11

9.2 LITERATURE REVIEW

This chapter aims to provide a comprehensive understanding of the perception and decision-making processes of autonomous cars, incorporating both technological and ethical considerations [1]. The objective is to redirect the sociotechnical discourse surrounding decision-making in AVs from the prevailing emphasis on "trolley framings" toward more profound ethical considerations. This paper examines the recent progress made in the field of deep learning as it pertains to the forecasting of traffic flow. The article encompasses a multitude of deep-learning architectures [2]. To effectively extract increasingly intricate features from unprocessed input, deep-learning models necessitate the inclusion of multiple layers of computing. In light of the complex structure of transport networks, we examine the latest deep-learning models developed to tackle this challenge. The suggested design incorporates a novel component known as the Trust Threshold Limit. This implementation aims to mitigate excessive usage and subsequently minimize waste in the production processes [3]. This study emphasizes the advantages of employing AI inside a decentralized blockchain framework, specifically in conjunction with smart contracts. Additionally, it explores the use of AI in establishing a company's trading rules and effectively managing market risk evaluations, particularly in times of financial crisis [3]. The model that was developed incorporated cost functions, evaluations of delivery time, and assessments of energy, with support from real-world scenarios [3]. The present study [4] constitutes an exploratory examination of AI in the operational activities of multiple case companies. The analysis of AI systems and their consequential benefits to organizations is undertaken. This study has highlighted multiple areas in which the utilization of AI in the supply chain might contribute to value generation [4]. In this paper [5], we perform a quick bibliometric study of the published research on AI in smart vehicles. Findings from a comprehensive literature review and analysis powered by AI. The purpose of this research [6] is a systematic literature review (SLR) on AI in smart cities. It demonstrates how AI and smart cities may interact in the near future. Through thematic analysis, we were able to identify overarching themes, more specific subthemes, and research gaps in more than a hundred different manuscripts. This paper [7] reviews recent papers on XAI research and deployment in domains like medicine, industry, transportation, and finance, and presents a comprehensive literature review of XAI approaches for a range of applications. The paper [8] provides an overview of the most popular AI-related technologies, evaluates their advantages, and highlights how they might help improve the efficiency and effectiveness of the renewable energy industry as a whole [8]. Furthermore, this work discusses the potential challenges of implementing large-scale integrations of renewable energy for a carbon-neutral transition, explains how to overcome them with the use of appropriate AI techniques, and recommends some interesting new possibilities for investigation [9].

When developing an SSC model using AI-based methodologies, researchers and practitioners can benefit from the study methodology proposed in the paper [9]. The scientific contribution to the fields of AI and SSC was assessed after a comprehensive review of the relevant literature. The purpose of this paper [10] was to synthesize the findings of a user-centric literature review on explainable AI. The particular objectives were to characterize the breadth of end-users' desire for explanations, to investigate how explanations change end-users' perspectives, and to identify knowledge gaps and suggest future research topics for XAI from the perspective of end-users. This paper [11] provides a for all existing technologies associated with AD and IVs, including a history of each, a synopsis of their primary achievements, and details on their viewpoints, ethics, and possible future research directions. This report [12] examines the current status and possible development of XAI technology for smart cities in great detail. According to the paper [13], simulations are a vital aspect of creating state-of-the-art AI systems. We begin by addressing the necessity of simulations in the creation of cutting-edge AI systems. Then, we dive into the problems that simulation-based AI systems face, as well as possible solutions. Finally, we wrap up by talking about areas for further research. This research [14] explores self-learning models, tracing applications and technological needs that inspired their development. The results of this academic study [15] indicate that uncertain predictions of the future of a moving object can cause incongruous actions with dire consequences. This proposes an approach to motion prediction uncertainty estimates based on deep ensembles. The primary focus of this investigation [16] is to examine techniques put to use to formulate the best possible policies for a wide range of challenging smart city scenarios. In this investigation, we dive deeper into how such methods have been put to use. The book [17] focuses on the applications of machine learning (ML) and intelligent systems in business and industry. Many existing systems may not be capable of processing data in real time, limiting the effectiveness of predictive models for real-time decision-making. By addressing these gaps, the automotive industry can move toward a future where smart vehicles leverage advanced AI predictive models for improved performance, safety, and user experience. Advocate for industry-wide standards for data formats and protocols to ensure seamless data sharing and collaboration among smart vehicles and infrastructure.

9.3 PREDICTIVE MAINTENANCE

To eliminate unscheduled downtime and predict when operating equipment may break, a data-driven strategy known as predictive maintenance is employed. With the assistance of on-board sensors, maintenance histories, and ML algorithms, it can determine the likelihood of a system or piece of equipment failing. AI then alerts the driver and automaker/maintainer that a part or system is failing and needs to be fixed or replaced. This method reduces the amount spent on unanticipated repairs while increasing the useful life of the component. Additionally, quality control systems driven by AI can be trained to identify potential problems in parts prior to installation, as described in the Figure 9.1.

Advantages of predictive maintenance include optimizing the production of replacement parts, managing inventory levels, maximizing product lifecycles, minimizing waste, and more. Fleet operators, especially those in the transportation and logistics industries where downtime may have devastating financial consequences, have made predictive maintenance a top priority. The automobile industry is just one of several sectors experiencing radical change due to the advent of AI.

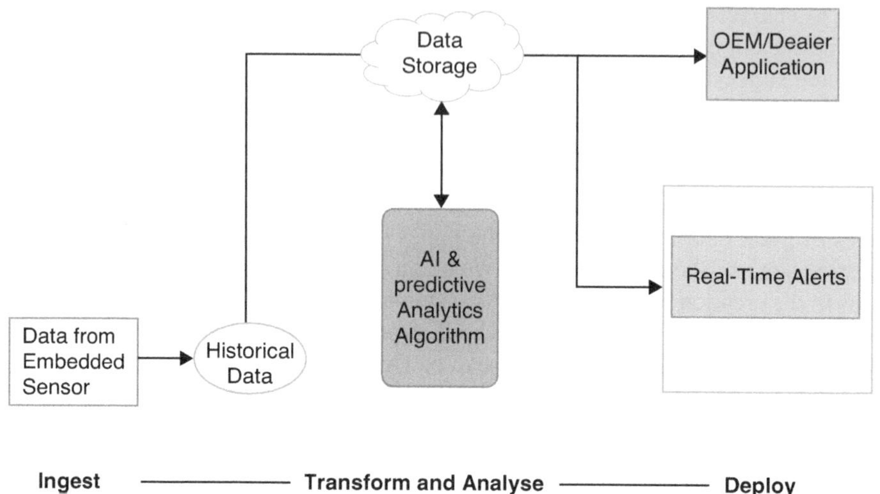

FIGURE 9.1   Predictive maintenance of car.

9.4   AUTOMOTIVE APPLICATIONS OF ARTIFICIAL INTELLIGENCE

9.4.1   STRENGTHENING PROTECTIONS

One of the most noticeable benefits of AI in the automotive sector is the improvement in passenger protection. Advances in AI have made several life-saving safety features possible.

9.4.2   CUSTOMIZED HIGHWAY TIME

AI allows cars to deliver a personalized ride for each passenger. The seat position, temperature, and informational features of a vehicle can all be tailored to the individual driver by using AI algorithms to study their habits, preferences, and trends. AI makes driving more enjoyable by learning the driver's preferences and responding accordingly.

9.4.3   UPKEEP PREDICTION

Predictive maintenance, made possible by AI, is revolutionizing the car industry by maximizing uptime and minimizing performance losses. AI algorithms can anticipate the need for repairs or upkeep by collecting data from various sensors and devices. Taking this preventive stance helps both vehicle owners and service providers avoid breakdowns and expensive repairs. AI harnesses the intelligence of the Internet of Things in vehicles, facilitating predictive maintenance. By analyzing the massive wealth of vehicle data, IoT devices allow managers to keep tabs on the current status of their fleet and know when maintenance is needed. When an Internet of Things sensor detects a problem, it notifies the car's administrators so that they can take preventive action before the issue escalates.

In addition to reducing pollutants and maximizing fuel efficiency, AI also enhances vehicle performance. When it comes to managing traffic, AI is crucial for increasing productivity and decreasing gridlock. AI algorithms are at the core of intelligent traffic management systems, using them to assess flow patterns, fine-tune light cycles, and notify drivers in real time. Cities can benefit from increased mobility and reduced commute times by using AI-powered smarter transportation networks.

9.4.4 MACHINE LEARNING AND INTERNET-CONNECTED VEHICLES

The use of AI-powered systems in connected vehicles is on the rise. Through AI, capabilities such as remote diagnostics, V2V communication, and V2I connectivity are made possible between cars and the surrounding infrastructure. Drivers and passengers alike can benefit from connected automobiles due to their superior navigation, constant access to information, and other convenient features.

9.4.5 THOUGHTS ABOUT ETHICS

The incorporation of AI into vehicles raises ethical questions. Considerable thought must be given to concerns like data privacy, liability in the event of accidents involving AVs, and decision-making under pressure. If we want people to feel safe using AI in their cars, we need to find a balance between advancing technology and ethical considerations.

9.5 PREDICTIVE MAINTENANCE AND MODELING'S VERSATILITY

Since AI utilizes the power of IoT in cars, it also aids the industry in planning repairs ahead of time. IoT systems help track how cars are performing in real time by analyzing a vast amount of data about them. This allows managers to know when maintenance is needed. When the IoT monitor senses a potential problem, it notifies the car's managers so they can take preventive steps before the issue worsens. AI also helps reduce emissions, enhance fuel economy, and overall improve vehicle performance.

The predictive maintenance market is expected to grow by an impressive 28%, with software experts playing a key role because the technology requires a successful combination of data science, thorough analysis, and ML to create reports predicting future events. The flexibility of predictive modeling lies in its ability to be integrated into a vehicle's computer system, allowing critical parts to be checked for performance. Software developers have also made the system user-friendly.

Data from sensors can be utilized to identify which parts of a car are likely to wear out or break prematurely, monitor them, and notify the owner through connected devices or in-dash alerts. It's essential to note that this feature differs from a check engine light on the dashboard. Predictive modeling, leveraging the complexity of ML, generates more detailed reports that predict when a car part will break. While shop manuals suggest when maintenance should be done, predictive modeling uses these suggestions in real time, providing owners and manufacturers a better understanding of the condition and performance of vehicle parts.

Predictive modeling is not exclusive to end-users or individual car owners; it can be invaluable in the car-making process. Digital Twin Technology (DT) further enhances the industry's efficiency. This resource is a digital model created and implemented by expert-level software developers, mimicking the stress and operation of a real machine or asset during normal tasks. Industry workers can monitor their machines and make proactive changes to ensure the car-making process runs smoothly. What makes DT intriguing is that its model ages like the machine or car part it's monitoring. Predictive analysis is crucial for understanding how the real car or manufacturing asset performs over time and in specific circumstances.

9.6 PREDICTIVE MAINTENANCE WITH ARTIFICIAL INTELLIGENCE

By considering traffic, weather, and the time of day, AI systems in automobiles analyze data to determine the optimal routes, ensuring the safety of both drivers and pedestrians. However, these new gadgets are capable of more than just enhancing the fun and safety of driving. It is possible to foresee mechanical issues and failures using AI and IoT, preventing accidents and saving money. In fact, according to a McKinsey report, manufacturers will save anywhere from $240 billion to $627 billion annually by using predictive maintenance by 2025.

Analytics and AI are combined in predictive maintenance solutions to identify potential issues with an automobile before they arise. Owners and drivers benefit financially from this, avoiding the need to pay for unexpected repairs. Additionally, AI-based predictive maintenance makes cars safer by alerting drivers to potential issues before they escalate. New instruments will facilitate the advancement of the mobility of the future. It will be interesting to observe how society, communities, and corporations utilize this path to a more eco-friendly future.

In phase 1, all sources of data must be looked at and used to put the advanced model into the business. This step is important for figuring out how well the asset is doing for the first time. During the proof-of-concept process, there aren't too many models to choose from for a more critical evaluation of the performance by figuring out how the project will fail and how long it will take to finish. Also, the last phase of the diagram shows how to predict the performance of an asset in real time, since the model in this phase needs to be changed and scaled all the time to give the best results.

9.7 SYSTEM METHODOLOGY

System uses a variation neural network (VNN), which takes inputs like maps and real-time traffic data and outputs like anticipated speed. There are three layers to the model: an encoder, a latent space, and a decoder. Map, traffic, and vehicle speed are all inputs that are transformed into latent space by the encoder layer. By adjusting the weights of a neural network, the latent space can learn the approximation of the input data's statistical distribution. The next kilometer's speed is predicted by the decoder layer based on the latent space's statistical distribution. We instrumented a Ford F-250 to record the speeds, traffic conditions, and location of multiple drivers and then used this information to progressively hone our VNN model. Previous work is described in detail.

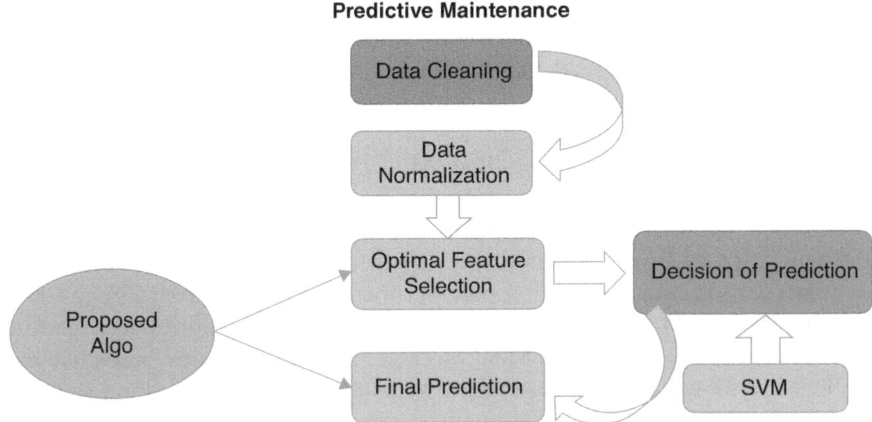

FIGURE 9.2 System workflow.

The PdM method is used to improve asset maintenance plans by using data-driven models to predict when assets will break. Using PdM helps to cut down on downtime and improve the quality of the output. Condition-based maintenance is another name for PdM. It tries to find failures and eventual deterioration by looking for trends in the conditions of parts using information from the past. Fast actions need to be taken into account. But PdM has some difficult jobs to do when it comes to predicting system failures and fixing parts to make them last longer. Also, this method depends on using devices to collect and change information. With continuous monitoring and inspection, there are two ways to find out what the situation is. PdM decision-making also requires putting together different types of information, such as maintenance records, causes, tracking data, knock-on effects of failures, and work orders. To improve FMM tasks, computer-based models are used instead of maintenance methods. Figure 9.2 shows how the suggested planning model for PdM is put together.

The suggested PdM planning model uses datasets of aircraft engines and lithium-ion batteries to get the best details about what will happen in the future. The proposed model has five key steps: (i) cleaning the data, (ii) normalizing the data, (iii) choosing the best features, (iv) letting the prediction network decide what to do, and (v) making a prediction. First, the datasets are cleaned up by looking for outliers and filling in any missing numbers. The cleaned data are then "normalized," which means that the data are put in a certain range (0–1). After normalization, duplicate data are taken out during the optimal feature selection step. Here, the suggested J-SLnO algorithm is used to choose the best features. ML systems are hard to make accurate because prediction values change within their ranges. In this case, an SVM is used to make a prediction, and the prediction must be made for the given range.

According to the literature review as author showed that business, the economy, and society can all benefit from using AI in the travel and transportation field. They pointed out that the current AI, deep-learning neural network can be more useful than the old methods.

The services related to travel and transportation are where the biggest gains can be found. For example, AI methods can be used to find the best and fastest way to get from one place to another on the road, which is helpful for drivers and delivery services. By looking at information from monitors on the roads, one European company has been able to figure out how trucks work and how their drivers act in real time. This cut fuel costs by 15% and reduced shipping time. AI techniques can also help airlines avoid the cost of canceling flights due to bad weather or traffic. Other perks include less traffic congestion, better scheduling for public transportation, safety for people riding in self-driving cars, and improved air quality. In general, for the travel industry, the value goes from 62% to 128%, and for transport and operations, it goes from 62% to 89%. In the future, the use of self-driving cars will lead to a rise in the value of deep learning.

Focusing on Long Short-Term Memory (LSTM) in smart vehicle predictive models offers several advantages that make it a suitable choice for handling sequential and time-series data in the context of smart vehicles. Focusing on LSTM in smart vehicle predictive models is driven by its capability to handle sequential data, capture long-term dependencies, adapt to dynamic conditions, handle irregular time intervals, automatically learn features, provide precise predictions, facilitate efficient training, and mitigate challenges like the vanishing gradient problem. These attributes contribute to building robust and effective predictive models for smart vehicles.

AI is employed in smart vehicles' predictive models across various domains, from processing sensor data and predicting maintenance needs to enhancing navigation, optimizing energy usage, personalizing user experiences, and contributing to advanced safety features. These applications collectively contribute to making smart vehicles more efficient, safe, and user-friendly. Many obstacles must be overcome before self-driving cars become commonplace. Creating AI systems that are trustworthy and secure enough for public use is a major obstacle. There are also regulatory, legal, and ethical concerns to think about, such as how to handle liability in the event of an accident and how to keep passengers and pedestrians safe. There is great potential for the application of AI in the auto sector in the future, but first, we must overcome a number of obstacles. Important issues include developing industry standards, addressing legal and ethical concerns, and ensuring the safety and reliability of AVs. It will be impossible to meet these challenges without the combined efforts of automakers, technological firms, and government agencies.

9.8 CONCLUSION

The use of AI in vehicles and transportation could significantly improve both processes. The automobile industry is primed for a dramatic AI-driven transformation. AI ushers in a future of ease, from self-driving cars and individualized driving experiences to enhanced security and traffic management. The entire potential of AI in the automobile industry has yet to be realized, however, due to the continued importance of ethical issues and overcoming difficulties. The potential of AI in the automobile industry is promising in the future. The development of AVs and connected cars will be influenced by future AI advancements like reinforcement learning and deep neural networks. Increased safety, greater energy economy, and a more pleasurable driving experience are all on the horizon as AI becomes more deeply embedded in the automobile ecosystem.

REFERENCES

1. M. Cunneen, M. Mullins, F. Murphy, D. Shannon, I. Furxhi, and C. Ryan, "Autonomous vehicles and avoiding the trolley (Dilemma): Vehicle perception, classification, and the challenges of framing decision ethics," *Cybernetics and Systems*, vol. 51, no. 1, 2020. [Online]. Available: https://doi.org/10.1080/01969722.2019.1660541
2. A. A. Kashyap et al., "Traffic flow prediction models: A review of deep learning techniques," *Cogent Engineering*, vol. 9, no. 1, 2022. [Online]. Available: https://doi.org/10.1080/23311916.2021.2010510
3. V. Manimuthu et al., "Design and development of automobile assembly model using federated artificial intelligence with smart contract," *International Journal of Production Research*, vol. 60, no. 1, 2022. [Online]. Available: https://doi.org/10.1080/00207543.2021.1988750
4. P. Helo and Y. Hao, "Artificial intelligence in operations management and supply chain management: An exploratory case study," *Production Planning & Control*, vol. 33, no. 16, 2022. [Online]. Available: https://doi.org/10.1080/09537287.2021.1882690
5. D. P. S. Samanta et al., "Analyzing bibliometrics of AI and smart vehicles," In A. Khang, V. Shah, Vrushank, and S. Rani (eds.), *Handbook of Research on AI-Based Technologies and Applications in the Era of the Metaverse*. IGI Global. https://doi.org/10.4018/978-1-6684-8851-cho19
6. A. K. M. B. Haque and A. N. R. Nusrat Chowdhury, "Artificial intelligence in smart city-systematic literature review of current knowledge and future research avenues," In M. A. Ahad, G. Casalino, and B. Bhushan (eds.), *Enabling Technologies for Effective Planning and Management in Sustainable Smart Cities*, pp. 53–77. Springer Nature, Berlin.
7. A. Saranya et al., "A systematic review of explainable artificial intelligence models and applications: Recent developments and future trends," *Decision Analytics Journal*, vol. 7, June 2023. [Online]. Available: https://doi.org/10.1016/j.dajour.2023.100230
8. Z. Liu et al., "Artificial intelligence powered large-scale renewable integrations in multi-energy systems for carbon neutrality transition: Challenges and future perspectives," *Energy and AI*, vol. 10, November 2022. [Online]. Available: https://doi.org/10.1016/j.egyai.2022.100195
9. F. Naz et al., "Reviewing the applications of artificial intelligence in sustainable supply chains: Exploring research propositions for future direction," *Business Strategy and the Environment*, vol. 31, no. 5, July 2022. [Online]. Available: https://doi.org/10.1002/bse.3034
10. A. K. M. B. Haque et al., "Explainable Artificial Intelligence (XAI) from a user perspective: A synthesis of prior literature and problematizing avenues for future research," *Technological Forecasting and Social Change*, vol. 186, Part A, January 2023. [Online]. Available: https://doi.org/10.1016/j.techfore.2021.120305
11. L. Chen et al., "Milestones in autonomous driving and intelligent vehicles: Survey of surveys," *IEEE Transactions on Intelligent Vehicles*, vol. 8, no. 2, February 2023. [Online]. Available: https://doi.org/10.1109/TIV.2022.3223131
12. A. R. Javed and W. Ahmed, "Survey of explainable artificial intelligence for smart cities," *Electronics*, vol. 12, no. 4, 2023. [Online]. Available: https://doi.org/10.3390/electronics12041020
13. L. Li et al., "Simulation driven AI: From artificial to actual and vice versa," *IEEE Intelligent Systems*, vol. 38, no. 1, January-February 2023. [Online]. Available: https://doi.org/10.1109/MIS.2023.3235676
14. L. Ismail and R. Buyya, "Artificial intelligence applications and self-learning 6g networks for smart cities digital ecosystems: Taxonomy, challenges, and future directions," *Sensors*, vol. 22, no. 15, 2022. [Online]. Available: https://doi.org/10.3390/s22155750

15. X. Tang et al., "Prediction-uncertainty-aware decision-making for autonomous vehicles," *IEEE Transactions on Intelligent Vehicles*, vol. 7, no. 4, December 2022. [Online]. Available: https://doi.org/10.1109/TIV.2022.3188662

16. A. G. Prawiyogi, S. Purnama, and L. Meria, "Smart cities using machine learning and intelligent applications," *International Transactions on Artificial Intelligence*, vol. 1, no. 1, 2022. [Online]. Available: https://doi.org/10.33050/italic.v1i1.204

17. P. R. Anisha et al., *Intelligent Systems and Machine Learning for Industry: Advancements, Challenges, and Practices*. CRC Press, Boca Raton. FL, December 21, 2022. ISBN 9781032261447

10 Digitising Sustainable Fashion Technology to Make Way for a Tech-Savvy, Eco-Friendly, and Responsible Society

Rishab Manocha

10.1 INTRODUCTION

Technology is being used by companies of all sizes in the fashion sector to increase their ability to recognise and foresee market demand, as well as to more quickly respond with appealing designs and customised trends. This is all being done to improve their competitive advantage. As stated in the conclusions of the EU Strategy on Sustainable and Circular Textiles, Russia's unlawful military intervention against Ukraine has resulted in an escalation of the energy crisis, a lack of raw materials, and a demonstration of the vulnerability of global supply chains. This was caused by Russia's unlawful military intervention against Ukraine. Artificial intelligence will fundamentally alter how companies approach the planning, design, and production of the things they sell, with a particular emphasis placed on the ability to anticipate what kinds of clothes consumers will want to buy in the foreseeable future. On the other hand, the participation of humans in the design process will continue to be absolutely necessary for the foreseeable future.

The fashion industry is dominated by internet platforms and digital marketing tactics that may encourage clients to use augmented or virtual reality technology. These technologies are becoming increasingly popular. According to Gazzola et al.'s [1] research, these platforms and methodologies make it possible to communicate directly with clients. Tests conducted by fashion companies have demonstrated one thing, and that is that the involvement of humans is necessary to make use of the insights supplied by AI and translate them into apparel that is both appealing and wearable. Aside from the fashion industry, businesses are already applying AI to develop out-of-the-box prototypes for a wide range of things, such as airplane components, golf equipment, and more [2]. Fashion is just one of the industries that is using AI. One of the fields that is currently making use of artificial intelligence is the fashion business. According to the findings of an analysis conducted by CB Insights' Industry Analyst Consensus, the market for generative design software is expected to

DOI: 10.1201/9781032664828-12

reach $44.5 billion by the year 2030. It is projected that by the year 2030, this number will have been achieved. Google's Project Muze [3] was an experiment that was conducted by Google in collaboration with Zalando, a fashion marketplace that is situated in Germany. The experiment was first introduced in 2016. This was Google's first foray into the realm of user-driven AI fashion design, and this particular project was the one that got the ball rolling for the entire initiative. Based on information from Google's Fashion Trends Report and Zalando's design and trend data, a neural network was trained to understand colours, textures, style preferences, and other "aesthetic parameters" for this study.

The neural network was trained with these "aesthetic parameters" in mind. After that, Project Muze used an algorithm to generate designs that catered to user tastes and were consistent with the network's established aesthetic norms. These designs were then developed and implemented. After that, these designs were incorporated into the product that was ultimately created. Additionally, Amazon is making substantial headway in the direction of technological advancement in this area. Amazon is working on a project that will utilise machine learning to determine whether or not a certain item can be classified as "stylish." This is one of the projects that they are working on (Figure 10.1).

The direction of this project would be handled by experts who are headquartered in Israel. Another one, which was developed by Amazon's Lab126 [4] Research and Development division in California, would take photographs to learn about a certain fashion trend and then make new images that are visually comparable to the originals. This one was able to do both simultaneously. If you're thinking that this sounds like "fast fashion by Amazon," then you're probably right to have that idea of what's being discussed here. In 2017, the e-commerce giant submitted a patent application for a manufacturing system that would enable the creation of clothing on demand. This would be made possible by the system's ability to make garments on demand. The technology was developed so that it might be used in the manufacturing of clothes. This technology might be put to use to help support Amazon's Essentials

FIGURE 10.1 Project muze. (Courtesy: Strategy Lead, Creative Works, Google.)

line of products or the suppliers in Amazon's logistics network. Both of these areas are possible applications for this technology. These two domains both present opportunities for the implementation of this technology.

It is to be expected that the designs generated by AI that do not include any input from humans will not always be ready for prime time on the runway. The majority of the designs that were generated for customers of Google's Project Muze were scribbles and scrawls that could not be worn, while several publications on the Amazon Lab126 initiative referred to the design outcomes as being "crude." Difficulties that are both legally and aesthetically problematic have arisen as a result of the utilisation of algorithms to generate garments. The implementation of these methods has resulted in the occurrence of various complications. A number of online T-shirt merchants, for instance, were found to be using bots to collect photos (under which users had written things like "I want this on a T-shirt") and submit them to marketplaces in 2019 so that they could be manufactured and sold on demand. This was done so that the online T-shirt retailers could make more money off the demand for their products. The desire that individuals had expressed for the T-shirt merchants' items was capitalised on by this method, and the online T-shirt merchants were successful as a result. This particular controversy was one of many that rocked the world of online retail in 2019, but it stood out as one of the most significant ones. As a direct and immediate result of this, claims of theft of intellectual property and breach of copyright were raised almost immediately. In spite of this, the distance between the designs generated by AI and those generated by humans is getting smaller.

An AI "designer" named DeepVogue [5] placed second overall and won the People's Choice Award in the International Fashion Design Innovation competition held in China in April 2019. This competition was part of the International Fashion Design Innovation Competition. Shenlan Technology [6], a Chinese business that specialises in information technology, developed a system that applies the concept of deep learning. Shenlan Technology is responsible for the development of this system. The photos, topics, and keywords that are entered into the system by human designers are the basis for the system's creative design output, which is produced based on the system's creative design output. As part of a project that they call "Algorithmic Couture," the design consultancy company Synflux [7], which has its offices in Tokyo, has also been utilising AI in order to develop one-of-a-kind ideas. They refer to this endeavour as "Algorithmic Couture."

A device that is capable of producing customised clothing by carrying out a set of actions in a specific order was created by a group that included both designers and software developers among its members. This group was responsible for the development of the device. The software starts by performing a scan in three dimensions of a person's body to gain an accurate measurement of the proportions of that person's physique. After that, the data are given to machine learning algorithms to be analysed, which results in the fabrication of clothing patterns with the intention of cutting down on the amount of fabric that is wasted during the manufacturing process. In the last stage of the process, designers develop fashion patterns that can be used to construct a variety of pieces of clothing by modelling these 2D patterns using computer-aided design (CAD) software. These patterns may then be utilised to create the garments themselves. After that, the patterns can be utilised to make the clothes in question.

10.2 REVIEW OF LITERATURE

The term "smart textiles" can refer to both an academic discipline and concrete products that increase textiles' practicality and utility. Fibres, filaments, and yarn are all examples of "smart textiles" that can communicate with their surroundings or the person wearing them. You can knit, weave, or make these fabrics without ever weaving at all. E-textiles are textiles that have been integrated with electronic components and could be used in the production of smart materials. These materials are referred to as "smart materials." The term "smart material" refers to this category of hybrid product in its more complete form. The term "smart textiles" refers to materials that have been designed with the assistance of contemporary technologies to confer additional advantages on the person who is wearing them. The body of research pertaining to intelligent textiles can be partitioned into a few distinct categories. In spite of this, one of the categories that is recognised by the vast majority of people is the classification that takes into account both the aesthetic and functional functions of the garments. The use of interactive fabrics has made it possible to increase the capabilities of the tactile interface.

A design for a textile interface that has potential should take into account applications that are typically used. Interactive textiles will likely serve as the new paradigm for the development of creative solutions to the issue of integrating electronics into fabrics or clothing. This might happen if interactive textiles become increasingly popular. They can respond and adjust their conduct in intelligent ways and present potential answers for challenges in a range of fields, such as the medical, athletic, automotive, and aerospace industries, amongst others. Additionally, they can respond and alter their behaviour in a manner that is consistent with their intelligence. They can adapt in response to the conditions or stimuli in their environment. It is feasible that one day, the method by which we interact with interactive textiles will take the place of the use of mobile devices in our daily lives. Because of a technology known as smart textiles, it is possible that one day the capabilities of a smartphone may be incorporated into the clothing that we wear on a daily basis. There is a growing movement towards the functionality of mobile devices being integrated into materials that may be worn. In fact, handheld electronic devices have become so ingrained in our routines that we frequently fail to notice the significance of the role they play in our lives. What do you believe the next step will be for electronic gadgets that can be carried around? What type of technology would be bold enough to replace the functions that are currently being carried out by smartphones and keep moving forward into the next stage of technological progress and cultural evolution? (Figure 10.2)

According to Adam Pruden [8] from Frog Design, the era after smartphones "will integrate wearable technology with leading companies such as DHL, Google, and Amazon." In the not-too-distant future, the incorporation of electronics into textiles will open up new avenues of communication, ushering in an era in which pervasive and ambient computing will be the norm. In the future that we are looking at, we will be encircled by textile interfaces that are designed to make the digital aspects of our lives easier. Integration of electronics into textiles allows for new communication possibilities, which can now be taken advantage of with these new waves. The term "digitalisation" has amassed a great deal of popularity in contemporary culture as a

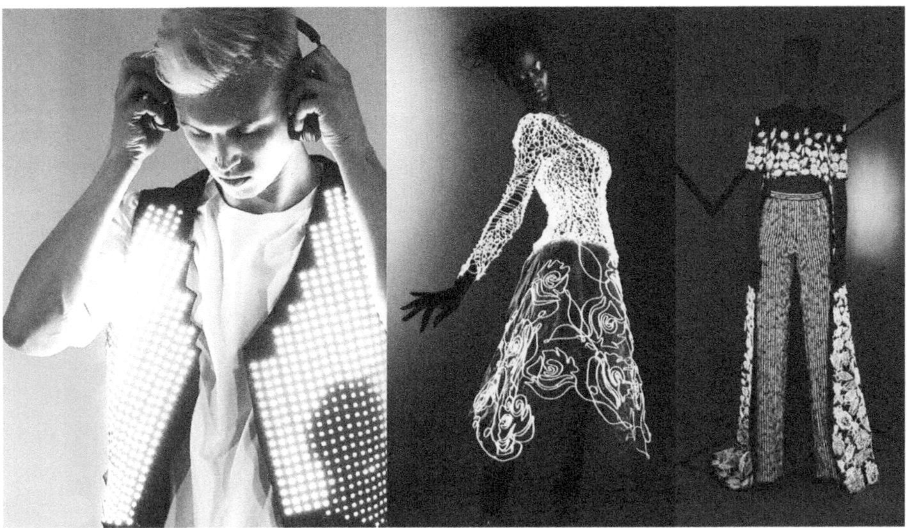

FIGURE 10.2 Smart textiles. (Courtesy: https://medium.com/@Geniemode/smart-textiles.)

result of its rapid spread into every facet of modern life in the second decade of the 21st century. Participation in the development of complex sociotechnical systems that can manage processes in the economic, social, and cultural sectors as well as in everyday life has become an integral component of the digitalisation proliferation sphere in recent years [9].

10.3 AESTHETIC SMART TEXTILES

Fabrics that have intelligent aesthetic features, such as the ability to glow in the dark, light up, and change colour, are increasingly being used in the fashion industry. One example of this trend is the use of glow-in-the-dark fabric. Textiles that are both aesthetically beautiful and intelligent have a number of applications, both in traditional and commercial settings, such as clothing that emits light.

10.4 PERFORMANCE SMART TEXTILES

Smart textiles can be placed into one of three categories according to the degree of intelligence they possess: passive, active, or extremely smart.

10.5 PASSIVE FABRICS

Passive intelligent textiles are the first generation of smart textiles that sense ambient conditions. They are referred to as "passive" for short. UV-protective clothing, conductive fibres, and other technologies that are analogous to these are some examples of passively intelligent textiles. Because they are only sensors, passively intelligent textiles can only gather information about the world around them and cannot actively

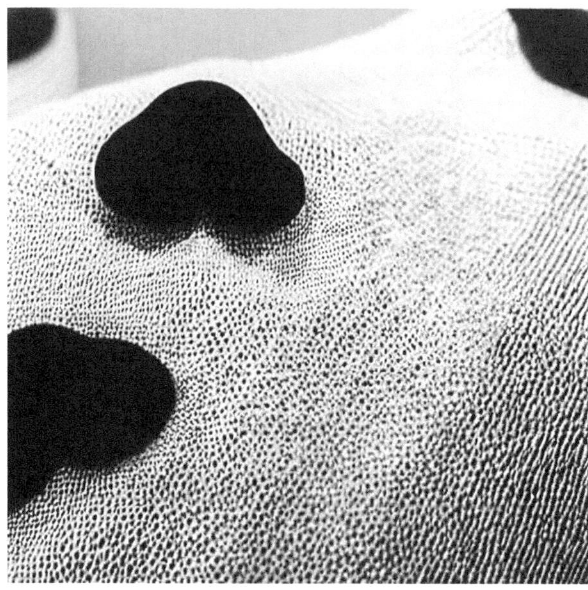

FIGURE 10.3 Passively intelligent fabrics (AI-generated image).

interact with it. The functioning of passively intelligent fabrics, which are commonly referred to as the "first generation" of intelligent textiles, is frequently alluded to as being superior to that of standard ones. On the other hand, it is essential to stress that passive materials, in general, do not adapt as a result of the information that they are exposed to. This is an extremely crucial distinction to make. To put it another way, even if the circumstances of the surroundings change, the fabric will not transform as a result of this. Using a cooling towel, for instance, may make it simpler to keep a healthy body temperature, but it does not actively produce cooler temperatures on its own (Figure 10.3).

The process of liquid evaporation is not significantly sped up by the construction of the cloth; rather, it merely makes the process slightly more noticeable. The same idea applies to anti-microbial, anti-static, and UV protection qualities that are included in various types of products besides just apparel, such as shoes and bags. Although process and product innovations have frequently been effective in mitigating the negative consequences of manufacturing and service operations, for the most part, they have been unable to achieve the critical mass required for widespread adoption [10].

10.6 ACTIVE FABRICS

Active and intelligent textiles can modify their functions in response to shifts in either the external environment or the input provided by humans. These shifts could involve things like mobility or changes in the weather. In addition to being able to do various functions, these textiles can alter their shape, retain heat, and exert control over them. Passive textiles, on the other hand, rely solely on their construction as their only support for the actuators and sensors that are embedded in active textiles. Active textiles also require the use of power. Actuators and temperature sensors were incorporated into the intelligent material, which enables it to respond to touches and determine its surrounding temperature. In addition to this, it can assess and make sense of a vast array of environmental data.

10.7 ULTRA-SMART FABRICS

Active smart textiles, in the same manner as super smart fabrics do, monitor their environments, adjust their qualities in response to what they observe, and learn from their experiences [11]. Ultra-smart textiles can detect, respond to, monitor, and adapt to a wide variety of environmental factors, including but not limited to thermal, mechanical, chemical, and magnetic stimulation, as well as other types of stimulation. These chemicals are also known as "smart" materials in some circles. A module embedded in an ultra-smart textile possesses the processing capability necessary to imitate the functions of the human brain in areas such as cognition, logic, and activation (Figure 10.4).

10.8 SMART TEXTILES IN FASHION

Possible future applications of smart textiles include the apparel industry. These are the kinds of materials that are increasingly being used to give clothing a more polished look and feel better functionality, and the ability to connect with its environment and other devices.

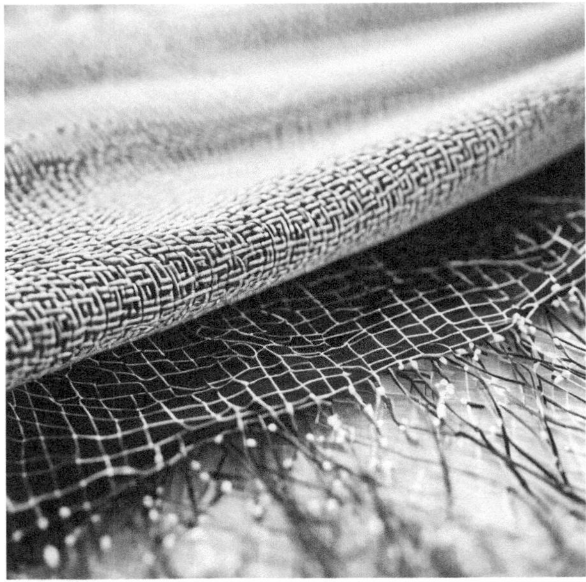

FIGURE 10.4 Ultra-smart fabrics (AI-generated image).

10.9 AESTHETIC VALUE

Smart fabrics can be designed with aesthetic considerations in mind, which may include making them light up and change colour, adding an interactive component, or adapting to their surroundings. Full collections have already been produced by fashion designers who have embraced the new technology and are using intelligent fabric. Textiles with photochromic, thermochromic, electrochromic, or solvatochromic properties can be used in a variety of ways in the fashion and interior design industries. For instance, CuteCircuit [12] creates a broad variety of distinctive garments that let the wearer express themselves in their own individual style. The Mirror Handbag, made from ultra-light aircraft aluminium and laser-etched acrylic mirrors, is a perfect illustration of this idea. These materials allow the light from the white LEDs to pass through, allowing for the creation of stunning animations in the form of display messages and tweets.

10.10 PERFORMANCE ENHANCEMENT

Actions taken to boost efficiency technologically sophisticated fabrics offer a one-of-a-kind feel that is specific to the purpose for which the fabric was designed. One way to achieve this is to prioritise the item's practicality over its form. This includes keeping the body at an appropriate temperature, protecting it from the elements (wind and water), blocking radiation, and keeping tabs on how hard the heart and muscles are working. Passive elements, like optical brighteners and UV absorbers, can be included in fibres to increase their capacity to shield the wearer from the sun. These materials, when used in conjunction with other technological elements, may be able to successfully protect the skin against UV damage and related illnesses.

These conditions are caused by prolonged exposure to the sun. A nickel-titanium alloy is a good illustration of a shape memory alloy that is used in the textile industry (Figure 10.5).

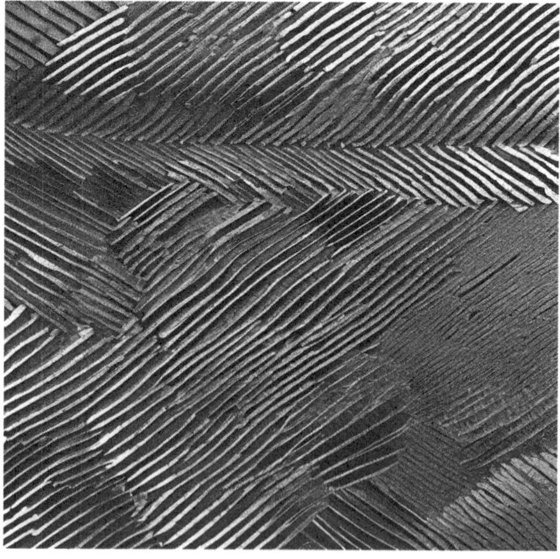

FIGURE 10.5 Nickel-titanium alloy (AI-generated image).

This particular alloy is used in the construction of protective gear that is worn in environments with high temperatures and the potential for fire. The amount of protection it offers is proportional to the temperature at which it is worn. A number of firms are now working on the development of intelligent clothing and accessories for sports. These clothes and accessories will be able to monitor and gather data based on the actions of the athlete. For instance, ReTiSense [13] came out with a high-tech insole that runners can stick inside their shoes to track their progress and get feedback on how they're doing. It is likely that the intelligent insole will assist runners in improving their form and lessening the likelihood of them becoming injured. In addition, businesses like WearableX [14] and Athos [15] have played a significant role in paving the way for high-performance athletic apparel that provides players with added functionality.

10.11 INTERACTION

Google's Jacquard [16] is a great example of how e-textiles may be used to maximise the potential of external devices. Information can also be shown on e-textiles. Jacquard was developed by Google as a wearable technology, and it is now being incorporated into a wide range of apparel and accessories. Google and Levi's collaborated on the creation of a smart jacket, a wearable electronic device. The jacket's capacitive touch grid interface allows the wearer to perform actions like answering calls, playing music, taking photos, and receiving directions with a single move, as the grid is woven directly into the fabric. It's a stylish jacket, for sure. E-textiles not only make it easier to operate electrical devices, but they also reduce battery drain (Figure 10.6).

10.12 THE PUSH FOR SUSTAINABILITY IN FASHION

It is generally agreed that innovative eco-design is necessary for the fashion industry to evolve into a more sustainable ecology [17]. Sustainability has emerged as a major new trend in recent years, influencing businesses across many sectors. This includes the retail sector. According to the NYU Stern Center for Sustainable Business, sustainability-marketed products were responsible for more than half of the consumer packaged goods industry's growth between 2015 and 2019. However, they only made

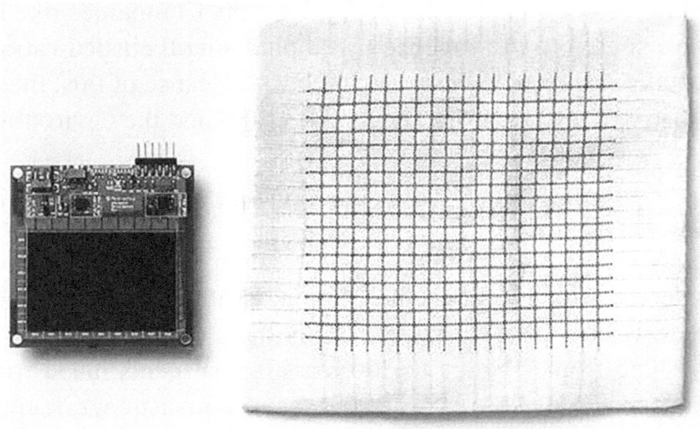

FIGURE 10.6 Google Jacquard. (Courtesy: Fortune.com.)

up 16% of the market in 2015. This occurred despite the fact that just 16% of sales are of sustainability-marketed products. More than $130 billion was made in retail sales in the United States in 2021 from environmentally friendly consumer packaged goods. This amounted to a 16.8% increase from the previous year. As consumers become more conscious of the issues associated with rapid fashion, the apparel industry is beginning to suffer the effects of the push towards eco-friendly shopping. The reason is that green purchasing is one of the best strategies for businesses to lessen their environmental footprint.

The notion of "slow fashion," which encourages the use of environmentally friendly materials and transparency about production techniques, is gaining popularity among consumers who are socially conscious. The "slow fashion" ideology places a strong emphasis on the utilisation of eco-friendly materials. The first three months of 2020 saw a 37% rise in the number of searches for phrases linked to sustainability compared to the same period in 2019. As a result, the average number of monthly searches for these phrases now exceeds 32,000, up from 27,000 in 2019. There was a decline in these keyword searches in 2019. A survey conducted by the Conference Board to gauge global consumer confidence found that 83% of American millennials have a positive view of companies that adopt programmes to enhance the environment. Especially the younger generations are becoming increasingly concerned with environmental sustainability. Furthermore, 75% of respondents are willing to adjust their buying habits to support eco-friendly enterprises.

Firstinsight's [18] research found that the demographic that includes Generation Z consumers is the most likely to pay a premium for sustainable goods. Emerging brands in the fashion sector are experiencing changes to accommodate the industry-wide trend toward catering to a more discerning clientele. It is only natural that further fashion industry transformation will take place online, as current shifts are founded on the benefits of digital advances, which are most apparent in a setting that necessitates the collaboration of specialists from a wide range of fields. The benefits of digital advances become most apparent in interdisciplinary settings [19]. When it comes to their fitness products, Girlfriend Collective prioritises honesty and openness above all else. Products include recycled polyester leggings, and the brand values transparency and honesty above all else. The Swiss-based firm, in founded in 2010, introduced a completely recyclable pair of shoes in the fall of 2021, together with a subscription service that served to further complete the recycling loop. Companies like Everlane and Reformation have soared to prominence by emphasising their dedication to socially and environmentally responsible business practices. Because of this, these businesses may now advertise to consumers that they value ethics and the environment.

10.13 RESALE PLATFORMS AS SUSTAINABLE ALTERNATIVES

Poshmark, ThredUP, and Depop are just a few examples of the thriving resale and consignment industry, which has benefited from consumers' growing concern for the planet's resources. Users of these sites can easily trade or resell gently used clothing. It can take anywhere from 20 years to 200 years for garments made from synthetic, non-biodegradable materials like polyester and spandex to disintegrate entirely in landfills. This is because these materials do not break down into smaller molecules as they break down naturally. Cotton can be broken down in as little as one week, whereas silk might take anything from one to four years to do the same thing. Producing synthetic textiles takes significantly longer. The act of reselling a garment extends its useful life and postpones the day on which it will be thrown away, but it does not eliminate the possibility that this day may eventually arrive. The trend is beginning to make its way into the consciousness of major retailers. The environmentally conscious department retailer Selfridges in London unveiled their Project Earth sustainability initiative in August 2020. This strategy incorporates the use of clothing that is less harmful to the environment, the establishment of a clothing rental business, and the opening of a secondhand store [20]. In addition, the brand Cos, which is sold by H&M, has started running its own reselling firm. Cos is a brand that H&M owns. In the meantime, well-known names in the fashion business, such as Anna Sui, Rodarte, and Christopher Raeburn, are targeting clients in the Gen Z age by leveraging the online secondhand marketplace Depop. The return scheme at Nike is called "Nike Refurbished," and it was started very recently. This initiative will collect previously owned footwear, sterilise it, and then put it up for sale at a reduced price after collecting, cleaning, and sanitising it. Under this plan, shoes can be recycled if they fall into one of three categories: brand new, shoes that have been worn only very lightly, or shoes that have cosmetic faults. As a direct result of the resale trend, white-label recommerce solutions have rapidly become one of the most sought-after commodities among fashion businesses. Startups such as Trove offer B2B services on both the back-end and front-end of their business operations. These services include the sorting of clothing and the establishment of branded resale websites. Trove's clientele consists of well-known brands such as Lululemon, Patagonia, and Levi's, among others.

10.14 NEW EARTH-FRIENDLY TEXTILES

Substitute materials, such as leather derived from plants and manufactured in a lab, may play a significant role in the fashion industry's efforts to improve its environmental footprint. Recent technical developments are inspiring many of the initiatives in this field. U.S.-based biotech firm Modern Meadow, which has raised over $180 million, is developing a leather substitute by fermenting special yeast strains to produce collagen, a key ingredient in conventional leather, and then processing the resulting protein. Traditional leather production comes with ethical and environmental difficulties that can be sidestepped using this method. As a result of this method, no animals are harmed in the production of the company's leather goods (Figure 10.7).

Bolt Threads, a startup, is creating a product called Mylo that gives the impression of being made of leather by extracting the proteins from mushrooms. Our company's Vice President of Product Development, Jamie Bainbridge, claims that our material feels identical to natural leather upon touch. This is what Jamie Bainbridge has said. If no one had told you, you would have sat there trying to guess whether or not it was made of leather.

The collections that Bolt has produced in collaboration with Stella McCartney, Adidas, and Lululemon have attracted the attention of a wide variety of fashion and retail businesses, which has resulted in Bolt receiving interest from a variety of these brands.

FIGURE 10.7   Modern Meadow. (Courtesy: https://www.usatoday.com/.)

10.15   RE-THINKING FASHION FROM END TO END TO ACHIEVE CIRCULARITY

In the special issue titled "Sustainable Chemistry for Circular Fashion" [21], an investigation into the role that green chemistry plays in the process of designing textile goods for use in circular fashion is conducted. The reality that achieving sustainability is a challenging goal is made abundantly obvious by the environmentally responsible actions that have been taken by fashion labels and technological companies. Circularity is essential for ensuring long-term sustainability; nevertheless, its implementation is difficult since it requires the adjustment of processes all along the supply chain. Long-term sustainability cannot be achieved without circularity. On the other hand, numerous fashion companies like Lululemon, Patagonia, and Nike are striving towards the objective of adopting some aspect of circularity in their businesses. This is often achieved by taking in used garments in exchange for credit, with the latter then being resold or the fabric used to make "new" garments.

The goal is to make it possible for clothes to have a shelf life that is as long as it is possible to make it since recycling worn textiles is favoured over producing new textiles. The term "circular fashion" refers to a business model in which items of clothing are reused and recycled until their value has been exhausted, and then they are disposed of in an environmentally responsible manner. Karel [22] identifies the following as important drivers of the closed fashion value chain: collaboration with partners, innovation, waste management, consumer connectivity, and altering usage models. To achieve circularity in the fashion industry, haute couture designer Yuima Nakazato is reimagining the garment-making process to create "garments for life." She really wants to get there. By using precise measurements of the wearer's body in conjunction with 3D printing, he reduces the amount of scrap that is thrown away during production. This is achieved by producing only the specific elements that are needed at any given time (Figure 10.8).

In addition, Nakazato eliminates the need to use needles and thread by manufacturing "modular" outfits with a 3D computational knitting machine. This process allows the clothes to be worn without any alterations. It is far simpler to repair and restyle existing clothing as opposed to purchasing fresh new garments. This is because

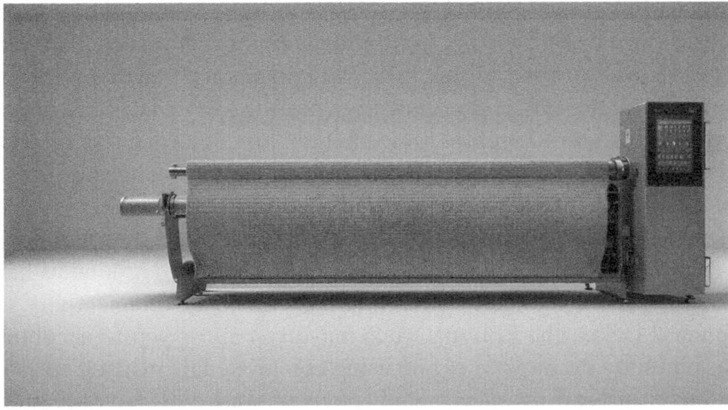

FIGURE 10.8 Computational knitting machine (AI-generated image).

various components of an article of clothing can be detached and reattached to other components of the garment. The idea that the designer has in mind is for clothes to become "permanent," which would do away with the necessity and the incentive to throw them away after each use. Nakazato has taken this process one step further and used it to manufacture a collection of ready-to-wear shirts. This is a step that Nakazato has taken. In the past, it had only been used for the creation of prototypes. There is a chance that in the future of fashion, unique fabrics that are made of materials of the next generation will also find their way into commercial use. It has been stated that professional football player Tom Brady sleeps in pyjamas that include a "bio-ceramic print" that, when paired with the athlete's body heat, generates far infrared radiation (FIR). FIR is a form of radiation that has the potential to lessen inflammation, improve circulation, and assist the body in more quickly recuperating from physically taxing activities. Alternating-coloured clothes are another fashion trend that could gain popularity in the years to come. Colour-changing T-shirts and mood rings were all the rage in the 1990s, but the most recent iterations of these products are substantially more complicated than their predecessors. Using a method that has been given the moniker ColorFab 3D and was created by researchers at MIT, it is possible to "print" products with "photochromic inks."

Upon being illuminated by certain wavelengths of UV light, the objects that contain these inks undergo a colour transformation. The first product that ColorFab developed utilising the technology was a ring that can be personalised to display any colour and can be pre-set to change into a variety of different colours. The ring was the first product that ColorFab used the technology to develop. The same group of individuals who are working on Google's Jacquard Threads are also creating a technology called Ebb, which is a method for colour-shifting fabric. Technology developed in collaboration with researchers at the University of California, Berkeley, has the potential to be programmed to respond to changes in the user's emotional state or the surrounding environment. Ivan Poupyrev, a researcher at Google, describes this as "moving away from the electronics" material. If you can weave sensors into fabric, you can incorporate them into apparel. To quote Google's Ivan Poupyrev, "moving away from the electronics" sums up what happens when you utilise this stuff. If you can effectively weave the sensors into the fabric, you can use them in clothing. It has been said that "by giving the components that make up our reality and the environment around us the ability to interact with one another," you... The Ebb materials have the same app connectivity as the Levi's Jacquard jacket, so they may one day allow you to replace your phone with colour signals for many of the functions you're used to using it for. For example, the Ebb materials may allow you to achieve many things that you currently do on your phone.

For instance, if you're wearing Ebb materials, the cuff on your wrist might turn a different colour to signal that you have a new call. The world around us has the ability to interact with one another, "you..." With the same app connectivity seen in the Levi's Jacquard jacket, the Ebb materials may even allow you to accomplish many things you presently do on your phone using colour signals instead, such as alerting you to an incoming call by altering the colour of your cuff. If you're wearing Ebb materials, for instance, your cuff may change colour to alert you to a new call. The fact that Harvard University scientists stated in January 2022 that they had created stretchy and soft temperature sensors that are suitable for use with intelligent clothes and soft robotics is even more incredible. This is a big advancement in the field of wearable electronics because sensors are typically rigid.

10.16 SENSORS IN TEXTILES

Temperature sensors, touch sensors, pressure sensors, optical sensors, chemical sensors, olfactory sensors, and many more types of sensors are among the numerous that can be included in textiles and used in a wide range of applications. Among the many senses that may be detected by various sensors, some of the more common ones include sight, hearing, taste, and smell. According to Oxman [23], the convergence of ecology, biology, nanotechnology, and the arts is providing designers with motivation to collaborate on the development of new ecological themes in design. Researchers have been concentrating their efforts on carbon-based nanomaterials such as carbon nanofibers, graphene, and carbon nanotubes (CNT) in recent years due to the promise that these materials hold as lightweight, flexible, and high-strain sensors. Nanofibers made of carbon are just one example of this category of nanomaterial (Figure 10.9).

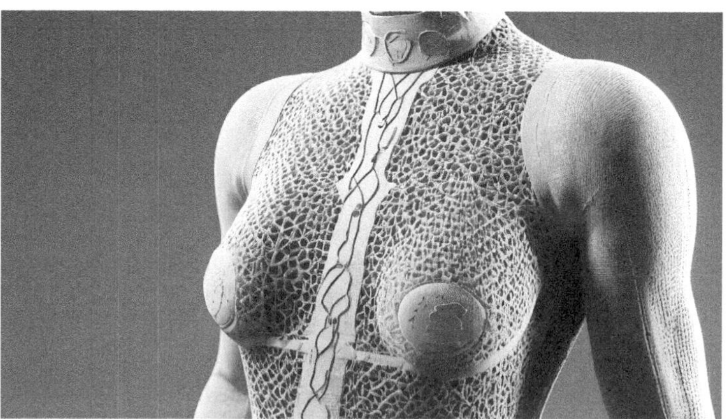

FIGURE 10.9 Algorithms and textiles (AI-generated image)

The development of intelligent clothing, the monitoring of human health, and the detection of human movements are all examples of possible future applications for these kinds of sensors. There are a few different approaches that have been developed for the synthesis of carbon-based nanoparticles, and these particles are currently being uniformly disseminated throughout polymers to employ them as strain sensors. Direct film casting and electrospinning were the two processes that were utilised in the production of strain sensors. In addition, it was possible to cast performers in films through direct casting. It has been discovered that high-performance strain sensors can benefit from the utilisation of carbon nanofibers as well as the materials that are produced by weaving nanofibers together. The results of the tests backed up this assertion. Human hairs that had been covered in graphene and used as the sensing element in strain gauges were also produced. In order to make strain sensors, in addition to spray painting, carbonisation, and stabilisation, silk and cotton fabrics were also utilised in the manufacturing process.

There are currently a variety of applications for which it is feasible to make use of sensors that are based on plasmons in intelligent textiles. It has been found that plasmonic sensors have a very high sensitivity when it comes to the detection of biological substances. Drawing techniques have the potential to be used in the production of multiple distinct varieties of plasmonic optical fibre sensors. Plasmon resonance is an important factor in the operation of a plasmonic fibre sensor, which is why it depends largely on this phenomenon. It is possible to stimulate, at a particular frequency, the phase-matching condition that exists between a mode that is guided by the core of an optical fibre and a surface plasmon mode at the interface between a metal and a dielectric. When a substance on a metal layer undergoes changes in its refractive index, the phase-matching requirement has been broken. Because of this, the spectral dip that usually appears at resonance is moved, and the change that this creates is recorded as a signal. In addition to the use of conventional single- or multimode optic fibres, the creation of a plasmonic sensor involves the implementation of a number of additional processes, such as etching, cladding, or polishing, and the subsequent deposition of several tens of metal nanolayers. These processes are followed by the latter step of depositing the nanolayers. It is necessary to make these modifications to achieve the desired results. However, the stack-and-draw process has the potential to provide the effective manufacture of a plasmonic fibre sensor of the highest possible quality. As a direct result of this, one might anticipate the most favourable conclusion. This is due to the fact that there are a great deal of challenges to conquer when applying these techniques to the production of plasmonic fibre sensors (Figure 10.10).

By embedding flexible capacitors into the fibre, it is also possible to produce fabrics that are sensitive to touch.

After the capacitor fibres had been woven into a one-dimensional sensor array using a Dobby loom, the array was subsequently encased in a wool matrix. A wool matrix served as the structure that held up the array. In the final design of the architecture of the touch sensor fabric, there were a total of fifteen capacitor fibres utilised. The voltage distribution and the local current both shift whenever a finger is brushed against one of these capacitor fibres. This shift is recorded, and then it is utilised to determine precisely where on the surface of the object the contact was made. These capacitors in the form of fibres can be joined with other conductive or battery fibres

FIGURE 10.10 Apparel with sensors (AI-generated image).

to produce wearable clothing that serves as an electrical circuit. This capability has a wide range of potential applications, some of which include programmable textiles, protective garments, and the fashion sector. Additionally, it is now possible to make pressure-sensitive fabrics, which can change their physical properties in reaction to an external force being applied to them. The dye-coating procedure was used to coat the fibres that were utilised in the fabrication of pressure sensors with organic conductive polymers such as poly (3,4-ethylenedioxythiophene) and poly (styrene-sulfonate), in addition to a layer of dielectric perfluoropolymer. These coatings were then covered with a dielectric perfluoropolymer. This was done to facilitate the construction of pressure sensors. After that, a layer of dielectric perfluoropolymer was deposited on top of these organic polymers. The treated fibres were used to weave the wefts and warps, and the unprocessed nylon fibres were used to fill in the matrix. The matrix that was created was made from nylon fibres. Capacitors were produced at the spots in the network where the individual fibres made new connections by coming into contact with one another.

After being subjected to a pressure of 4.9 N/cm², the capacitance of the fabric grew from 0.22 to 0.63 pF. This resulted in a sensitivity range that extended from 0.98 to 9.80 N/cm². In a similar fashion, temperature and humidity sensors are woven right into the fabric of the material. This procedure is very similar to the one that was described earlier. For the production of the sensors that are woven into fabrics, contemporary manufacturing methods such as photolithography and inkjet printing have been utilised. The creation of sensors that can be woven into textiles is made possible through the use of these approaches.

10.17 CONCLUSION

The contribution made by the fashion business is 2% of the whole value of the world economy, which comes to a total of three trillion dollars. This means that the fashion sector is responsible for a total of six hundred billion dollars. In addition to this, the United Nations concluded that this industry does not care about social and environmental concerns, such as the generation of CO_2 emissions and the wastage that occurs in the consumption and manufacturing of resources. This result was reached because the United Nations came to the opinion that this industry does not care about social and environmental concerns. Because the United Nations came to the conclusion that this industry does not care about social and environmental issues, this result was reached as a consequence of that decision. However, to accomplish the United Nations Sustainable Development Goals that are pertinent to the fashion sector, it is essential to integrate digitalised technologies such as the Internet of Things (IoT), Artificial Intelligence (AI), blockchain, Augmented Reality (AR), and Virtual Reality (VR). These technologies will help ensure that the goals are accomplished. As a direct result of the aforementioned reality, the investigation into the development of digitalised technology within the fashion sector was the primary objective of this particular piece of research. During the exploration, the study focused on digitalising technologies in the fashion industry for smart clothing, forecasting fashion trends, making dress recommendations based on environmental circumstances, predicting health, the real-time supply chain, as well as fashion and the shopping experience. Other topics that were investigated included fashion and the shopping experience. In addition to such subjects, we looked into fashion as well as the overall shopping experience. Based on the findings of the expedition, this report detailed the constraints and offered recommendations for how they should be addressed in the future. In the fashion supply chain, widespread adoption of blockchain technology is one of these proposals, along with breakthroughs in energy storage for smart fabric, integration of the IoT, AI, and edge computing, and a framework for smart clothing-based rescue operations.

REFERENCES

1. Gazzola, P., Pavione, E., Pezzetti, R. and Grechi, D. Trends in the fashion industry. The perception of sustainability and circular economy: A gender/generation quantitative approach. *Sustainability*, 12(7), 2809 (2020). https://doi.org/10.3390/su12072809
2. Pierson, H.A. and Gashler, M.S. Deep learning in robotics: A review of recent research. *Advanced Robotics*, 31, 821–835 (2017).
3. Project Muze. Stink Studios. https://www.stinkstudios.com/work/zalando-project-muze (2016).
4. Amazon Lab126. Amazon.Jobs. https://amazon.jobs/en/teams/lab126/ (2004).
5. Golub, A. DeepVogue AI: The Era of AI Design – Farewell Human Karl Lagerfeld? ELSE Research by ELSE Corp. https://blog.else-corp.com/2019/05/deepvogue-ai-the-era-of-ai-design-farewell-human-karl-lagerfeld/ (2019).
6. JCTC 胜蓝股份官网-连接创造价值 https://jctc.com.cn/ (2020).
7. Synflux. A Speculative Fashion Laboratory based in Tokyo. Synflux: A Speculative Fashion Laboratory Based in Tokyo. https://synflux.io/ (2018).
8. Adam Pruden. Frog, Part of Capgemini Invent. https://www.frog.co/authors/adam-pruden (2022).
9. Hund, A., Wagner, H.-T., Beimborn, D. and Weitzel, T. Digital innovation: Review and novel perspective. *The Journal of Strategic Information Systems*, 30(4), 101695 (2021). https://doi.org/10.1016/j.jsis.2021.101695
10. Shahi, C. and Sinha, M. Digital transformation: Challenges faced by organizations and their potential solutions. *International Journal of Innovative Science*, 13, 17–33 (2020).

11. Fernandez, C.T.M. and Fraga-Lamas, P. Towards the internet of smart clothing: A review on IoT wearables and garments for creating intelligent connected e-textiles. *Electronics*, 7(12), 405 (2018).

12. Cutecircuit. https://cutecircuit.com/ (2004).

13. ReTiSense. Stridalyzer Sensor Insoles – Sensor Insoles for Gait Research, Physiotherapy, Runners and Rehab. https://www.retisense.com/ (2014).

14. Wearable, X. Fashion Technology Company Building Future of Clothing. Wearable X. https://www.wearablex.com/ (2013).

15. Athos. Wearables.com. https://wearables.com/collections/athos (2012).

16. Jacquard by Google - Home. Jacquard by Google. https://atap.google.com/jacquard/ (2015).

17. UN Alliance. For Sustainable Fashion. https://unfashionalliance.org/ (2021).

18. Firstinsight. Gen Z Shoppers Demand Sustainable Retail. Retrieved from https://www.firstinsight.com/white-papers-posts/gen-z-shoppers-demand-sustainability (2020).

19. Nambisan, S., Lyytinen, K. and Yoo, Y. Digital innovation: Towards a transdisciplinary perspective. In Nambisan, S., Lyytinen, K., and Yoo, Y. (eds.), *Handbook of Digital Innovation*. Edward Elgar Publishing, Cheltenham, pp. 2–12 (2020). https://doi.org/10.4337/9781788119986.00008

20. Park, H. and Kim, D.Y. Feasibility and user experience of virtual reality fashion stores. *Fashion and Textiles*, 5, 32–17 (2018).

21. European Commission. EU Strategy for Sustainable and Circular Textiles. https://eur-lex.europa.eu/resource.html?uri=cellar:9d2e47d1-b0f3-11ec-83e1-01aa75ed71a1.0001.02/DOC_2&format=PDF (2022).

22. Karel, E. and Niinimäki, K. *Design for Circularity: The Case of circular.fashion. Sustainable Fashion in a Circular Economy*. Aalto University, Espoo, Finland, pp. 96–127 (2018).

23. Oxman, N. Design at the Intersection of Technology and Biology. Mit Media Lab. https://www.media.mit.edu/articles/design-at-the-intersection-of-technology-and-biology-2/ (2015).

Section C

Educational Technologies

11 Intelligent Tutor Systems and the Personalization of Pedagogical Objects in Adaptive e-Learning

Lamya Anoir, Mohamed Khaldi,
and Mohamed Erradi

11.1 INTRODUCTION

The rapid evolution of information technologies has profoundly transformed the educational landscape, offering new opportunities to personalize the learning experience. The emergence of intelligent tutoring systems (ITS) and adaptive e-learning represents a significant advance in this field, capitalizing on the capabilities of artificial intelligence (AI) to provide individualized and dynamic teaching [1]. These systems are designed to adapt in real time to the learner's knowledge, affective state, and behavior, offering an unprecedented level of personalized learning [2].

ITS represent a cutting-edge approach in the realm of educational technology, harnessing the power of artificial intelligence to enhance the learning experience for students. These systems go beyond traditional teaching methods by dynamically adapting to the unique needs and capabilities of individual learners. One of the key elements that contribute to the effectiveness of ITS is the personalization of pedagogical objects in adaptive e-learning environments.

In the rapidly evolving landscape of education, the demand for personalized learning experiences has gained prominence. Adaptive e-learning, facilitated by ITS, tailors instructional content, pacing, and assessment to suit the specific requirements and learning styles of each student. This personalization is achieved through the utilization of sophisticated algorithms and data-driven insights, allowing the system to continuously assess a learner's progress and dynamically adjust the educational content accordingly.

Adaptive e-learning has revolutionized education by giving learners access to tailor-made learning resources, adapted to their individual needs. ITS plays a crucial role in this transformation, offering personalized support to learners.

This manuscript aims to explore in depth the crucial importance of personalized learning objects within adaptive e-learning platforms. By integrating advances in AI, we aim to illuminate the underlying mechanisms that enable this personalization and assess its impact on learning effectiveness. Through an in-depth analysis of these systems, we aim to provide essential food for thought for researchers, educators,

DOI: 10.1201/9781032664828-14

and e-learning platform designers, to optimize digital educational environments and foster more effective learning adapted to each learner.

11.2 THEORETICAL AND METHODOLOGICAL FRAMEWORK

In this section, we present our theoretical and methodological framework used to frame the subject of our study.

11.2.1 THEORETICAL FRAMEWORK

11.2.1.1 Intelligent Tutoring Systems

ITS are a fundamental pillar of personalized e-learning. They leverage advances in AI to provide individualized, adaptive support for each learner [1]. ITS are distinguished by their ability to analyze the learner's interactions with educational content, making it possible to personalize the learning path according to his or her skill level, learning preferences, and previous performance [2]. This approach promotes more effective learning by focusing on the specific needs of each individual [3].

A notable aspect of ITS is its ability to precisely calibrate pedagogical content. They enable learners to progress at their own pace, which is essential to ensure optimal assimilation of knowledge [4]. Furthermore, ITS aim to optimize learner motivation by offering activities and challenges tailored to their skill level, helping to maintain their engagement throughout the learning process [5].

In sum, ITS represent a major advance in online education, offering personalized and adaptive teaching through the judicious use of AI. They play a crucial role in optimizing learning efficiency by targeting the specific needs of each learner and fostering their engagement and motivation throughout the educational process.

11.2.1.2 Adaptive e-Learning

Adaptive learning, sometimes called intelligent teaching, is a pedagogical method that uses technology as a teaching tool and is responsible for organizing human resources and learning materials according to the unique needs of each learner. An adaptive learning environment is one in which it is possible to monitor learners' activities, interpret them according to the domain-specific model, and deduce learners' requirements and preferences from the activities explained [6].

Adaptive e-learning is based on progressive assessments carried out throughout the learning process. These assessments make it possible to target the learner's shortcomings and create personalized learning objects containing additional explanations and application exercises.

11.2.1.3 Learning Object

A learning object, also known as a learning resource, is defined as "any element or support used to facilitate learning" [7]. These elements can take various forms, such as texts, images, videos, interactive simulations, exercises, presentations, quizzes, and computer applications, among others. The main purpose of a pedagogical object is to provide information, illustrate concepts, encourage reflection, or engage learners in learning activities.

Learning objects are often used in formal educational contexts, such as classrooms, online courses, or vocational training [8]. They are carefully selected and designed to meet specific pedagogical objectives and learner needs. Increasingly, with the advent of digital technologies, learning objects are available in digital formats and can be integrated into e-learning platforms [9].

In short, a pedagogical object is a tool or support designed to facilitate learning by providing information, illustrating concepts, or engaging learners in educational activities. These resources play a crucial role in creating an effective and stimulating learning environment [7].

11.2.1.4 Personalization

The term personalized is often encountered in the literature and refers to the provision of information specifically adapted to the needs of an individual or group of individuals [10].

Personalized learning is a systematic learning design that focuses on adapting teaching to learners. Personalized learning offers flexibility and supports what, how, when, and where learners learn and demonstrate their mastery of knowledge. Specifically, these flexibilities and supports are designed in terms of pedagogical approaches, content, activities, objectives, and learning outcomes. Personalized learning systems often exploit technology to improve access to personalized, adaptive learning for all learners [11,12].

Following definitions of our theoretical and scientific framework, we will present our methodology, which is based on a systematic approach to personalized learning for learners.

11.2.2 METHODOLOGY

This part of the methodology integrates learner profiling, domain modeling, pedagogical modeling, and interface design to create an adaptive and effective learning environment. Each step of the methodology contributes to providing tailor-made teaching that optimizes learner engagement, effectiveness, and motivation.

11.2.2.1 Learning Data Collection

The first step in our methodology is to collect learning data from users of the e-learning platform. This includes information such as responses to assessments, time spent on each module, interactions with content, and other relevant indicators. For this, we opted to use our learning platform made with Edx [13] which goes by the name TEN (Abbreviation of: Tetouan Ecole Numérique) [14]. This platform was made to offer learners an optimal learning experience by offering MOOCs [15], to different learner profiles according to their learning need based on their learning preference and pace [16].

11.2.2.2 Learner Profiling

Using the data collected, we will proceed to profile each learner. This involves analyzing individual characteristics such as skill level, learning preferences, identified gaps, and other relevant factors [2]. Learner profiling will enable us to create an accurate representation of their needs and preferences.

In the context of adaptive e-learning and ITS, learner profiling involves the continuous collection and analysis of data related to a learner's interactions with educational content. These data may include performance on assessments, time spent on different types of activities, responses to feedback, and more. Machine learning algorithms often play a crucial role in synthesizing this information to create a dynamic profile that evolves as the learner progresses.

11.2.2.3 Domain Modeling

Once the learner profile has been established, we turn to modeling the learning domain. This involves creating a model describing the structure of the knowledge to be taught. This model will serve as the basis for personalized learning content tailored to the specific needs of each learner [4].

Domain modeling refers to the process of creating a representation or abstraction of a specific knowledge domain, often within the context of software development, AI, or educational technology. In the realm of adaptive e-learning and ITS, domain modeling plays a crucial role in understanding the subject matter that is being taught and tailoring instructional content to the specific needs of learners.

11.2.2.4 Pedagogical Modeling

In parallel, we will develop a pedagogical model that defines teaching strategies and methods of interaction with the learner. This model takes the form of a learning scenario for each activity presented to the learner. The role of a scenario is to take into account the learning preferences of each individual, as well as best pedagogical practices to maximize teaching effectiveness.

11.2.2.5 Adapting the Platform Interface

Finally, we adopt a learner-friendly interface, taking into account each learner's interaction preferences. The interface will be designed to facilitate access to personalized educational content and to encourage learner engagement. With this in mind, we developed personalized models based on the data collected. These models made it possible to recommend specific pedagogical objects, such as videos, quizzes, and readings, according to learners' individual preferences and needs.

Our study aims to create a comprehensive system for personalized e-learning, taking into account the specificities of each learner. We aim to provide an optimized learning experience that promotes learner engagement, effectiveness, and motivation. This approach also aims to provide teachers with useful recommendations for adapting their teaching to the individual needs of each learner.

11.3 RESULTS AND DISCUSSION

The integration of personalized learning objects into adaptive e-learning platforms has a significant impact on learning efficiency. Learners benefit from content that is better adapted to their individual needs, boosting engagement and motivation.

11.3.1 Learning Data Collection

In this first phase, we gathered crucial information from learners on the TEN learning platform.

TABLE 11.1

Example of Data Collected from a MOOC

Learners	Time Spent (in minutes)	Ratings (out of 10)	Interactions (clicks)
Learner A	45	8	120
Learner B	60	7	150
Learner C	30	9	100
Learner D	55	6	130

TABLE 11.2

Interaction of Different Learner Profiles in the Chosen MOOC

Learners	Predominant Kolb Style	Motivation Level (out of 10)	Preferred Content	Time Spent (in minutes)	Score Evaluations (out of 100)
Learner A	Active	8	Videos	120	85
Learner B	Reflective	7	Articles	150	70
Learner C	Pragmatic	9	Simulations	100	90
Learner D	Experimental	6	Interactive exercises	130	75
Learner E	Active	7	Videos	110	80
Learner F	Reflective	8	Articles	140	65
Learner G	Pragmatic	9	Simulations	120	95
Learner H	Experimental	6	Interactive exercises	125	70
Learner I	Active	8	Videos	115	75
Learner J	Reflective	7	Articles	135	85

These data include time spent in the MOOC entitled "Pedagogical Approaches," ratings awarded by learners, and the number of interactions with the content (Table 11.1). ITS played a key role in analyzing these data to understand each learner's learning preferences and habits. This provides a solid foundation for personalized learning.

The information gathered, such as time spent on the MOOC, ratings, and interactions, will give us a detailed insight into the behavior of the learners we selected. This was crucial in understanding their preferences, performance, and learning habits.

11.3.2 LEARNER PROFILING

The learner profiling process was highly beneficial.

By analyzing the data collected, we were able to identify specific patterns for each learner (Table 11.2). This enabled us to understand their strengths, weaknesses, and learning preferences. This information was essential for effectively personalized learning paths. The learner profiling stage was enriched by the integration of Kolb's learning style model. This model, developed by David A. Kolb, identifies four distinct learning styles: active, reflective, pragmatic, and experiential. These profiles have been integrated by ITS to adapt learning content and activities to the specific needs of each learner. Each individual has a preference for one of these styles, which influences the way they assimilate and use information.

Table 11.2 presents fictitious learner profiling data, including predominant learning style according to Kolb's model, level of motivation, preferred content type, time spent on the platform, and assessment scores.

These data were used to personalize learning paths on the TEN platform, taking into account the specific preferences and needs of each learner. This contributed to a more adapted and efficient learning experience.

11.3.3 Pedagogical Modeling

Pedagogical modeling was central to the design of tailor-made learning scenarios. These scenarios were specifically designed to meet learners' individual needs and preferences. They provided objects, pedagogical activities, and resources that were in perfect alignment with each learner's profile.

Active Style: Activity scenarios in the form of hands-on simulations
- **Objective:** To enable the learner to put into practice the concepts learned in an interactive virtual environment.
- **Description:** The learner will be placed in a simulation where he/she will have to make decisions and solve problems using the knowledge acquired. He will be able to experience the consequences of his actions.

Reflective Style: Scenario activity in the form of analysis and discussion
- **Objective:** To encourage in-depth reflection and critical analysis of the concepts studied.
- **Description:** The learner will be invited to read an article or watch a video, and then write a thoughtful analysis. They will then take part in a group discussion to share their thoughts and hear those of others.

Pragmatic Style: Activity scenario in the form of a practical case study
- **Objective:** To emphasize the practical application of knowledge in real-life situations.
- **Description:** The learner will look at real-life case studies and propose practical solutions using the principles taught. They will be encouraged to identify and implement best practices.

Experimental Style: Activity scenario in the form of an exploration project
- **Objective:** To encourage experimentation and the exploration of new ideas.
- **Description:** The learner will be given the opportunity to choose an exploration project related to the subject studied. They will design and implement an experiment or project, and then analyze the results obtained.

In this step, we showed features of tailor-made learning scenarios taking into account learner profiles and pedagogical content available on the TEN platform, and also based on our previous work [6], these scenarios were designed to maximize teaching efficiency by proposing specific activities adapted to each learning style.

Using Kolb's learning style model, we identified the individual learning preferences of each learner on the TEN platform. ITS used these models with the scenarios to recommend relevant content adapted to each learner's learning style. This enabled us to categorize learners according to their predominant style, whether active, reflective, pragmatic, or experiential.

This personalized approach aims to maximize learning efficiency by aligning activities with each individual's learning preferences.

For active learners, we have developed hands-on simulations that challenge them to directly apply the concepts taught.

For reflective learners, we offer analysis and discussion activities, encouraging deeper reflection and critical analysis.

Pragmatic learners were challenged through practical case studies, encouraging them to apply their knowledge in real-life situations.

Finally, for experimental learners, we encouraged the exploration of new ideas through experimentation projects.

This tailored approach optimized learning engagement and effectiveness for each individual, taking into account their specific learning preferences and strengths. ITS contributed by dynamically adapting learning paths according to each learner's preferences, helping to create a more enriching learning experience tailored to each learner on the TEN platform.

11.3.4 Adapting the Platform Interface

In this step, we personalized the TEN platform interface to meet the specific needs of each learner according to their learning profile. Here are the results of this adaptation:

Personalized Recommended Content: Based on each learner's learning preferences, the platform interface was configured to recommend content that matches their learning style (Figure 11.1). For example, active-style learners are prioritized for hands-on simulations, while reflective learners receive recommendations for articles and analyses.

Intuitive, Interaction-Facilitating Interface: For active-style learners, the interface has been optimized to allow direct interaction with simulations and hands-on projects (Figure 11.2). For reflective learners, note-taking and reflection features have been integrated to encourage reflection and analysis.

Personalized Progress Tracking: Each learner can track his or her own progress and easily access recommended resources based on his or her learning style (Figure 11.3). Personalized dashboards provide specific information for each learner.

Adapted Feedback and Encouragement: The TEN platform interface has lbeen designed to provide feedback specific to each learning style (Figure 11.4). For example, active learners receive encouragement for their active engagement, while reflective learners receive feedback highlighting their deep reflection.

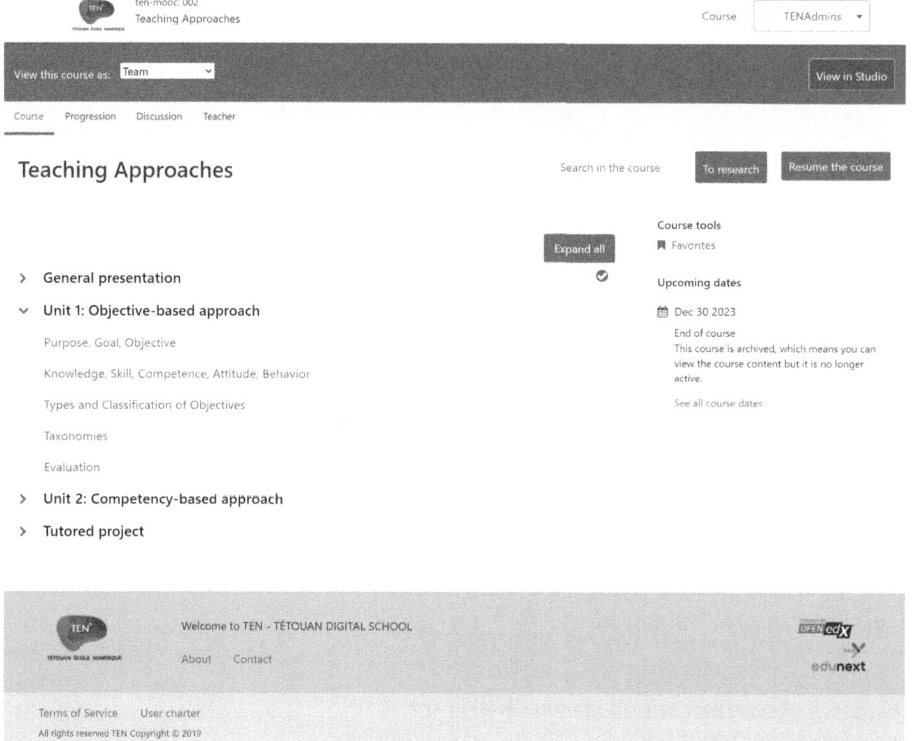

FIGURE 11.1   Recommended course interface.

By combining interface personalization with learning scenarios adapted to each style, we have created a learning environment that maximizes effectiveness and engagement for every learner on the TEN platform. This tailor-made approach has optimized the learning experience and fostered greater assimilation of knowledge.

The results of our study highlight the effectiveness of the personalized ITS-based approach to adaptive e-learning. By combining learning data collection, learner profiling, domain modeling, pedagogical modeling, and interface adaptation, we have created a dynamic learning environment adapted to each individual on the TEN platform.

The integration of Kolb's learning style model enabled learners to be categorized according to their learning preferences, facilitating the design of tailored learning scenarios. These scenarios were specifically adapted to maximize learning effectiveness, taking into account the strengths of each learning style.

In addition, ITS played a crucial role in analyzing learning data in real time and adapting pedagogical content according to each user's learning profile. This has delivered a highly personalized learning experience and led to improved knowledge retention.

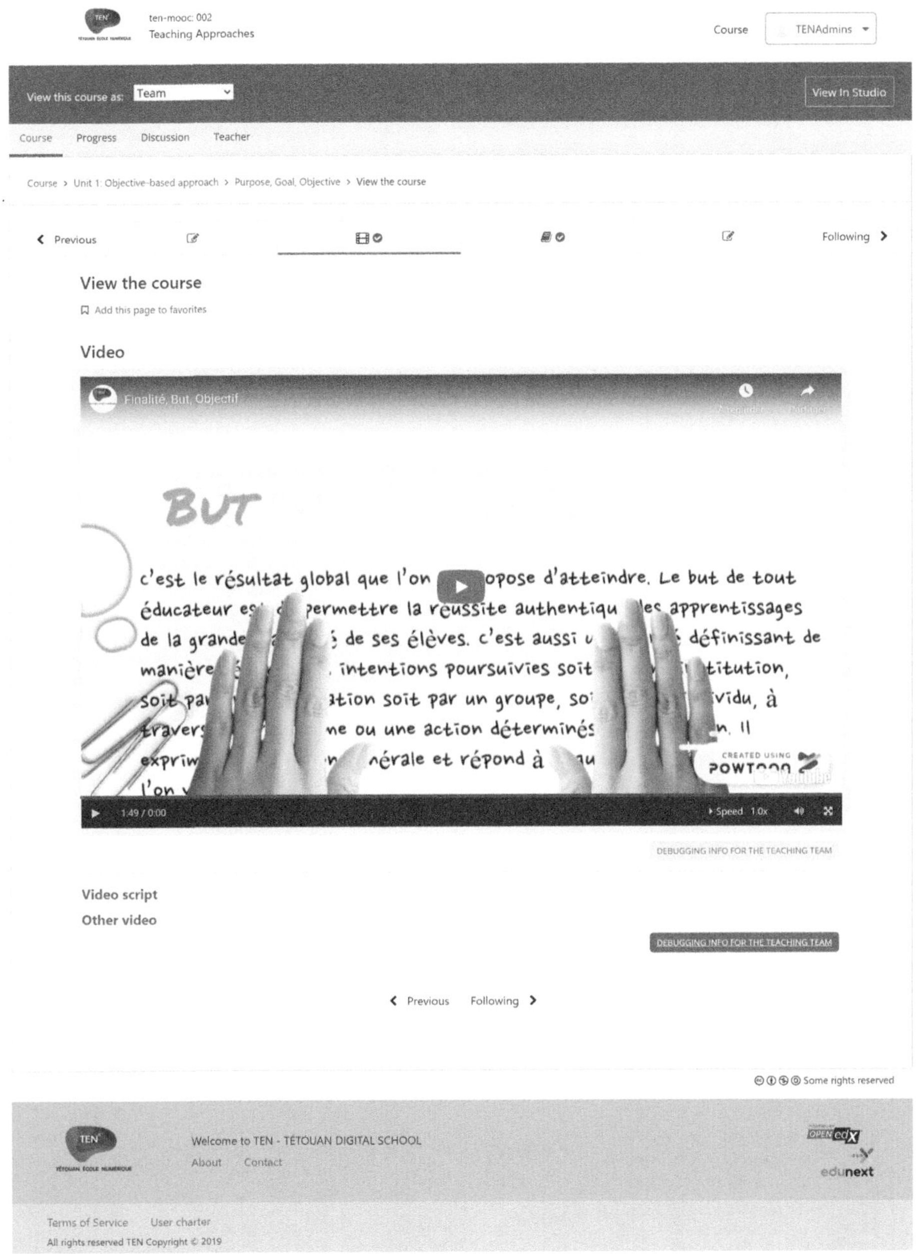

FIGURE 11.2 Intuitive, interaction-facilitating interface.

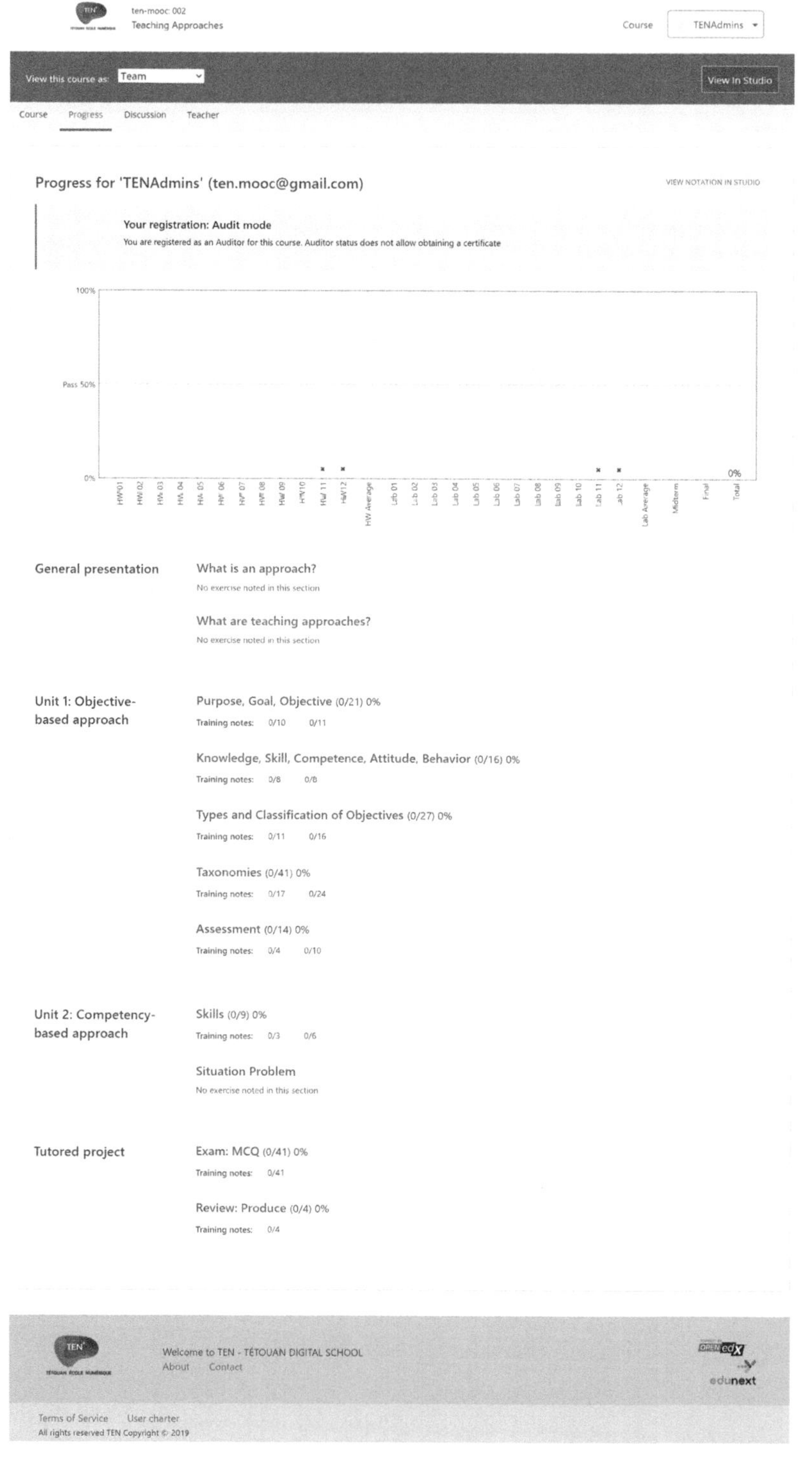

FIGURE 11.3 Personalized progress tracking.

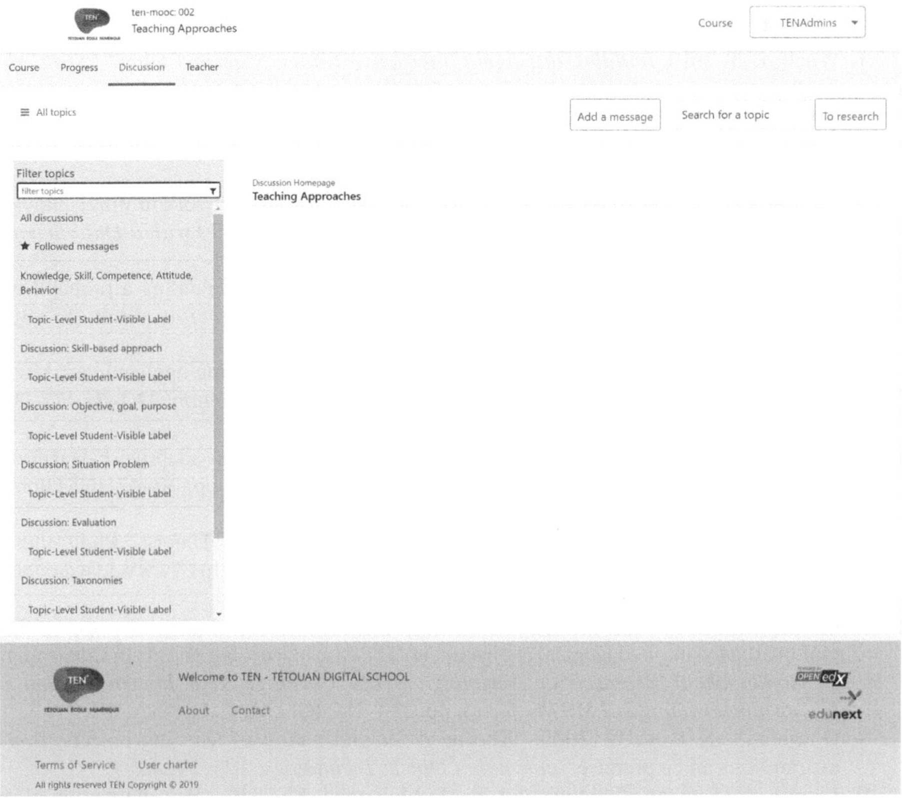

FIGURE 11.4 Feedback and encouragement.

As a concluding guide, our approach demonstrated that personalizing pedagogical objects using ITS has a significant impact on the effectiveness of adaptive e-learning. This methodology offers considerable potential for improving learner engagement, motivation, and success on the TEN platform. It also paves the way for new perspectives in personalized and adaptive online education.

11.4 CONCLUSION

In this chapter, we explored recent advances in the field of ITS and their ability to personalize educational objects, thus promoting a more effective and engaging learning experience.

In conclusion, the personalization of pedagogical objects in ITS and adaptive e-learning environments is an essential element in improving learning efficiency. The use of AI to personalize educational content according to the specific needs of each learner opens up new perspectives for online education. This approach makes learning more effective, motivating, and adapted to each individual, contributing to the advancement of digital education.

REFERENCES

1. Woolf, B. P. 2010. *Building Intelligent Interactive Tutors: Student-Centered Strategies for Revolutionizing e-Learning.* Morgan Kaufmann, Cambridge, MA.
2. Brusilovsky, P. 2000. "Adaptive hypermedia: From intelligent tutoring systems to web-based education." In *Intelligent Tutoring Systems*, pp. 1–7. Springer, Berlin, Heidelberg.
3. Banubakode, A., Gosavi, C. S., & Satpute, M. 2022. "Social network mining, analysis, and research trends: A case study." In *Designing User Interfaces with a Data Science Approach*, pp. 210–226. IGI Global, Hershey, PE.
4. Conati, C. & Heather, M. 2009. "Empirically building and evaluating a probabilistic model of user affect." *User Modeling and User-Adapted Interaction,* 19(3): 267–303. https://doi.org/10.1007/s11257-009-9062-8.
5. Vassileva, J. 2012. "Motivating participation in social computing applications: A user modeling perspective." *User Modeling and User-Adapted Interaction,* 22(1–2): 177–201. https://doi.org/10.1007/s11257-011-9109-5.
6. Anoir, L., Khaldi, M., & Erradi, M. 2022. "Personalization in adaptive E-learning." In *Designing User Interfaces with a Data Science Approach*, pp. 40–67. IGI Global, Hershey, PE.
7. Wiley, I. I. 2000. "Learning Object Design and Sequencing Theory." Ph.D. thesis, Brigham Young University. Retrieved July 10, 2024 from https://www.learntechlib. org/p/120055/.
8. Chelliq, I., Erradi, M., & Khaldi, M. 2023. "Mediatization and pedagogical mediation in a multimedia techno-pedagogical environment: The case of learning objects in E-learning." In *Handbook of Research on Scripting, Media Coverage, and Implementation of E-Learning Training in LMS Platforms*, pp. 289–314. IGI Global, Hershey, PE.
9. Downes, S. & Kim, W. 2000. "The need for and nature of learning objects: Some assumptions and a premise." *Journal of Object Technology,* 1(1): 29–40.
10. Lamya, A., Kawtar, Z., Mohamed, E., & Mohamed, K. 2020. "Personalization of an educational scenario of a learning activity according to the learning styles model david kolb." *Global Journal of Engineering and Technology Advances,* 5(3): 99–108. https://doi.org/10.30574/gjeta.2020.5.3.0114.
11. Walkington, C., & Bernacki, M. L. 2020. "Appraising research on personalized learning: Definitions, theoretical alignment, advancements, and future directions." *Journal of Research on Technology in Education*, 52: 235–252.
12. Anoir, L., Chelliq, I., Erradi, M., & Khaldi, M. 2023. "Adapting personalized learning activities to the learner's profile according to kolb." *International Journal on Engineering Applications*, 11(4): 240.
13. Open edX. 2018. "Accueil - Ouvrir edX." December 4, 2018. https://openedx.org/fr/.
14. Edunext.Io. 2023. "TEN (Tétouan Ecole Numérique)." Accessed November 6, 2023. https://ten-mooc.edunext.io/.
15. Lamya, A., Mohamed, K., & Mohamed, E. 2022. "Personalization between pedagogy and adaptive hypermedia system." In *Proceedings of the 5th International Conference on Big Data and Internet of Things,* ENSIAS, Rabat, Morocco, pp. 223–234. Springer International Publishing, Cham.
16. Reich, J. 2015. "Rebooting MOOC research." *Science*, 347(6217): 34–35.

12 A New Pedagogy Technique for Research-Based Application-Oriented Case Study Presentation

Shridhar Kedar and Himani Kadam

12.1 INTRODUCTION

While big lecture classes have mostly focused on peer discussion, research in engineering studies connected to the use of active learning methods has been more focused on instructions in lab sessions. The 'think-pair-share' is an active learning approach used in classroom settings. Think-pair-share is a group activity that was implemented in the undergraduate mechanical engineering program by Anand K. Bewoor [1]. The study reveals that group activities improve active engagement and boost student confidence. In contrast to the traditional lecture style, the new teaching paradigm integrates peer learning, faculty-facilitated learning, practical experimentation, group discussions, and web-enabled discovery. Sandhu Jaspal et al. [2] say that a set of web-based learning modules is a key component of this educational reform. Now, Outcome-Based Education is the main focus of engineering education rather than result-based education. For graduating students, attaining graduate qualities is very important because it speaks to their talents in communication, teamwork, self-learning problem-solving, topic understanding, etc. The classroom environment and student participation can both be enhanced by using active and cooperative learning strategies. It is difficult to get final-year students to participate actively in the learning process, which underlines the value of using active learning techniques with undergraduate final-year students. The use of classroom assessment techniques has been advocated as a way to close the learning-teaching gap. Formative assessment approaches are classified as classroom assessment procedures Sandeep Desai [3]. The one-minute paper, idea maps, muddiest point, and research-based learning are a few examples of classroom assessment techniques by Budiyanto Cucuk et al. [4,5]. In this article, the authors try to discuss two active learning strategies: (i). research paper collection for a given allocated topic and (ii). a 5–10-min subject presentation.

DOI: 10.1201/9781032664828-15

12.1.1 Gap Analysis

It can be observed that most of the studies from the literature review for various pedagogical practitioners used pedagogy techniques to enhance active learning skills for engineering students. Most of the researcher mainly focuses on their impact analysis of pedagogy technique. The present work mainly focuses on research-based pedagogy techniques for the subject of renewable energy sources (RES). It enhances student technical skills and current learning knowledge for related areas and motivation for various aspects and opportunities in the renewable energy sector.

12.2 LITERATURE SURVEY

Students in the last year of automotive engineering used active learning strategies in the "Hydraulic and Pneumatics" course created by Sandeep Desai et al. [3]. The jigsaw technique, muddiest point technique, concept mapping, case study, and team-based learning were a few of them. The various methods are applied to various sections of the "Hydraulics and Pneumatics" course that is taught to final-year automobile engineering students during their last semester of study. He concluded that the use of active learning strategies had a direct impact on passing rates. He noted that passing rates had risen by 21%.

The potential benefit of experiential learning to Indonesian primary school pupils' comprehension of renewable energy is evaluated by Çoker Bünyamin et al. [6]. The beginning points for the analysis are Bloom's elements of learning, specifically cognitive and affective. The authors highlight that researchers and actual teachers need to pay more attention to how elementary school pupils learn.

Three case studies for rural and urban communities in India have been described by Christian E. Casillas [5]. He had incorporated short-term learning projects that included theoretical and practical learning opportunities for the design, installation, and maintenance of renewable energy systems. He concluded that drawing on the pedagogical models of project-based learning and mentoring, the workshops demonstrated how complicated technical topics (i.e., current, voltage, energy, power, and electrical system design) can be learned quickly by a diverse group of children who had limited exposure to the topics and only basic skills in math and science. He also concluded that project-based learning (PBL) and mentoring have a great potential to empower students. PBL allows students to gain a practical understanding of technology that is not typically presented in schools and is often only handled by a small group of professionals. The paper written by Çoker Bünyamin et al. [6] aimed to investigate Turkish primary and secondary students' knowledge about RES. One hundred and seven primary and secondary school students participated in the study. Open-ended questions are used to determine students' knowledge about the topic. Students' answers were analyzed using the descriptive analysis technique. They concluded that much more learning outcomes related to RES must be added to the primary and secondary education curricula with environmental issues, which could be faced in daily life context. The research-based teaching idea can improve the quality of teaching and learning. Hongiun Yang [7] concluded that research-based teaching changes students to active participants and positive responders. The research-based teaching promotes the students' scientific learning method. S.A. Kedar et al. [8–16] mainly worked on a hybrid solar groundwater desalination system. They have developed various experimental setups to obtain fresh water from groundwater which is useful to society. Students select the topic of solar desalination systems for their study.

To overcome the logistical challenges, step-by-step introduction of in-class active learning techniques is recommended by leading engineering educators in a guide provided by Richard M. Felder. Active participation of students in the classroom plays a significant role in the outcome-based education model. Rujuta Agevakar et al. [17] implemented the muddiest point and one-minute paper for third-year mechanical students for the subject. They concluded that the techniques implemented improved the overall understanding of the course content. They also mentioned that active learning strategies help students develop interest in the course content. As per Sasikumar [18], a good teacher must have technical competency, but along with that, the teacher must employ various techniques to involve the students in learning process and make the learning a fruitful experience for both slow and advanced learners. Sasikumar has discussed various active learning techniques such as think-pair-share, class discussion, student debate, etc. He has also summarized the typical problems encountered while implementing these techniques and remedial measures for the same Richard M. Felder et al. [19]. From the literature review and discussion with various experts in the engineering domain, various studies have been conducted on pedagogical practices to enhance the knowledge and skill sets of engineering graduates. In this paper, a new pedagogy technique was adopted for presentation of research-based, application-oriented case studies.

12.3 METHODOLOGY

The RES is offered as open elective course for final year engineering students. This course has five units in the syllabus: (i) solar energy, (ii) wind energy, (iii) biomass technology, (iv) ocean, tidal, geothermal energy, and (v) hybrid energy systems. 14 students from Electronics and Telecommunication Engineering and 38 students from the Instrumentation and Control Engineering department had registered for this course. 73 students from the Mechanical Engineering department had registered for this course. A new pedagogy research-based application-oriented case study presentation has been implemented for students registered from the Electronics and Telecommunication Engineering and Instrumentation and Control Engineering departments. A total of 52 students were registered from these two departments. The subject teacher had discussed the active learning method research-based application-oriented case study presentation at the start of the semester around February 2022 with students. The active learning strategy had been implemented for all 52 registered students (Figure 12.1).

12.4 IMPLEMENTATION

The active learning strategies, viz. research paper-based presentation for hybrid system, is implemented for courses of the final-year open elective subject for Electronics and Telecommunications Engineering and Instrumentation and Control Engineering

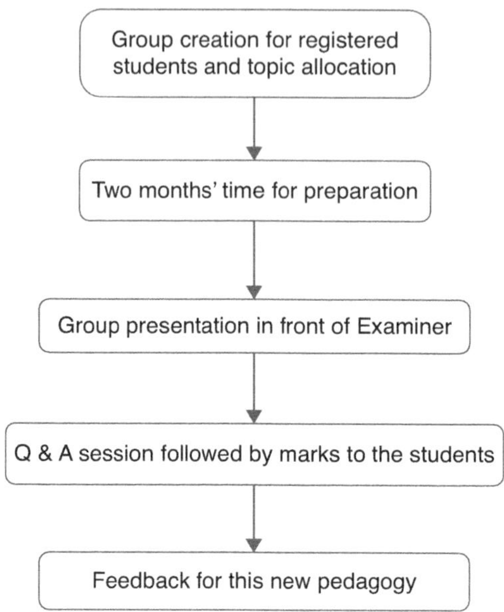

FIGURE 12.1   Procedural steps for the new pedagogy.

undergraduate programs. The course is RES which is an open elective subject offered at the institute level. This activity has been divided into three parts.

Part 1: Group creation based on the number of registered students and allocation of topic presentation. Part 2: Research paper searching and PowerPoint presentation, in the presence of senior faculty. Part 3: Feedback on this research-based active learning pedagogy. Part 4: Freedom was given to the students to choose their own team members. Topic allocation to the groups was done by the subject teacher. This is done three weeks after the start of the course teaching-learning process. Part 4: Two months for preparation were given to the student for research paper selection for their respective topic. Two weeks before the end of teaching-learning of the course, the presentation was conducted in the classroom. Senior faculty from the Mechanical department who has already taught this subject was the examiner for this group presentation and evaluation. The question & answer session was followed after the presentation. Ten groups were created. Each group was given a different topic for this activity. The following are topics given to the students: (i) solar thermal energy + photovoltaic, (ii) solar energy with wind energy, (iii) conventional energy and nonconventional energy, (iv) solar + wind + biomass, three-way hybrid system, (v) electric vehicle solar car hybrid (petrol/diesel, battery charged and solar PV), (vi) solar/wind, biomass, and tidal, (vii) biomass tidal geothermal and wind energy, (viii) four-way hybrid system (solar, wind, biomass geothermal tidal), (ix) electricity generation using hybrid energy system, and (x) solar energy application solar tree. Table 12.1 shows different topics allocated to groups.

12.4.1 SELECTION SAMPLING METHOD

When you conduct a research study about a group of students, it's rarely possible to collect data from every group of students. The sample should be selected from the group of students who actually participate in the research-based activity. To draw a valid conclusion we have to carefully decide on the selection of a sample of the group. In our research student activity probability sampling method was used which involves the random selection of a maximum group of students about the entire group activity. Table 12.1 shows a sample of group creation with topic names for their research-based case study presentation.

TABLE 12.1

Sample of Group Creation and Topic Name

Sr. No.	Department	Division	Roll No	C Number	Group No.	Topic
01	Instrumentation & Control	A	4509	C22018331509	1	Solar thermal energy + P V
02	Instrumentation & Control	A	4531	C22018331510	1	
03	Instrumentation & Control	A	4558	C22018331539	1	
04	Instrumentation & Control	A	4557	C22018331538	1	
05	Instrumentation & Control	A	4561	C22018331543	1	
06	Instrumentation & Control	A	4516	C22018331514	2	Solar energy with wind energy
07	Instrumentation & Control	A	4547	C22018331532	2	
08	Instrumentation & Control	A	4527	C22018331521	2	
09	Instrumentation & Control	A	4518	C22019332507	2	
10	Instrumentation & Control	A	4502	C22018331503	3	Conventional energy + nonconventional energy
11	Instrumentation & Control	A	4506	C22018331508	3	
12	Instrumentation & Control	A	4506	C22018331547	3	
13	Instrumentation & Control	A	4555	C22018331537	3	
14	Instrumentation & Control	A	4565	C22018331548	3	

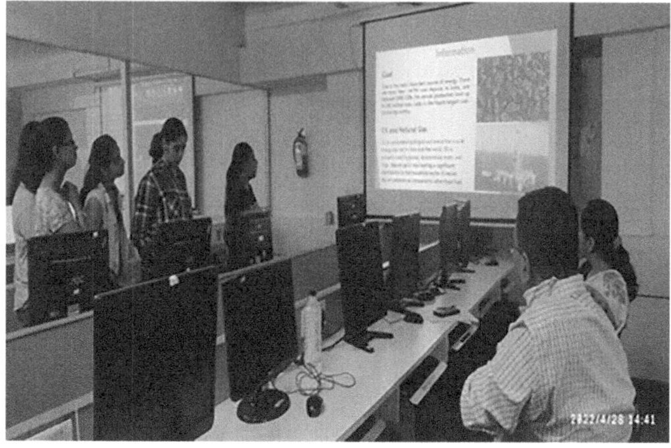

FIGURE 12.2   A group presentation was carried out in front of the examiner.

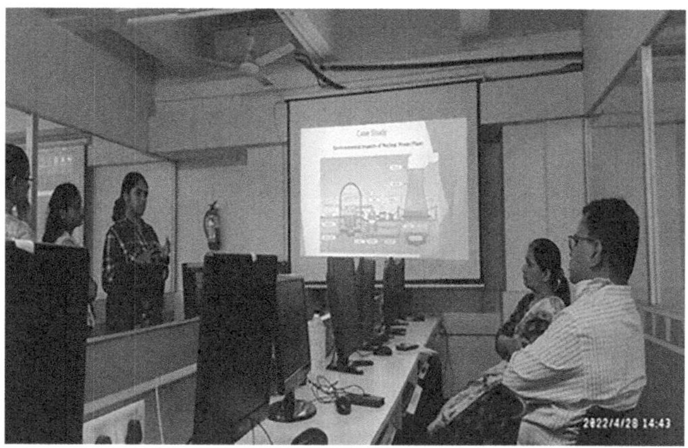

FIGURE 12.3   Group presentation in front of the examiner.

Figures 12.2 and 12.3 shows some of the photos of the same.

Figures 12.2 and 12.3 mainly show the group activity students' topic presentation in front of the examiner. The entire group prepration of (4–5) students presented the entire topic with their case study. Students inbuilt self-confidence and good learning ability for the said RES course.

Figure 12.4 shows the creation of Google Classroom for the final-year B.Tech. students for the RES course. Google Classroom is a useful tool for online platforms for students as well as faculty members to give assignments, study material, and new announcements for the RES course (Figure 12.5).

Figure 12.6 mainly shows feedback for research-based assignments for the RES course. It shows that more than 90% of students are satisfied with getting the actual

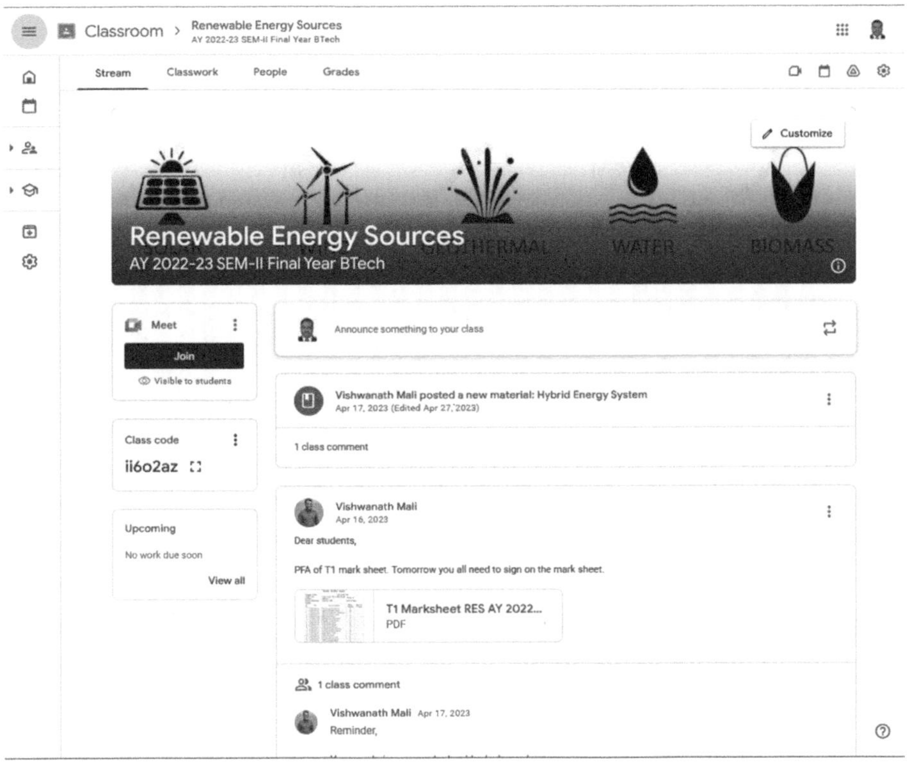

FIGURE 12.4   Screenshot for creation of Google Classroom for RES course.

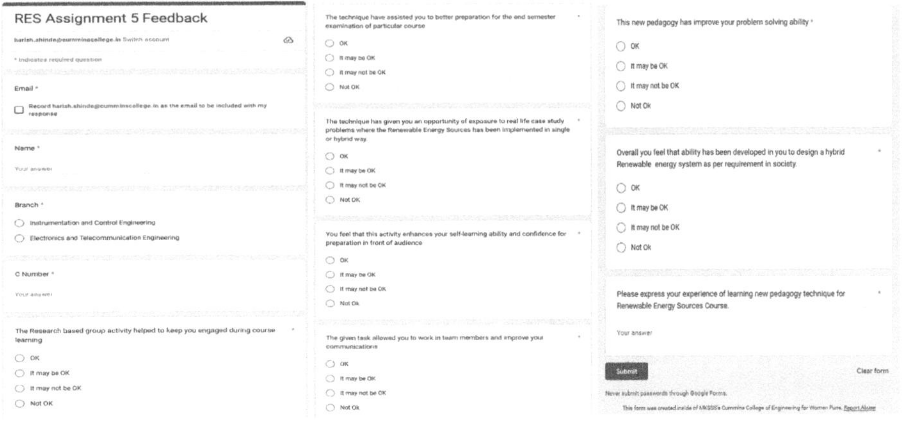

FIGURE 12.5   Screenshot for creation of Google form for RES course.

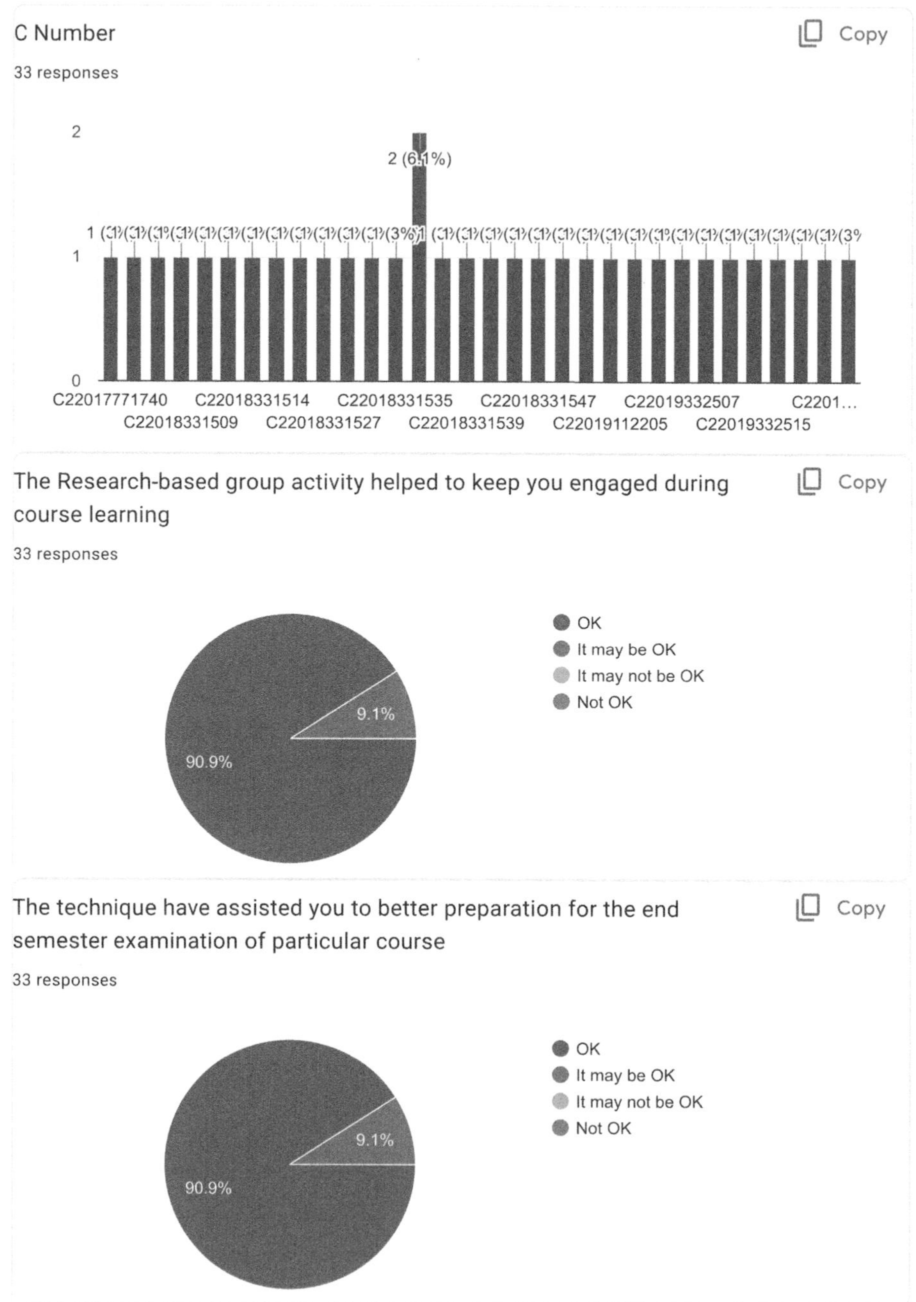

FIGURE 12.6   Screenshot for research-based assignment feedback for RES course.

implementation of the pedagogy technique—research-based case study presentation. Students can build their skillsets for understanding problems and defining case study-oriented presentations in front of an audience.

TABLE 12.2
Active Learning Survey Questionnaire

Sr. No.	Statement	OK	It May Be OK	It May Not Be OK	Not OK
1	The research-based group activity helped to keep you engaged during course learning	30	3	0	0
2	The technique has assisted you to better prepare for the end-semester examination due to thorough understanding of the course	30	3	0	0
3	The technique has given you an opportunity of exposure to real-life case studies where the RES has been implemented in a single or hybrid way	3	2	0	0
4	You feel that this activity enhances your self-learning and confidence	3	2	0	0
5	The given task allowed you to work in a team and improve your communications	29	4	0	0
6	This new pedagogy has improved your problem-solving ability	30	3	0	0
7	Overall you feel that ability has been developed in you to design a hybrid Renewable energy system as per the requirement of mankind	30	3	0	0
8	Please express your experience of learning new pedagogy techniques for RES course				

12.5 SURVEY QUESTIONNAIRE

To study and analyze the effectiveness of active learning strategies, viz. research paper-based presentation, a survey questionnaire was prepared. During the process of preparing the questionnaire, the survey questions related to active learning strategies prepared by Sandeep Desai [3] were referred to. The objectives of the survey were to measure (i) the enhancement of students' engagement in the teaching-learning process to measure an opportunity for exposure to real-life case studies where the RES have been implemented in a single or hybrid way and (ii) to measure team spirit, self-learning, and confidence. The survey was conducted using Google Forms. There were ten questions in the questionnaire. Students' responses were recorded on a four-point scale: (A) Ok, this is equivalent to strongly agree; (B) It may be ok, this is equivalent to agree; (C) It may not ok, this is equivalent to disagree; and (D) Not ok, this is equivalent to strongly disagree. The survey questionnaire is shown in Table 12.2. (33 students participated in the survey.) The survey was conducted for female students aged 21–22 years in the final-year engineering.

12.5.1 SOME OF THE COMMENTS FROM STUDENTS ARE AS FOLLOWS

"It was very helpful." "It was nice." "I understood how to read research papers and extract information from that. And i enjoyed working with a team." "My experience was very nice it gave me confidence and it was very useful." "This technique helps a lot to us to find new way of learning things." "I got to learn new interesting things."

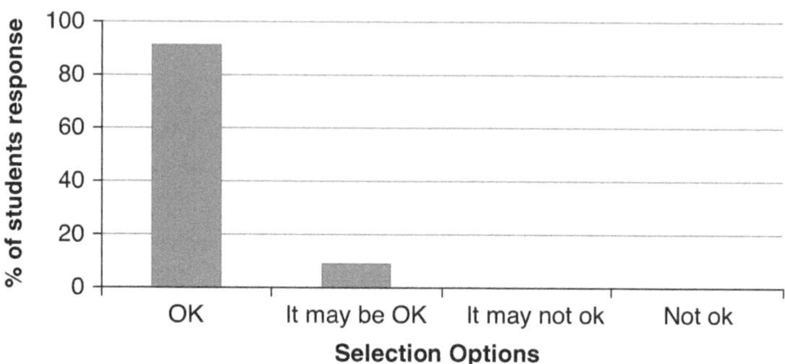

FIGURE 12.7　Engagement during course learning.

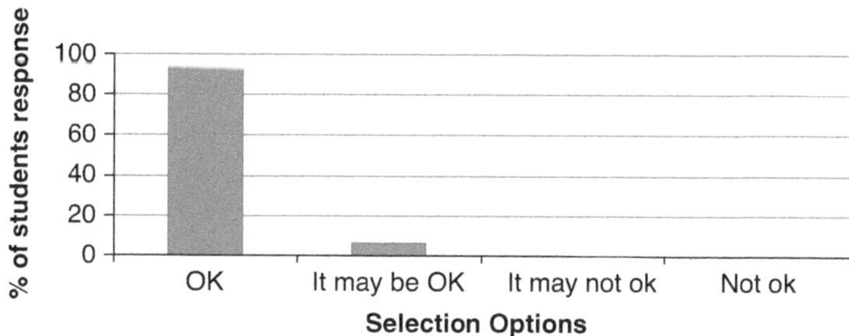

FIGURE 12.8　Exposure to a real-life case study where RES have been implemented in a single/hybrid way.

12.6　RESULTS AND DISCUSSION

In this section, the students' responses to the survey questionnaire are presented. The responses are analyzed to understand students' perception of the implementation and effectiveness of the active learning strategies viz. Research-based application-oriented case study presentation.

The analysis of the students' responses mainly shows that there is a better understanding subject of RES. Figure 12.7 shows that 90.9% of students feel that research-based group activity keeps them engaged during course learning, and because of this engagement during course learning, students get more interested.

Figure 12.8 mainly shows that 93.9% of students feel that the technique has given them an opportunity of exposure to real-life case study problems where RES have been implemented in a single or hybrid way. Additionally, it helped them understand unit 5: Hybrid Energy System and motivated them. Since they understood the importance of a hybrid energy system, it was implemented in remote areas and villages where the population is <500.

Figure 12.9 mainly shows that 90.9% of students feel that this new technique assisted them for better preparation for the end-semester examination (ESE, 50 marks) of the RES course. Since the presentation was in front of an external

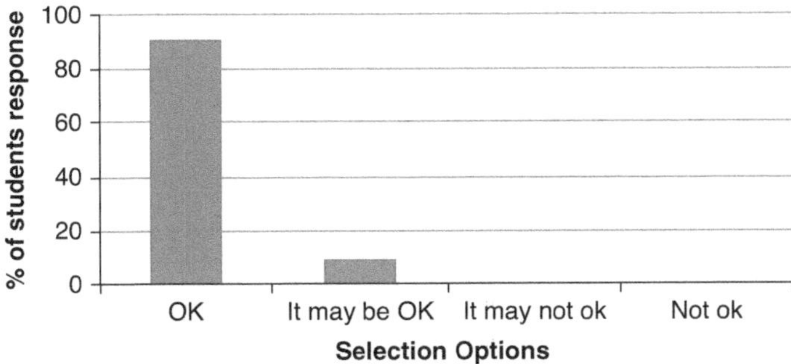

FIGURE 12.9 Assistance for the preparation of end-semester exam during course learning.

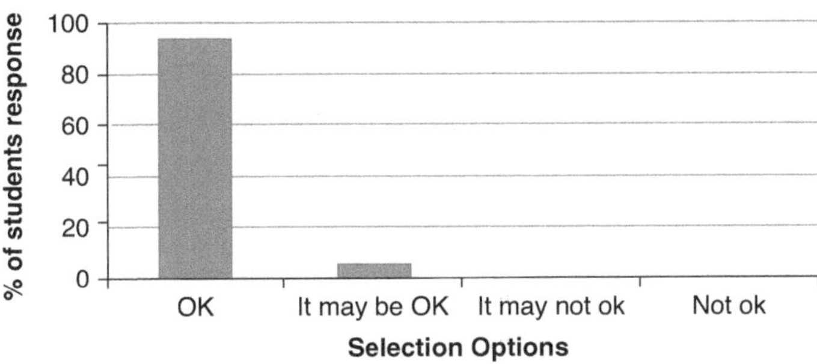

FIGURE 12.10 Enhancement in self-learning ability and confidence.

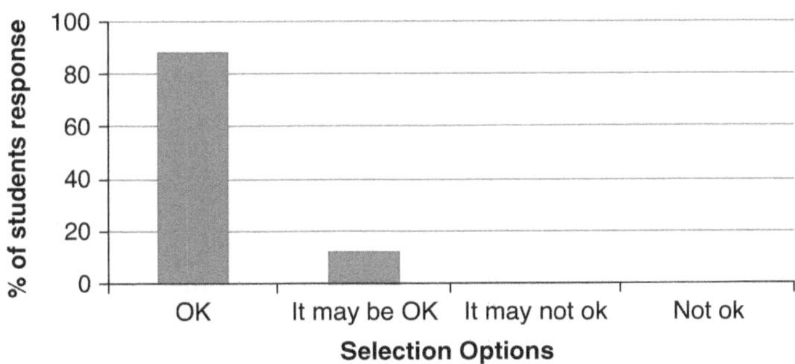

FIGURE 12.11 Working in a team and improvement in communication.

examiner from another department, the students feel that this activity enhanced their self-learning ability and confidence for preparation in front of an audience.

This is reflected in Figure 12.10 which mainly shows that 93.9% agree that self-learning ability and confidence have increased.

Figure 12.11 mainly shows that 87.9% of students feel that the given task allowed them to work with team members and improve their communication. This team working will benefit the students in their professional life while working in companies or Industries.

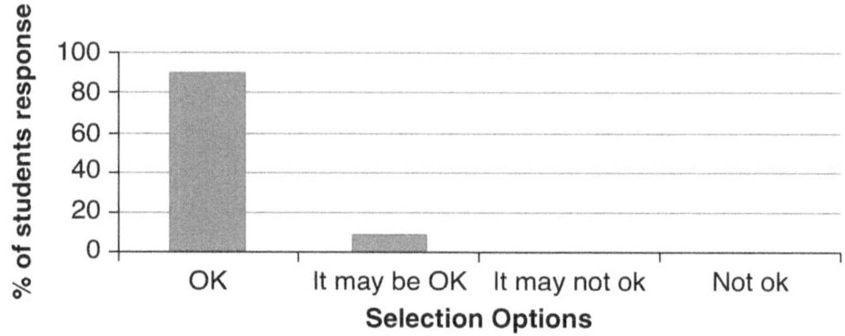

FIGURE 12.12   Improvement in problem-solving ability.

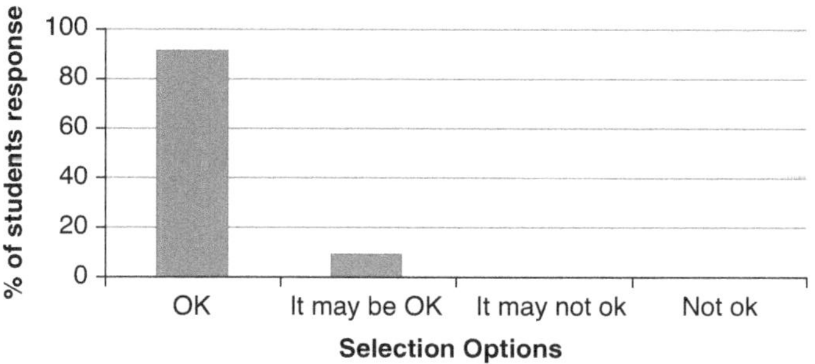

FIGURE 12.13   Task ability to design hybrid renewable energy.

Figure 12.12 mainly shows that 90.9% of the students feel that this new pedagogy has improved their problem-solving ability.

Figure 12.13 mainly shows that 90.9% of students feel that their ability has developed to design a hybrid renewable energy system as per requirement in society.

12.6.1 Valuable Contribution in the Field of Engineering Education

- A good engineering education is linked to better outcomes for students and the world. Engineering graduates make significant contributions to health care, national security, and many other fields. As per the National Academy of Engineering, disciplines have been the key to our nation India's growth. Engineering disciplines integrate scientific principles with a practically oriented approach/research providing the system for creating new knowledge. The entire pedagogy techniques gives vital role foe industrial innovation.

- National Education Policy (NEP 2020) is the golden step of the Government of India for the education system. Students can adopt said pedalogy techniques for upgrade their bright career. This pedagogy technique makes a valuable contribution to the field of engineering education for the benefit of students while working in various companies or research institutes.

12.7 FUTURE SCOPE

The future scope of this work is to be conducted for students from various background in open elective RES. Students have other branches such as Computer, Information Technology and Electronics & Telecommunication for selecting topic releated to Machine Learning (ML), Artificial Intelligence (AI) tools implemented in RES course. While selecting the topic from students for the case hybrid combination-related topic to be selected. Students should select topic – IoT based hybrid solar desalination system.

12.8 CONCLUSION

Active learning strategy, viz. research paper-based application-oriented case study presentation, was implemented for RES in the final-year undergraduate Electronics and Telecommunication and Instrumentation and Control Engineering programs. Feedback was received from students based on the analysis of the students' responses to the survey questionnaire. The following conclusions are drawn:

- The students enjoyed this activity and understood how to extract the information from a research paper. This active learning technique helps students develop thinking skills so that they can convey their conceptual understanding of the respective case study in the presentation.
- Due to real-life case studies of hybrid energy systems, students immensely benefited from the exposure to different technologies used in hybrid systems in different parts of the world.
- The active learning strategy helps students develop interest in the course content. Students refer to research papers related to their topic, which increases their self-learning ability and confidence. It leads to a better understanding of the concepts. The new pedalogy technique activity within the team. It enhances the team spirit and communication.
- Research paper-based, application-oriented case study presentations can be implemented for other courses in the final year for different engineering branches. This mode of teaching encourages and ensures active learning during the session.
- Due to this active learning technique, students can build up confidence for writing the answer in the end-semester examination.
- Application-based detailed studies will be more appropriate for setting good knowledge for the students.
- The major outcome of the said pedagogy technique is enhancement of students skillsets and better interaction of students to the industry.

ACKNOWLEDGMENTS

The authors acknowledge the cooperation and support of the final-year Instrumentation and Control Engineering and Electronics and Telecommunications Engineering students who played a key role in the implementation of the active learning techniques in the classroom. The authors also express their sincere thanks to the director and management of the MKSSS's Cummins College of Engineering for Women, Karvenagar, Pune for creating a conducive atmosphere for the implementation of such techniques and for providing necessary support and facilities.

REFERENCES

1. Bewoor, A. K. Use of think-pair-share (TPS) as an active learning strategy to teach metrology course. *Cummins College Digest of Engineering Education*, 1(1), 5–12, 2020.
2. Jaspal, S., E. Bamberg, J.-W. Hong, and M. Boyce. Active engagement pedagogy for an introductory solid mechanics course. *2002 Annual Conference*, 2002, 7–143, 2002.
3. Desai, S. R. Impact of active learning methods on students' learning and course results. *Journal of Engineering Education Transformations*, 35(3), 133–142, 2022.
4. Budiyanto, C. and I. Widiastuti. Hands-on learning on renewable energy: A proposed approach for technology dissemination. *AIP Conference Proceedings*, 1977(1), 060017, 2018.
5. Casillas, C. E. Renewable energy and experimental learning for development. In *Conference: Impact and Effectiveness of Developmental Relationships,* Chennai, January 2013.
6. Çoker, B., H. Çatlıoğlu, and O. Birgin. Conceptions of students about renewable energy sources: A need to teach based on contextual approaches. *Procedia-Social and Behavioral Sciences*, 2(2), 1488–1492, 2010.
7. Yang, H. Study on research-based teaching for mechanical engineering control basis. *In 2014 International Conference on Education Reform and Modern Management (ERMM-14)* Phuket, Thailand, pp. 90–93. Atlantis Press, Amsterdam, Netherlands, 2014.
8. Kedar, S. A., A. K. Bewoor, and K. Arul Raj. Design and analysis of solar desalination system using compound parabolic concentrator. *IOP Conference Series Material Science and Engineering*, 455, 12063, 2018.
9. Kedar, S. A., A. K. Bewoor, G. Murali, R. Kumar, M. Sadeghzadeh, and A. Issakhov. Effect of reflecting material on CPC to improve the performance of hybrid groundwater solar desalination system. *International Journal of Photoenergy*, 2021(6675236), 13, 2021
10. Kedar, S. A., A. K. Bewoor, and S. Madhusudan. Solar desalination system using evacuated tube collector & compound parabolic concentrator-theoretical approach. In *National Conference Advance Electrical Engineering Energy Science,* NIT Hamirpur India, June 2019.
11. Kedar, S. A., G. Murali, and A. K. Bewoor. Mathematical modeling and analysis of solar desalination system using evacuated tube collector & compound parabolic concentrator. *Mathematical Modeling of Engineering Problems*, 08(01), 45–51, 2021.
12. Kedar, S. A., K. Arul Raj, and A. K. Bewoor. Performance analysis of solar desalination system using ETC and CPC. *SN Applied Science*, 1, 965, 2019.
13. Kedar, S. A., G. Murali, and A. K. Bewoor. Effective hybrid solar groundwater desalination in rural areas. *International Transaction Journal of Engineering, Management, & Applied Sciences & Technologies*, 12(3), 1–10, 2021. https://TUENGR.COM/V12/12A3M.pdf
14. Kedar, S. A., G. Murali, and A. K. Bewoor. Thermal analysis of solar desalination system using evacuated tube collector. *AIP Conference Proceedings*, 2039, 020061, 2018. https://doi.org/10.1063/1.5079020

15. Kedar, S. A., G. Vijay More, D. S. Watvisave, and H. M. Shinde. A critical review on the various techniques for the thermal performance improvement of solar air heaters. *Energy Sources, Part A: Recovery, Utilization, and Environmental Effects*, 45(4), 11819–11852, 2023.
16. Kedar, S. A. and A. K. Bewoor. Patent on solar desalination system Ref.no 201921032482. *The Patent Office Journal*, 07/2021, 7767, 2019.
17. Agavekar, R., P. Bhore, H. Kadam, and M. Moharir. Effective application of one minute paper and muddiest point technique to enhance students' active engagement: A case study. *Journal of Engineering Education Transformations*, 3(6), 8–17, 2023.
18. Sasikumar, N. Impact of active learning strategies to enhance student performance. *Innovative Journal of Education*, 2(1), 206–214, 2014.
19. Felder, R. M. and R. Brent. *Teaching and Learning Stem: A Practical Guide*. John Wiley & Sons, Hoboken, NJ, 2019.

13 Cyber Forensic Accounting in Public-Sector Accounting Education for Sustainable Learning Commitment in Education 4.0

Evidence and Conceptualization of Behavioral Intention

Pham Quang Huy and Vu Kien Phuc

13.1 INTRODUCTION AND CONTEXT

According to commentators, this indicates that it is becoming more and more important for accounting graduates to receive a solid education in this area [1,2]. The significance of public-sector accounting education is paramount in relation to the dependability of public-sector financial disclosure, the proficiency of public officials [3], and the development of involved citizens [4]. It is important for a sustainable future to educate responsible managers in the public sector by methodically incorporating perspectives on Sustainable Development Goals (SDGs) and the common good [5]. Building on the perspectives of Ref. [6], higher education institution (HEI) plays a significant role in sustainable learning commitment generation (SLCO). The 2030 Agenda for SGD 4 calls for improving educational quality. In order for learners to succeed in demanding and complex situations, HEIs must provide them with the knowledge and abilities of sustainable learning in education [7].

In particular, digitalization is drastically altering public services by changing how citizens and public entities interact as well as how knowledge and information are produced, shared, and used online rather than in person. This presents new competencies and skill requirements for public managers and civil servants [8].

DOI: 10.1201/9781032664828-16

Digital technology, on the one hand, automates and facilitates a large portion of accounting activities and gives accountants real-time access to information that was previously unavailable, improving data quality, decision-making processes, accuracy, and timeliness [9]. But, in order to accomplish these objectives, accountants require specialized knowledge and abilities, particularly in the areas of cloud computing, data analytics, and more recent and cutting-edge technologies like blockchain, sophisticated business intelligence software, and programming [8].

There appears to be a disconnect between public-sector accounting practice and education since some academics worry that there will not be many graduates with the necessary training to handle the complexity of public-sector accounting reforms in the near future [10]. As a result, it forces HEIs and professional associations to update their training programs and curricula to better reflect the real-world issues that public entities confront now and, in the future [11,12].

The necessity for forensic accounting education development and extension has been acknowledged on a global scale [13]. To effectively detect and prevent cyber fraud, forensic accountants must adjust their investigative tactics and tools in light of the constantly changing nature of cyber threats and the intelligence of hackers [14]. Cyber forensic accounting entails looking at financial crimes committed online and calls for expertise in data analysis, digital forensics, and cybercrime investigation [15]. Different levels of support and failure have been linked to attempts to advance forensic accounting education and practice [16]. Although there have been previous evaluations of the forensic accounting literature [17], it is crucial to comprehend the variables that influence forensic accounting education [18], particularly cyber forensic accounting education.

As a result, the goal of this study is to pinpoint the critical success factors that influence instructors' behavioral intention to cyber forensic accounting implementation into public-sector accounting education (BITI) in order to fulfill the commitment to sustainable learning in the context of Education 4.0.

RQ1: What are the critical success factors of BITI?
RQ2: How far do they impact BITI?

Rested on the analysis of the overall results and chief observations, the present study offers several contributions in the theoretical aspect. The scholarly literature on accounting education in the public sector is extremely scarce [19]. Furthermore, a number of studies contend that there remains a disconnect between public-sector accounting education and research and practice, and that the latter has lagged behind what is required to understand the public sector's most recent advancements [11]. The aforementioned disparity implies that public-sector organizations frequently lack the capacity to address contemporary issues, and accounting graduates' employment prospects are restricted [11,12]. According to Ref. [20], there has been a current push for additional research on the state of public-sector accounting education delivery, instructional methods, and active learning resources. The culmination of the research lies in enriching the current frontier of knowledge on public-sector accounting education from a new angle in the lens of Education 4.0. More importantly, forensic accounting education must be included in accounting and auditing curricula since it is crucial to preventing and identifying fraud. According to Ref. [21], forensic accounting is crucial for spotting fraud early on, which is why undergraduate and graduate degree courses must include forensic accounting. Although it has been discovered that forensic accounting approaches are useful in identifying and stopping fraud practices in developing economies, little is still understood about the specific circumstances of these economies [22]. Against this backdrop, it fills a study void left by earlier studies by generating empirical data on the vital success elements of BITI. To the best of the researchers' knowledge, this may be one of the few scholarly works that provide novel and crucial insights on how to realize SLCO by highlighting the extraordinarily distinctive effects of potential factors, such as lecturers' intelligence (LI), technology readiness (TR), and environmental readiness (ER).

On the practical significance, the acquired observations of this study will offer managerial and operational pointers and suggest a roadmap, at the policy level, regarding the implementation of cyber forensic accounting education to managers, administrators, and educators of higher institutions, in addition to enlarging academicians' horizons in terms of the behavioral intention of instructors with regard to implementing cyber forensic accounting education.

The entire research process proceeds as follows. The review of literature including theoretical foundation and conceptualization is demonstrated in Section 13.2. Section 13.3 lays forth the research model interpretation and the formulation of hypotheses. Research methodology is covered in Section 13.4. Section 13.5 provides an illumination of the obtained observations. In conclusion, Section 13.6 offers useful implications and agenda for upcoming academic works.

13.2   LITERATURE REVIEW AND THEORETICAL BASIS

13.2.1   CAPABILITY, OPPORTUNITY, MOTIVATION, AND BEHAVIOR MODEL

Refs [23] and [24] established theoretical paradigm of capability, opportunity, motivation, and behavior model, which is employed in the current study to visualize behavioral intention to adopt. This paradigm has gained recognition as a dynamic and systematic approach to behavioral shifts. It may be useful in identifying factors that facilitate or impede actor behavior, as well as reasons why interference methods have been successful or unsuccessful [23,24]. According to Ref [24], this paradigm states that a behavior change can only occur when one or more capabilities, opportunities, and motives to carry out the action interact. Accordingly, the competence aspects represent a person's mental and physical ability to engage in relevant activities [24]. Opportunity, as defined by Ref. [24], is all external factors that cause behavior to occur or intensify. On the other hand, motivation, which is considered the foundation of capability, opportunity, motivation, and behavior framework, refers to mental processes that guide behaviour [24].

13.2.2   TECHNOLOGY-ORGANIZATION-ENVIRONMENT (TOE) FRAMEWORK

The TOE framework was first established by [25], and it has been widely used in the adoption of information systems. The TOE model has been widely employed by academics to elucidate the adoption of technology within organizational contexts. Its capacity to elucidate the adoption of diverse technologies in an organizational context has been validated by numerous research [26]. One of TOE's primary advantages over other theories explaining technology adoption is that it considers both internal and external factors at the same time, unlike resource-based view, institutional theory, knowledge-based view, and technology structuration theory [27]. The term "technological context" describes the technology associated with the institution, whether it be internal or external [28]. The features of an organization that either encourage or prevent the adoption of technological innovation are referred to as the organizational context [29]. The environment context, which includes rules, rivals, resource availability, and connections to governmental entities, is the setting in which a company operates [29].

13.2.3 CYBER FORENSIC ACCOUNTING

As stated by Ref. [30], there are many different kinds of cybercrimes that might happen, including phishing, identity theft, hacking, and cyberstalking. Digital forensics, financial investigations, and data analytics are some of the methods used to track down the origins of cyberattacks and retrieve funds that have been pilfered. In this regard, cyber forensic accounting is the application of forensic accounting principles to the investigation and prevention of cybercrimes [30].

13.3 RESEARCH MODEL DEVELOPMENT AND HYPOTHESIS FORMULATION

An expanding corpus of studies indicate that integrating technology into the classroom may boost student interest [31]. According to Ref. [32], there were instances in which the use of technology provided additional chances for students to interact with the course material. When technology-based components offered several learning opportunities spread out over time, they seemed to be very successful in fostering extra engagement with the subject and retrieval practice [33]. When technology was included in the classroom, students' motivation to study and their level of participation increased [34]. In the end, a meta-analysis revealed that the integration of technology into the classroom improved student involvement in a variety of learning environments [35]. Students who made use of these tools reported feeling better about their contacts with teachers and were more inclined to contribute to class discussions. Cogitating all the above, the first hypothesis in this research is postulated as follows.

HYPOTHESIS 1 (H1): TR INDUCES AN IMPACT ON SLCO IN A SIGNIFICANT AND POSITIVE MANNER

One of the most frequently mentioned crucial elements for the effectiveness of reform is human potential [36]. Faculty members who specialize in accounting might be considered ardent advocates for the inclusion of accounting in the curriculum [37].

Public-sector accounting lecturers are essential for giving students access to this material and ensuring that it continues to be covered. Because of this, adding the necessary information to the curriculum requires hiring academic personnel who are adequately competent in public-sector accounting [38].

According to Ref. [39], the foundation of organizational digitalization is intelligence competencies. As one of the most important skills in the globalized world, creativity is widely considered to play a major role in all domains [40]. Building on the perspective of cognitive psychology, creativity was associated with the domain of problem-solving and broadly defined as the ability to generate original and beneficial concepts and solutions that were appropriate, useful, correct, and functional [41]. When it comes to instances when students need effective options and solutions to achieve behavioral, emotional, cognitive, and agentic engagement in their learning, creative intelligence would allow higher education lecturers to come up with fresh and creative ideas. Thus far, the adversity quotient has been conceptualized as a form of intelligence that serves as the foundation for an individual's success in managing adversity [42]. According to Ref. [43], adversity quotient is the ability to put forth effort in interpreting complexities into problems that need to be solved and then turning those problems into opportunities for improved performance. A lecturer with a high adversity quotient would undoubtedly be able to handle any issues pertaining to experiential learning and win over the students.

Based on the suggestions made by Ref. [44], the research on the digital intelligence of lecturers focused on those who were able to manage their personal lives in an online setting in order to achieve work-life balance, handle large amounts of information, share ideas and perspectives with an online audience, and interact meaningfully within a digital environment. According to Ref. [45], the term "digital intelligence quotient" refers to a broad range of digital competencies that are derived from shared moral principles and that people can use, oversee, and create in order to advance humankind through technology. As an alternative, it was thought that the digital intelligence quotient would be associated with the alignment of business and information technology strategy, as well as the final administration and management of information technology projects [46]. Cogitating all the above, the second hypothesis in this research is postulated as follows.

HYPOTHESIS 2 (H2): LI INDUCES AN IMPACT ON SLCO IN A SIGNIFICANT AND POSITIVE MANNER

According to Ref. [47], public organizations typically prioritize government laws and other sector-specific criteria. Furthermore, pressure on public institutions may still come from community expectations and consumer readiness. When it comes to specifics, the job system is highly influential, especially for institutions and undergraduate programs. Authorities in charge of hiring need employees who can perform operational tasks and who possess specific, in-depth applied knowledge and abilities. Because of this, these authorities anticipate relevant knowledge about public-sector accounting that focuses on current accounting practices. Further financial assistance for schools can be given by the government and other education stakeholders, together with an increase in professional development activities for teachers and the creation of learning platforms that facilitate unrestricted access to high-quality educational resources [48]. Government pressure and the imposition of advantageous regulations would thereby increase HEIs' willingness to support the BITI process. On the other hand, even while HEI management deemed BITI to be beneficial and compatible, they may decide to postpone implementation in the absence of external and governmental backing. Cogitating all the above, the third hypothesis in this research is postulated as follows.

Hypothesis 3 (H3): ER Induces an Impact on SLCO in a Significant and Positive Manner

In addition to seeking to create value for stakeholders, managers with a sustainability emphasis take accountability for their work and its effects on the environment, society, and economy. As a result, students must be able to question assumptions and raise pertinent issues more often. They also need to comprehend that doing business as usual is not always the greatest method to provide value to a company [49]. Furthermore, one must possess the following qualities: an open mind, cultural awareness, the capacity to handle complexity, a willingness to learn and listen, the ability to spot opportunities, effective communication, and teamwork. A more sustainable future can be attained by students with sustainable knowledge, skills, and talents because education is a crucial predictor of human development [50]. According to Ref. [51], sustainable development depends on sustainable accounting education; hence, it would be challenging to adopt and disclose sustainable business practices without sustainable accounting education. In order to provide students with the information and abilities to comprehend contemporary concerns and collaborate toward more sustainable futures, education for sustainability in higher education aims to establish transformative learning environments [52]. Building on the perspectives of Ref. [53], a process of continual development toward sustainability through individual and collective reflection requires a formal understanding of sustainability issues, critical thinking, and soft skills. Therefore, incorporating sustainability concerns into education involves more than just giving definitions and topics that are important to business. Alternatively, it involves acquiring a number of crucial abilities that managers and future managers might use [53]. Cogitating all the above, the fourth hypothesis in this research is postulated as follows.

Hypothesis 4 (H4): SLCO Induces an Impact on BITI in a Significant and Positive Manner

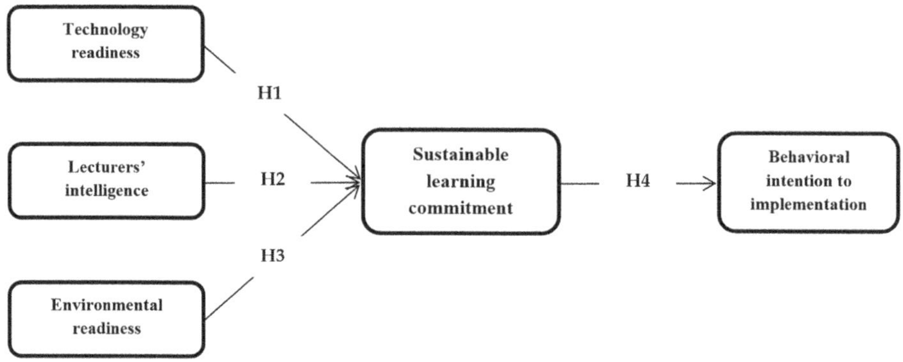

FIGURE 13.1   Hypothesized model.

13.4   MATERIALS AND METHOD

13.4.1   Operationalization of Research Variables

The hypothetico-deductive research strategy is taken advantage in the current research. Building on the viewpoints of Ref. [54], surveys are the most appropriate approach for this particular assignment because they are commonly leveraged in theory testing and non-experimental design. Creating a theoretical model, developing hypotheses, reviewing relevant literature, and drawing logical conclusions from study results are all part of this methodology. Gauging criteria for each topic in the current study are selected following an inclusive review of the literature. The comprehensive back-translation process is employed to fastidiously translate the first English version of the questionnaire into Vietnamese, guaranteeing a high degree of similarity between the source and target languages. A questionnaire using a five-point Likert scale is used to gauge the respondents' points of view. Furthermore, the pre-tested methodology involving seven specialists is leveraged to reduce unforeseen intricacy. Items that were not obviously measuring anything are taken out of the structure after professional advice. Additionally, the measurement instruments and questionnaire are assessed in a small-scale pilot study involving 30 individuals. The internal consistency of each pilot questionnaire concept is evaluated using Cronbach's alpha, which has to be greater than or equal to 0.7 in order to get the correct reliability coefficient. Cronbach's alpha score of the small-scale pilot test is more than 0.7, showing consistent completion of the questionnaire with rational answers.

13.4.1.1   Sustainable Learning Commitment

The SLOC construct is conceptualized as a second-order aggregate of three first-order micro-foundational constructs – sustainable learning opportunities, sustainable learning capabilities, and sustainable action-oriented. The criteria applied to evaluate first-order constructs are subsequently constituted employing the contributions of Refs. [52,55–58] and are pre-tested with academicians and practitioners.

13.4.1.2 Environmental Readiness

The items for measuring ER in the current investigation are drawn from the academic works of [47] as well as [48] and are pre-tested with academicians as well as practitioners.

13.4.1.3 Lecturers' Intelligence

The items for measuring LI in the current investigation are drawn from the academic works of Refs. [59,60] and are pre-tested with academicians as well as practitioners.

13.4.1.4 Technology Readiness

The items for measuring TR in the current investigation are drawn from the academic works of Refs. [61–63] and are pre-tested with academicians as well as practitioners.

Behavioral Intention to Implementation

The items for measuring BITI in the current investigation are drawn from the academic works of Ref. [64] and are pre-tested with academicians as well as practitioners.

13.4.2 TARGET POPULATION AND SAMPLING PROCEDURE

For Partial Least Square-Structural Equation Modeling (PLS-SEM), Ref. [65] recommended that the minimum sample size should be ten times greater than or equal to the structural path shown in any structural model. In terms of ethical considerations, participants receive sufficient information about the study's goals to justify the privacy of their information. Between March 2023 and October 2023, 450 questionnaires were delivered using a convenience and snowball sampling strategy to academics employed in various accounting faculties across the vast array of HEIs in Vietnam, after receiving the informed consent of the informants. After the questionnaires were screened and evaluated, a total sample size of 383 cases with a 14.89% data loss rate remained. In this study, statistical data analysis was conducted using SPSS 29.0 and SmartPLS 4.0.9.6. The utilization of PLS-SEM was deemed appropriate for this manuscript due to its ability to effectively manage intricate models containing numerous structures, all while maintaining statistical power [66]. In terms of academic ability, almost all of the participants earned at least a post-graduate degree. Table 13.1 delineates comprehensive demographic data captured from this investigation.

TABLE 13.1
Demographic Information

Respondent's Demographic Profile	Variables	Usable Responses	Weight (%)
Gender of respondent	Male	116	30.29
	Female	267	69.71
Age of respondent	30–under 40	74	19.32
	40–under 50	291	75.98
	Over 50	18	4.70
Experience of respondent (years)	10–under 20	85	22.19
	20–under 30	285	74.42
	30–under 40	13	3.39

TABLE 13.2

Results Summary for the Content Validity

	BITI	ER	LI	SAO	SLCA	SLO	TR
BITI1	0.869	0.074	0.109	0.286	0.243	0.206	0.122
BITI2	0.864	0.167	0.192	0.243	0.220	0.318	0.197
BITI3	0.820	0.145	0.091	0.270	0.164	0.191	0.104
ER1	0.098	0.857	0.247	0.079	0.255	0.247	0.163
ER2	0.176	0.841	0.113	0.113	0.195	0.301	0.142
ER3	0.117	0.888	0.228	0.206	0.241	0.273	0.127
LI1	0.127	0.177	0.816	0.139	0.251	0.182	0.137
LI2	0.148	0.201	0.817	0.098	0.193	0.173	0.189
LI3	0.030	0.156	0.863	0.082	0.267	0.199	0.159
LI4	0.199	0.225	0.870	0.186	0.340	0.282	0.242
SAO1	0.294	0.133	0.136	0.868	0.172	0.118	0.252
SAO2	0.277	0.162	0.125	0.856	0.066	0.106	0.147
SAO3	0.237	0.121	0.151	0.879	0.125	0.080	0.160
SLCA1	0.138	0.235	0.227	0.173	0.851	0.046	0.129
SLCA2	0.280	0.204	0.300	0.135	0.874	0.083	0.223
SLCA3	0.214	0.245	0.301	0.051	0.831	0.118	0.204
SLO1	0.261	0.266	0.243	0.112	0.085	0.903	0.252
SLO2	0.236	0.271	0.208	0.126	0.077	0.851	0.193
SLO3	0.261	0.310	0.236	0.076	0.094	0.911	0.240
TR1	0.151	0.078	0.236	0.172	0.194	0.264	0.871
TR2	0.207	0.178	0.139	0.163	0.177	0.265	0.820
TR3	0.063	0.120	0.169	0.134	0.164	0.116	0.783
TR4	0.117	0.174	0.190	0.243	0.185	0.180	0.843

13.5 INFERENTIAL STATISTICS

13.5.1 Assessment of Measurement Model

13.5.1.1 Content Validity

Hinged on the standpoints of Ref. [67], when cross-loading is taken advantage to gauge the questionnaire's content validity, the items are integrated into particular ideas and have meanings that are equivalent. The assessment of a construct has to have a value higher than others in the same rows and columns, as stated by [68]. Complete content validity is attained by the study's measuring scale, according to the values of all evaluated constructs displayed in Table 13.2.

13.5.1.2 Convergent validity evaluation

The measuring model is assessed using the standards set out by Ref. [66] before the structural model is analyzed and the hypotheses are tested. Based on indicator reliability, convergent validity, discriminant validity, and internal consistency reliability, the measuring model is assessed. Most outer loadings are more than the suggested indication reliability value of 0.708 [66]. The calculated values of Cronbach's alpha, rho_A, and composite reliability show that the scale items have strong internal consistency and lie within the acceptable range of 0.70 [66]. Furthermore, it is established that all of the constructs' AVE values are above the 0.50 permissible limit [66]. The accuracy of the indicators, the measurement model's convergent validity, and its internal consistency are highlighted in the specifics of these outputs shown in Table 13.3.

Correlations and discriminant validity. The heterotrait–monotrait ratio (HTMT) and Fornell and Larcker criterion are used to evaluate the measurement model's discriminant validity. Table 13.4 illustrates that the discriminant validity meets the Fornell–Larcker criterion, which states that the square root of each construct's AVE must be greater than the correlation between that specific construct and all other constructs [69].

TABLE 13.3
Results Summary of Measurement Model Assessment

Constructs and Operationalization		Convergent Validity		Construct Reliability			
		Factor Loadings	AVE	Cronbach's Alpha	Composite Reliability	PA	Result
Technology readiness	TR	0.783–0.871	0.689	0.850	0.898	0.865	Retained
Lecturers' intelligence	LI	0.816–0.870	0.709	0.866	0.907	0.905	Retained
Environmental readiness	ER	0.841–0.888	0.743	0.828	0.897	0.838	Retained
Sustainable learning commitment	SLCO						
Sustainable learning opportunities	SLO	0.851–0.911	0.790	0.867	0.919	0.868	Retained
Sustainable learning capabilities	SLCA	0.831–0.874	0.726	0.811	0.888	0.813	Retained
Sustainable action-oriented	SAO	0.856–0.879	0.753	0.836	0.901	0.838	Retained
Behavioral intention to implementation	BITI	0.820–0.869	0.725	0.811	0.888	0.821	Retained

TABLE 13.4
Results Summary of Discriminant Validity Using Fornell–Larcker Process

	BITI	ER	LI	SAO	SLCA	SLO	TR
BITI	0.852						
ER	0.151	0.862					
LI	0.157	0.228	0.842				
SAO	0.311	0.159	0.158	0.867			
SLCA	0.248	0.267	0.324	0.142	0.852		
SLO	0.285	0.318	0.258	0.117	0.096	0.889	
TR	0.169	0.166	0.222	0.218	0.218	0.257	0.830

Furthermore, as Table 13.5 shows, all of the constructs' HTMT ratios considerably fall below the maximum allowable limit of 0.85, indicating the discriminant validity of the constructs [70].

13.5.2 ASSESSMENT OF STRUCTURAL MODEL

Using SmartPLS 4, the second stage evaluated the significance and relevance of the structural model interlinks. Figure 13.2 demonstrates the structural model that outcomes from employing the PLS algorithm in SmartPLS 4.

In order to investigate the significance of the path coefficients, the bootstrapping procedure with 10,000 resamples is employed. Building on the bootstrap results in Table 13.6, TR is corroborated to demonstrate a significant and positive effect on SLCO ($\beta=0.256$; t-value$=5.320$; p-value$=0.000$). In the same vein, SLCO is verified to be impacted by LI ($\beta=0.262$; t-value$=4.559$; p-value$=0.000$) and ER

TABLE 13.5

Results Summary for Discriminant Validity on Heterotrait–Monotrait Ratio

	BITI	ER	LI	SAO	SLCA	SLO	TR
BITI							
ER	0.185						
LI	0.175	0.264					
SAO	0.379	0.185	0.176				
SLCA	0.301	0.327	0.372	0.172			
SLO	0.333	0.375	0.286	0.138	0.115		
TR	0.191	0.199	0.249	0.251	0.261	0.288	

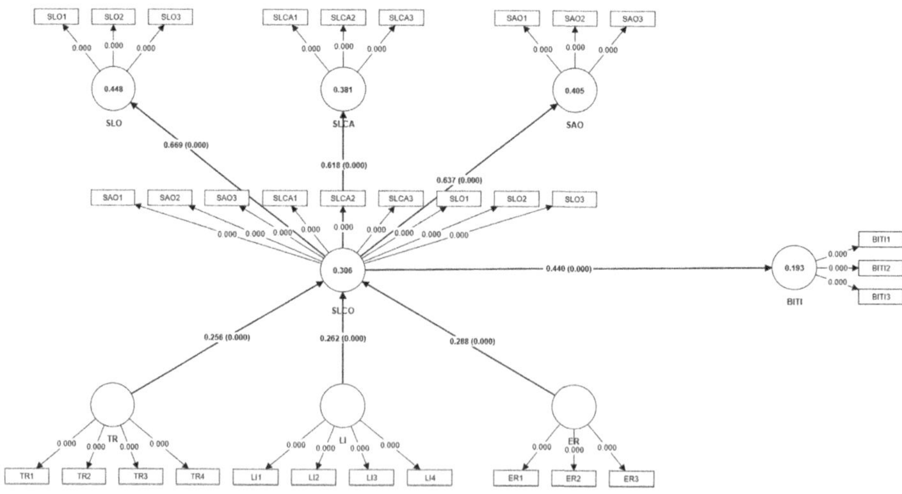

FIGURE 13.2 Structural model.

($\beta=0.288$; t-value $=4.947$; p-value $=0.000$). In line with the expectation of the researchers, SLCO is validated to induce a significant and positive effect on BITI ($\beta=0.440$; t-value $=8.509$; p-value $=0.000$). Thus, H1-H4 are accepted. The R^2 is 0.306 for SLCO and 0.193 for BITI. The score of f^2 in Table 13.6 puts the accent on the fact that TR and LI illustrate a small impact on SLCO (0.089 and 0.090, respectively). In the same vein, ER illustrates a small impact on SLCO (0.111) and SLCO illustrates a medium impact on BITI (0.239).

When compared to a linear benchmark model, the model illustrate perfect out-of-sample prediction capabilities for all construct indicators. The PLSpredict [71] results in Table 13.7 outlaid that the RMSE acquired by the linear model (LM) is much higher than the RMSE acquired by PLS-SEM for almost all items.

TABLE 13.6
Results Summary of Hypotheses Acceptance

Relevant Path	Path Coefficient	SE	95% Confidence Interval	VIF	t-Value	p-Value	Result
TR → SLCO	0.256	0.048	[0.160–0.348]	1.068	5.320	0.000	Accepted
LI → SLCO	0.262	0.057	[0.145–0.367]	1.095	4.559	0.000	Accepted
ER → SLCO	0.288	0.058	[0.166–0.394]	1.071	4.947	0.000	Accepted
SLCO → BITI	0.440	0.052	[0.335–0.535]	1.000	8.509	0.000	Accepted
R^2	$R^2_{\text{SLCO}} = 0.306;\ R^2_{\text{BITI}} = 0.193$						
f^2	$f^2_{\text{TR} \Rightarrow \text{SLCO}} = 0.089;\ f^2_{\text{LI} \Rightarrow \text{SLCO}} = 0.090;\ f^2_{\text{ER} \Rightarrow \text{SLCO}} = 0.111;\ f^2_{\text{SLCO} \Rightarrow \text{BITI}} = 0.239$						

TABLE 13.7
Results of Mediating Effect of SDT

Construct	Item	Q^2_{predict}	RMSE PLS-SEM	LM RMSE	IA Average Loss Difference (p-Value)	LM Average Loss Difference (p-Value)
			PLS$_{\text{predict}}$		CVPAT	
SLO	SLO1	0.149	1.346	1.350	−0.224 (0.002)	−0.062 (0.204)
	SLO2		1.246	1.285		
	SLO3		1.298	1.327		
SLCA	SLCA1	0.139	1.077	1.118	−0.131 (0.001)	−0.068 (0.004)
	SLCA2		1.013	1.045		
	SLCA3		1.139	1.159		
SAO	SAO1	0.045	1.431	1.462	−0.071 (0.325)	−0.059 (0.267)
	SAO2		1.467	1.485		
	SAO3		1.404	1.418		
BITI	BITI1	0.045	1.444	1.468	−0.066 (0.145)	−0.068 (0.203)
	BITI2		1.406	1.425		
	BITI3		1.436	1.464		
Overall					−0.131 (0.000)	−0.063 (0.010)

Moreover, when employing a cross-validated predictive ability test (CVPAT) [72], the result analysis discloses that the PLS-SEM can outperform the naive indicator average (IA) prediction benchmark. To this end, the model acquires some predictive capability that enables it to pass the IA test but not a more conservative LM benchmark.

13.6 FINAL DELIBERATIONS

13.6.1 PRACTICAL IMPLICATIONS

Expanding upon the managerial perspectives, the observations made during this study provide a multitude of practical takeaways. Every HEI leader should develop their managerial cognitive ability and focus more on cyber forensic accounting in accounting curricula. Therefore, in order to get the best results in cyber forensic accounting education, management educators were advised to focus their concerns on structural changes regarding the supply of infrastructure and digital platforms that were mandatory. A number of training and development programs have been suggested to offer lecturers in order to enhance their knowledge, skills, and capabilities to carry out their duties, as the training programs would serve as a first step in enabling the academic staff to do so effectively and efficiently. Alternatively, the focus of HEIs should be on providing suitable courses that provide guidance on enhancing the diversity of intelligence that their instructors can use in cyber forensic accounting. Given the critical role that modern information technologies play in improving and developing the efficacy and efficiency of information technology implementation across all organizations, legislators and other influential members of the government should create and implement laws and policies pertaining to the adoption of information technology, as well as take specific action regarding budget allocation, infrastructure reinforcement, and deployment planning. Additionally, the results of this study were crucial for standard-setters and policymakers to have when establishing and implementing measures and regulations pertaining to accounting practice and accounting education.

13.6.2 LIMITATIONS AND FUTURE SCOPES

Even though every effort is made to ensure the integrity of this research, there are a number of intrinsic limits that may arise when analyzing the study's results. On the plus side, these constraints will allow for new directions for this research stream's future work. The first significant issue that disallows explicit assertions on the result analysis is the cross-sectional data. Put differently, the current study is limited to providing static viewpoints on fit, which means that inferences can only be made regarding the shared relationships among the suggested variables. Therefore, it is recommended that future efforts include a thorough longitudinal analysis as well as more contemporary statistical approaches. The second drawback is the very small sample size, which in many instances will not have been sufficient to support firm findings. Hence, increasing the sample size is only advised as a function of effort and time. The sample's makeup is the third important constraint. The majority of the current study is based on samples, namely on the limited number of informants from a single dominant major. This will make it impossible to say for sure whether the observations made from this manuscript might be applied to lecturers of other majors. Therefore, relevant stakeholders and instructors in different majors will gain more insightful information. Fourthly, the results of this study cannot be generalized until similar studies from other nations are conducted, as it only examines data sets with Vietnamese origins. It will be beneficial to ground test the suggested model with a bigger sample size at the same time. Fifth, even though multiple processes are carried out to maximize the participants' impartiality, there is still a chance that self-serving bias originating from the informants may surface as a result of the survey questionnaire deployment [73]. In order to overcome this constraint, it is recommended that secondary-based data be used in future studies. Finally, given that correlation, regression, and structural equation modeling have lower predictive strengths, more research should consider different statistical techniques [74]. This will undoubtedly be easily overcome and will not provide a problem going forward.

ACKNOWLEDGMENTS

This chapter was funded by the University of Economics Ho Chi Minh City (UEH) in Vietnam.

REFERENCES

1. Elzain, E. (2021). Should forensic accounting be included in curricula of accounting departments: The case of Saudi Arabia. *International Journal of Higher Education*, *10*(4), 258–267.
2. Alshurafat, H., Beattie, C., Jones, G., & Sands, J. (2019). Forensic accounting core and interdisciplinary curricula components in Australian universities: Analysis of websites. *Journal of Forensic and Investigative Accounting*, *11*(2), 353–365.
3. Budding, T., de Jong, G., & Smit, M. (2022). New development: Bridging the gap-analysis of required competencies for management accountants in the public sector. *Public Money & Management*, *42*(7), 565–568. doi: 10.1080/09540962.2022.2068862
4. Karatzimas, S. (2020). The beneficial role of government accounting literacy in developing participatory citizens. *Accounting Education*, *29*(3), 229–246. doi: 10.1080/09639284.2020.1737547
5. Lorson, P.C., & Haustein, E. (2023). Debate: How to give university public sector accounting education the relevance it truly deserves. *Public Money & Management*, *43*(7), 729–730. doi: 10.1080/09540962.2023.2247709
6. Leal Filho, W., Tripathi, S.K., Andrade Guerra, J.B.S.O.D., Giné-Garriga, R., Orlovic Lovren, V., & Willats, J. (2018). Using the sustainable development goals towards a better understanding of sustainability challenges. *International Journal of Sustainable Development & World Ecology*, *22*, 1–12. doi: 10.1080/13504509.2018.1505674
7. Ben-Eliyahu, A. (2021). Sustainable learning in education. *Sustainability*, *13*(4250), 1–10. doi: 10.3390/su13084250
8. Tsiligiris, V., & Bowyer, D. (2021). Exploring the impact of 4IR on skills and personal qualities for future accountants: A proposed conceptual framework for university accounting education. *Accounting Education*, *30*(6), 621–649. doi: 10.1080/09639284.2021.1938616

9. Agostino, D., Saliterer, I., & Steccolini, I. (2022). Digitalization, accounting and accountability: A literature review and reflections on future research in public services. *Financial Accountability & Management, 38*(2), 152–176. doi: 10.1111/faam.12301

10. Adam, B., Brusca, I., Caperchione, E., Heiling, J., Jorge, S., & Manes-Rossi, F. (2020). Are higher education institutions in Europe preparing students for IPSAS? *International Journal of Public Sector Management, 33*(2/3), 363–378. doi: 10.1108/IJPSM-12-2018-0270

11. Heiling, J. (2020). Time to rethink public sector accounting education? A practitioner's perspective. *Journal of Public Budgeting, Accounting & Financial Management, 32*(3), 505–509. doi: 10.1108/JPBAFM-05-2020-0059

12. Sinervo, L.-M., Kork, A.-A., & Hasanen, K. (2021). Challenges in curriculum development process aimed at revising the capabilities of future public financial managers. *Teaching Public Administration, 41*(2), 243–256. doi: 10.1177/01447394211051878

13. Alshurafat, H., Beattie, C., Jones, G., & Sands, J. (2020). Perceptions of the usefulness of various teaching methods in forensic accounting education. *Accounting Education, 29*, 1–28. doi: 10.1080/09639284.2020.1719425

14. Akinbowale, O.E., Klingelhöfer, H.E., & Zerihun, M.F. (2020). An innovative approach in combating economic crime using forensic accounting techniques. *Journal of Financial Crime, 27*(4), 1253–1271. doi: 10.1108/JFC-04-2020-0053

15. Hossain, M.Z. (2023). Emerging trends in forensic accounting: Data analytics, cyber forensic accounting, cryptocurrencies, and blockchain technology for fraud investigation and prevention. *SSRN Electronic Journal, 27*, 1–30. doi: 10.2139/ssrn.4450488

16. Yang, C.-H., & Lee, K.-C. (2020). Developing a strategy map for forensic accounting with fraud risk management: An integrated balanced scorecard-based decision model. *Evaluation and Program Planning, 80*, 1–10. doi: 10.1016/j.evalprogplan.2020.101780

17. Botes, V., & Saadeh, A. (2018). Exploring evidence to develop a nomenclature for forensic accounting. *Pacific Accounting Review, 30*(2), 135–154. doi: 10.1108/par-12-2016-0117

18. Alshurafat, H., Al Shbail, M.O., & Mansour, E. (2021). Strengths and weaknesses of forensic accounting: An implication on the socio-economic development. *Journal of Business and Socio-economic Development, 1*(2), 135–148. doi: 10.1108/JBSED-03-2021-0026

19. Karatzimasa, S., Heiling, J., & Aggestam-Pontoppidan, C. (2022). Public sector accounting education: A structured literature review. *Public Money & Management, 42*(7), 543–550. doi: 10.1080/09540962.2022.2066356.

20. Heiling, J., Jorge, S., Karatzimas, S., & Aggestam-Pontoppidan, C. (2022). Editorial. *Public Money & Management, 42*(7), 538–540. doi: 10.1080/09540962.2022.2099143

21. Kaur, B., Sood, K., & Grima, S. (2023). A systematic review on forensic accounting and its contribution towards fraud detection and prevention. *Journal of Financial Regulation and Compliance, 31*(1), 60–95. doi: 10.1108/JFRC-02-2022-0015

22. Alharasis, E.E., Haddad, H., Alhadab, M., Shehadeh, M., & Hasan, E.F. (2023). Integrating forensic accounting in education and practices to detect and prevent fraud and misstatement: Case study of Jordanian public sector. *Journal of Financial Reporting and Accounting, 1*, 1–28. doi: 10.1108/JFRA-04-2023-0177

23. Michie, S., Atkins, L., & West, R. (2014). *The Behaviour Change Wheel: A Guide to Designing Interventions.* Silverback Publishing, London, UK.

24. Michie, S., van Stralen, M.M., & West, R. (2011). The behaviour change wheel: A new method for characterising and designing behaviour change interventions. *Implementation Science, 6*(1), 1–12. doi: 10.1186/1748-5908-6-42

25. Tornatzky, L., & Fleischer, M. (1990). *The Process of Technology Innovation.* Lexington Books, Lexington, MA.

26. El-Haddadeh, R., Osmani, M., Hindi, N., & Fadlalla, A. (2021). Value creation for realising the sustainable development goals: Fostering organisational adoption of big data analytics. *Journal of Business Research, 131*, 402–410. doi: 10.1016/j.jbusres.2020.10.066

27. Xu, W., Ou, P., & Fan, W. (2017). Antecedents of ERP assimilation and its impact on ERP value: A TOE-based model and empirical test. *Information Systems Frontiers, 19*(1), 13–30. doi: 10.1007/s10796-015-9583-0

28. Zhu, K., Rraemer, K.L., & Xu, S. (2003). E-business adoption by European firms: A cross-country assessment of the facilitators and inhibitors. *European Journal of Information Systems 12*(4), 251–268.

29. Yoon, T.E., & George, J.F. (2013). Why aren't organizations adopting virtual worlds? *Computers in Human Behavior, 29*(3), 772–790. doi: 10.1016/j.chb.2012.12.003

30. Xanthopoulou, A., Kalantonis, P., Arsenos, P., & Kallandranis, C. (2023). Forensic accounting: A strategic tool to strengthen corporate governance against fraud. In E. Karger & A. Kostyuk (Eds.), *Corporate Governance: An Interdisciplinary Outlook* (pp. 33–38). Virtus Interpress, Boston, MA.

31. Malik, R. (2023). Impact of technology-based education on student learning outcomes and engagement. *2023 10th International Conference on Computing for Sustainable Global Development (INDIACom)* (pp. 784–788), New Delhi, India.

32. Turan, Z., Meral, E., & Sahin, I.F. (2018). The impact of mobile augmented reality in geography education: Achievements, cognitive loads and views of university students. *Journal of Geography in Higher Education, 42*(3), 427–441. doi: 10.1080/03098265.2018.1455174

33. Hsiao, C.C., Huang, J.C.H., Huang, A.Y.Q., Lu, O.H.T., Yin, C.J., & Yang, S.J.H. (2019). Exploring the effects of online learning behaviors on short-term and long-term learning outcomes in flipped classrooms. *Interactive Learning Environments, 27*, 1160–1177. doi: 10.1080/10494820.2018.1522651

34. Mayer, R.E., & Moreno, R. (2002). Animation as an aid to multimedia learning. *Educational Psychology Review, 14*(1), 87–99. doi: 10.1023/a:1013184611077

35. Wu, W.-H., Jim Wu, Y.-C., Chen, C.-Y., Kao, H.-Y., Lin, C.-H., & Huang, S.-H. (2012). Review of trends from mobile learning studies: A meta-analysis. *Computers & Education, 59*(2), 817–827. doi: 10.1016/j.compedu.2012.03.016

36. Polzer, T., Adhikari, P., Nguyen, C.P., & Gårseth-Nesbakk, L. (2023). Adoption of the international public sector accounting standards in emerging economies and low-income countries: A structured literature review. *Journal of Public Budgeting, Accounting & Financial Management, 35*(3), 309–332. doi: 10.1108/JPBAFM-01-2021-0016

37. Goduscheit, R.C. (2014). Innovation promoters: A multiple case study. *Industrial Marketing Management, 43*(3), 525–534. doi: 10.1016/j.indmarman.2013.12.020

38. Reicharda, C., Küchler-Stahn, N., & Siegel, J. (2023). Education in public sector accounting at higher education institutions in Germany. *Public Money & Management, 43*(7), 741–749. doi: 10.1080/09540962.2023.2238913

39. Lenka, S., Parida, V., & Wincent, J. (2016). Digitalization capabilities as enablers of value co-creation in servitizing firms. *Psychology & Marketing, 34*(1), 92–100. doi: 10.1002/mar.20975

40. Voogt, J., & Roblin, N.P. (2012). A comparative analysis of international frameworks for 21stcentury competences: Implications for national curriculum policies. *Journal of Curriculum Studies, 44*(3), 299–321. doi: 10.1080/00220272.2012.668938

41. Walia, C. (2019). A dynamic definition of creativity. *Creativity Research Journal, 31*, 1–11. doi: 10.1080/10400419.2019.1641787

42. Muztaba, S., Bahri, S., & Farizal, F. (2020). The effects of adversity quotient and spiritual quotient on teacher performance. *Asian Journal of Science Education, 2*(1), 64–70.

43. Hidayat, W., & Prabawanto, S. (2018). Improving students' creative mathematical reasoning ability students through adversity quotient and argument driven inquiry learning. *Journal of Physics: Conference Series, 948*, 1–6. doi: 10.1088/1742-6596/948/1/012005

44. Marnewick, C., & Marnewick, A. (2021). Digital intelligence: A must-have for project managers. *Project Leadership and Society, 2*, 1–12. doi: 10.1016/j.plas.2021.100026

45. Park, Y. (2019). *DQ Global Standards Report 2019*. DQ Institute, United States of America.

46. Mithas, S., & McFarlan, F.W. (2017). What is digital intelligence? *IT Professional, 19*(4), 3–6. doi: 10.1109/mitp.2017.3051329

47. Neumanna, O., Guirguis, K., & Steiner, R. (2022). Exploring artificial intelligence adoption in public organizations: A comparative case study. *Public Management Review, 26*, 1–28. doi: 10.1080/14719037.2022.2048685

48. Batac, K.I.T., Baquiran, J.A., & Agaton, C.B. (2021). Qualitative content analysis of teachers' perceptions and experiences in using blended learning during the COVID-19 pandemic. *International Journal of Learning, Teaching and Educational Research, 20*(6), 225–243. doi: 10.26803/ijlter.20.6.12

49. Wamsler, C. (2020). Education for sustainability: Fostering a more conscious society and transformation towards sustainability. *International Journal of Sustainability in Higher Education, 21*(1), 112–130. doi: 10.1108/IJSHE-04-2019-0152

50. Dias, B., Onevetch, R., Santos, J., & Lopes, G. (2022). Competences for sustainable development goals: The challenge in business administration education. *Journal of Teacher Education for Sustainability, 24*(1), 73–86. doi: 10.2478/jtes-2022–0006

51. Sharma, U., & Kelly, M. (2014). Students' perceptions of education for sustainable development in the accounting and business curriculum at a business school in New Zealand. *Meditari Accountancy Research, 22*(2), 130–148. doi: 10.1108/MEDAR-12-2012-0042

52. Noy, S., Capetola, T., & Patrick, R. (2021). The wheel of fortune as a novel support for constructive alignment and transformative sustainability learning in higher education. *International Journal of Sustainability in Higher Education, 22*(4), 854–869. doi: 10.1108/IJSHE-08-2020-0289

53. Dzhengiz, T., & Niesten, E. (2020). Competences for environmental sustainability: A systematic review on the impact of absorptive capacity and capabilities. *Journal of Business Ethics, 162*(4), 881–906. doi: 10.1007/s10551-019-04360-z

54. Elkaseh, A.M., Wong, K.W., & Fung, C.C. (2016). Perceived ease of use and perceived usefulness of social media for e-Learning in libyan higher education: A structural equation modeling analysis. *International Journal of Information and Education Technology, 6*, 192–199. doi: 10.7763/IJIET.2016.V6.683

55. Hill, L.M., & Wang, D. (2018). Integrating sustainability learning outcomes into a university curriculum. *International Journal of Sustainability in Higher Education, 19*(4), 699–720. doi: 10.1108/ijshe-06-2017-0087

56. Claro, P.B., & Esteves, N.R. (2021). Teaching sustainability-oriented capabilities using active learning approach. *International Journal of Sustainability in Higher Education, 22*(6), 1246–1265. doi: 10.1108/IJSHE-07–2020–0263

57. Duarte, F. de P. (2017). Sustainability learning challenges in a Brazilian government organization. *International Journal of Organizational Analysis, 25*(4), 562–576. doi: 10.1108/ijoa-02–2015–0842

58. Hadi, N.U., & Abdel-Razzaq, A.I. (2023). Promoting sustainable learning among accounting students: Evidence from field experimental design. *Higher Education, Skills and Work-Based Learning, 58*, 1–13. doi: 10.1108/HESWBL-03–2023–0058

59. Lehmkuhl, G., Gresse von wangenheim, G., Helena Martins-Pacheco, L., Borgatto, A.F., Da Cruz Alves, N. (2021). SCORE: A model for the self-assessment of creativity skills in the context of computing education in K-12. *Informatics in Education, 20*(2), 231–254. doi: 10.15388/infedu.2021.11

60. Matore, M.E.E.M, Khairani, A.Z., & Razak, N.A. (2020). Development and psychometric properties of the adversity quotient scale: An analysis using rasch model and confirmatory factor analysis. *Revista Argentina de Clínica Psicológica, XXIX*(5), 574–591.

61. Wekerle, C., Daumiller, M., & Kollar, I. (2020). Using digital technology to promote higher education learning: The importance of different learning activities and their relations to learning outcomes. *Journal of Research on Technology in Education*, *54*(1), 1–17. doi: 10.1080/15391523.2020.1799455

62. Yeung, K.L., Carpenter, S.K., & Corral, D. (2021). A comprehensive review of educational technology on objective learning outcomes in academic contexts. *Educational Psychology Review*, *33*, 1583–1630. doi: 10.1007/s10648-020-09592-4

63. Alshammary, F.M., & Alhalafawy, W.S. (2023). Digital platforms and the improvement of learning outcomes: Evidence extracted from meta-analysis. *Sustainability*, *15*(1305), 1–21. doi: 10.3390/su15021305

64. Venkatesh, V., & Davis, F.D. (2000). A theoretical extension of the technology acceptance model: Four longitudinal field studies. *Management Science*, *46*(2), 186–204. doi: 10.1287/mnsc.46.2.186.11926

65. Hair, J., Ringle, C., & Sarstedt, M. (2011). PLS-SEM: Indeed a silver bullet. *Journal of Marketing Theory and Practice*, *19*, 139–151. doi: 10.2753/MTP1069-6679190202

66. Hair, J.F., Hult, G.T.M., Ringle, C.M., & Sarstedt, M. (2022). *A Primer on Partial Least Squares Structural Equation Modeling (PLS-SEM)* (3rd ed.). Sage, Thousand Oaks, CA.

67. Ur Rehman, S., Bhatti, A., & Chaudhry, N.I. (2019). Mediating effect of innovative culture and organizational learning between leadership styles at third-order and organizational performance in Malaysian SMEs. *Journal of Global Entrepreneurship Research*, *9*(1), 1–24. doi: 10.1186/s40497-019-0159-1

68. Hair, Jr, J.F., Anderson, R.E., Tatham, R.L., & Black, W.C. (2010). *Multivariate Data Analysis* (7th ed.). Pearson, Hoboken, NJ.

69. Fornell, C., & Larcker, D.F. (1981). Evaluating structural equation models with unobservable variables and measurement error. *Journal of Marketing Research*, *18*(1), 39–50. doi: 10.1177/002224378101800104

70. Henseler, J., Ringle, C.M., & Sarstedt, M. (2015). A new criterion for assessing discriminant validity in variance-based structural equation modeling. *Journal of the Academy of Marketing Science*, *43*(1), 115–135. doi: 10.1007/s11747-014-0403-8

71. Shmueli, G., Sarstedt, M., Hair, J.F., Cheah, J.-H., Ting, H., Vaithilingam, S., & Ringle, C.M. (2019). Predictive model assessment in PLS-SEM: Guidelines for using PLS predict. *European Journal of Marketing*, *53*(11), 2322–2347. doi: 10.1108/EJM-02-2019-0189

72. Sharma, P.N., Liengaard, B.D., Hair, J.F., Sarstedt, M., & Ringle, C.M. (2023). Predictive model assessment and selection in composite-based modeling using PLS-SEM: Extensions and guidelines for using CVPAT. *European Journal of Marketing*, *57*(6), 1662–1677. doi: 10.1108/EJM-08-2020-0636

73. Oduro, S., & Haylemariam, L.G. (2019). Market orientation, CSR and financial and marketing performance in manufacturing firms in Ghana and Ethiopia. *Sustainability Accounting, Management and Policy Journal*, *10*(3), 398–426. doi: 10.1108/sampj-11-2018-0309

74. Abubakar, A.M., Behravesh, E., Rezapouraghdam, H., & Yildiz, S.B. (2019). Applying artificial intelligence technique to predict knowledge hiding behavior. *International Journal of Information Management*, *49*, 45–57. doi: 10.1016/j.ijinfomgt.2019.02.006

14 A Shelter of Love
Assistive Technologies for Education amongst Students with Learning Disabilities in Uganda

Racheal Ddungu Mugabi and Rosemary Nakijoba

14.1 INTRODUCTION: BACKGROUND TO ASSISTIVE TECHNOLOGY AND DISABILITY

The global need for Assistive Technology (AT) is high. The Global Report on AT jointly published by the World Health Organization [1] and United Nations Children's Fund (UNICEF) estimated 2.5 billion people in need of assistive products in 2019. In 2022, the report acknowledged that 1 in 10 people in the world need assistive products [2]. By 2050, it is estimated that 3.5 billion people will require assistive products [2]. However, as of 2016, about 90% of those who required ATs did not access them [3]. The reality therefore is that most of us – globally – will require AT at some point in our lifespan. With this in mind, considering the acknowledgement that access to AT is a right and a means to access, and realize other rights [4], it is critical to note that as a field, we rarely discuss issues related to Learning Disabilities (LD) when it comes to AT usage and provision [5]. Researchers have noted that there is still an enormous gap between the potential of AT and how much it actually helps [6]. While studies demonstrate the ideal approaches for developing AT policy [7], we are less inclined to discuss the policy gaps for improvement, AT products used, and their impact on students with neurodisability in Uganda [8]. With limited studies in place to address this literature and AT for the education gap, this study explored the experiences of university students with Learning Disability on AT for education in Uganda. It also explored how access to AT can be improved in Uganda by interrogating different government initiatives and their gaps. This was in lieu of not only the complex learning needs they face, but also the difficulties they encounter in acquiring new skills in inclusive university education with traditional pedagogies.

14.1.1 DEFINING ASSISTIVE TECHNOLOGIES (AT)

AT refers to the development and application of organized knowledge, skills, procedures, and policies relevant to the provision, use, and assessment of Assistive

DOI: 10.1201/9781032664828-17

Products (AP) [9]. In consonance with the global report, AT is an umbrella term for AP and its related systems and services [2]. APs include devices, equipment, instruments, and software that are specifically designed and produced to maintain and/or improve functioning [9]. ATs are thus assistive and rehabilitative devices, objects or systems that are utilized to help disabled people [10] and mitigate the effects of disabilities [11,12]. They are tools for communication, social interaction and physical access to resources [13]. In Uganda, Persons with Disabilities Act, 2019 [14] uses the word AT and assistive devices interchangeably to mean equipment or devices that support persons with disabilities to participate effectively in all aspects of life. The selection of AT, along with appropriate services and supports, is of vital importance to ensure that students not only learn and communicate with peers and parents but also are welcomed as part of a mainstream classroom, with similar expectations and considerations as their peers. While some universities and parents embrace AT to support students with disabilities, others are yet to see how the technology enhances inclusive education for these students specifically for this study of LDs.

14.1.2 Defining Learning Disability

LD is a group of neurodevelopmental disorders that can significantly hamper a person's ability to learn new things [15]. Examples include some of the common types of LDs espoused by Sanjana [16] as individuals having dysgraphia (difficulty producing legible handwriting), dyscalculia (difficulty understanding and using math concepts and symbols), dyspraxia (difficulty in language comprehension and does not match language production. She/he may mix up words and sentences while talking), nonverbal learning disorder (demonstrated by below-average motor coordination, visual-spatial organization and social skills), and dyslexia (may mix up letters within words and words within sentences while reading. May also have difficulty spelling words correctly while writing and letter reversals are common. Some individuals with dyslexia may also have a difficult time with navigating and route finding using right/left and/ or compass directions). The LDs can, however, be bypassed through the application of ATs. Such individuals confine significant cognitive and adaptive behavioural impairment and are prone to experiencing traumatic grief symptoms such as difficulty or inability to find meaning and communication barriers [17]. The LDs also represent some of the personal difficulties students face in university classroom environments.

14.2 CONTEXT

As part of the COVID-19 lockdown regulations, all institutions across Uganda were required to shift to Open Distance and eLearning (ODeL). Before an institution was cleared to offer remote teaching and learning, they needed to show that they adhered to guidelines that were issued in July 2020 by the National Council for Higher Education (NCHE). One of the requirements was an interactive learning management system (LMS) that effectively supports eLearning, which should provide for student-to-student interactions, student and instructor interactions, and evaluation of interaction. After the lockdowns, institutions were required to continue in a blended learning mode. However, institutions were not required to ensure that all physical and remote learning activities, and in particular the LMS and its content, were accessible to students with disabilities, even though 12.4% of the population have a disability [18]. The situation for students with LD is even worse because they are a neglected group regarding AT.

In Uganda, the Ministry of Health [19] clearly indicated that the government has limited awareness of the need for and importance of AT, lacks enforceable guidelines, lacks mainstream AT schemes. There is no national strategy for AT, the government plays no or very limited role in ensuring availability and access to ATs, its contribution is ad-hoc, and there is no monitoring and evaluation plan and indicators of AT impacts. The usage of AT data in Uganda is not available attributed to lack of a central place for registration, distribution of AT, low coordination and lack of contemporary information management systems which highly contributes to insufficient data collection, and lack of money to finance ATs. The Uganda Functional Difficulties Survey (UFDS) 2017 [20] indicated that over 62% of persons with disabilities who needed AT did not have such devices. Moreover, data of the number of disabled students with LD enrolled in universities in Uganda are not readily available [21]. Albeit the uncertainty of only students with LD statistics in the selected universities, primary data collection in Kyambogo showed 121 students with disabilities, but only seven (five male and two female) were living with LDs. The argument of more male than female students was attributed to gendered cultural practices that favour a boy child rather than a girl child to access education that is experienced by many girls on top of their disability to join university. Other universities did not have clear disaggregated data on disability and thus could not even specify the exact number of students with LDs attributed to the invisible nature of the difficulty.

Students with LDs face barriers that limit their learning and achievement in different activities that take place in the classroom setting [22]. LDs tend to create deficits in adaptive behaviours that can interfere with individuals' learning progress. For example, reduced functioning and the deficiency of independence among students with such disabilities can hinder their involvement in the learning process [23]. In some cases, students may exhibit below-average intellectual functioning that impacts their academic progress [24]. Moreover, insufficient communication and social skills are additional key markers that such students can experience, and limitations that are associated with mental functioning can contribute to slowing students' learning progress. While concerns such as reliance on others and a lack of self-confidence are often experienced by students with disabilities [25], appropriate support can help them deal with a range of challenges [23].

AT can thus offer opportunities to overcome barriers and engage in learning activities, access information, participate in academic activities, complete key academic assignments, and foster a sense of inclusion that encourages them to keep pace with their nondisabled peers [23]. Additionally, it is worth noting that AT complements accommodations and individualized instruction. ATs thus may help students learn how to complete the task and it can help to bypass an area of difficulty. For example, when a student decides to listen to a digital version of a book, they are bypassing an area of difficulty. It is thus critical that students with LDs have access to the same opportunities to participate as their peers. In this regard, evidence has shown that ATs (computers, laptops and mobile devices) have changed many students' lives with disabilities [26]. Despite these changes affecting education, little attention has been paid to the experiences of students with LDs and how they incorporate technologies into their daily lives. The recent situation analysis done in Makerere, Kyambogo and Kabale by Sikoyo et al. [27] and a subsequent peer-reviewed paper [28] focused on the preparedness of Uganda's public universities to provide education to students with visual impairment and staff capacities for inclusive teaching and learning of students with visual impairment in public universities respectively. This is not surprising, given that existing research often excludes this sector of the student population (LD). To ensure that equality and equity have been embraced, the Government of Uganda has created an enabling environment by enacting policies and regulations from the international to national levels over the past decades.

14.3 POLICY ENVIRONMENT

International and national policy instruments mandate access, equity and quality of educational services for persons with disability, and Uganda is a signatory. This section reviews key international and national policies that promote education accessible to students with disabilities.

14.3.1 INTERNATIONAL EXISTING POLICY ENVIRONMENT

At the global level, Uganda has committed herself to the global disability agenda. We mainly focused on three policy documents that cover the rights of adults with disabilities: the 2030 Sustainable Development Goals, the UN and The Universal Declaration of Human Rights.

14.3.2 COMMITMENT TO THE GLOBAL AGENDA

The 2030 Sustainable Development Goal 4 is "To ensure equitable and inclusive quality education and promote lifelong learning opportunities for all" [29]. In September 2015, while Uganda was holding the UN presidency, the 192 member states of the UN adopted a resolution committing themselves to the 2030 Agenda for Sustainable Development. The 2030 agenda and its associated principle of leaving no one behind is not only a holistic approach, but also the most people centred drive to achieving sustainable development for all. This drive is also connoted "persons with disabilities as agents and beneficiaries of development" [30, p. 9]. Along with this drive was the obligation for lecturers and university systems to plan for, teach and assess all students in a manner that is appropriate to meet their learning needs, irrespective of their circumstances [31].

14.3.3 THE UNIVERSAL DECLARATION ON HUMAN RIGHTS (1948)

This instrument recognises education as a universal right [32]. Denying any person a right to education is dubiously condemning such a person to a limitation in the enjoyment of fundamental rights. Article 26(1) desperadoes' discrimination in education at all levels and comprehensively requires states to make educational services available, accessible, acceptable and adaptable [32]. The requirements apply to persons with disabilities as well as the principle of equality and non-discrimination which is the basis of the human rights law (article 1(3)). This is enshrined in the philosophy of inherent dignity, and of the equal and inalienable rights of all human beings.

14.3.4 Convention on the Rights of People with Disabilities (2006)

On 13th December 2006, the United Nations during its General Assembly adopted the resolution drafted by the International Convention on the Rights of People with Disabilities (CRPD) [33]. The resolution was to promote measures for research and development of disability.The measures included friendly technologies, their availability and use such as specific technical devices designed to improve the daily lives of people with disabilities. Article 24 of the CRPD in 2006 [34] encourages public/private partnerships to ensure the provision of ATs with reasonable accommodation in education. In 2016, Uganda had a dialogue with the CRPD Committee regarding conditions for people with disabilities. Delegates from Disabled Persons Organizations demanded the need to translate the CRPD into amendments to national law, including the 2006 Persons with Disabilities Act, and the need for budgetary allocations to disability programmes particularly in education. These programmes included Disability Rights Fund [35] and Uganda DPOs Present Critical Rights Issues to CRPD Committee [36]. In 2018, Uganda endorsed the CRPD which mandates all states to protect, respect and fulfil the right to education without discrimination.

14.3.5 Regional Level

At the regional level, the benefits arising from integrating AT in learning to accommodate students with LDs in the general classroom compelled policy changes to promote its use and stimulate a positive attitude in teachers. For instance, the East Africa Community Protocol adopted in 2012 recognises the need to empower Persons with Disabilities with education as one of the special emphasis [37].

14.3.6 National Level

At the national level, we focused on policies that embrace rights of the people with disabilities.

14.3.7 The Constitution of the Republic of Uganda (1995)

Articles 21(1&2), 32(1) and 35(1&2) guarantee that all persons have the right to education [38].

14.3.8 Persons with Disability Act (2019)

The Ministry of Gender, Labour and Social Development in conjuction with its partners worked on the development of disability Act [39]. The Act was assented to by the President in September 2019 and gazetted in December 2019. The Act is the first legal document in Uganda to interpret 'assistive devices'. The Act [Section 6(5)] guides the government to provide assistive devices to all learners with disabilities. The Act [Section 7(7)] continues to specifically guide the government that persons with disabilities should be provided with assistive devices at no cost or subsidized prices.

14.3.9 The Education Policy Draft (2011)

It stipulates that all children with special learning needs – including those with disabilities – should access learning in an inclusive environment [40]. This policy is guided by the Ministry of Education and Sports (MoES). The MoES established a department for Special Needs Education (SNE) to ensure; disability inclusive planning and budgeting; development of policies and laws for learners with disabilities that are in tandem with the constitution, and other national and international laws. Similarly, SNE is catered for and adequately programmed and managed right from curriculum development at the National Curriculum Development Centre (NCDC), the Directorate of Educational Standards (DES), and the Uganda National Examinations Board (UNEB).

14.3.10 The Universities and Other Tertiary Institutions Act (2003)

It was established by the NCHE of Uganda and details its mandates [41]. The Act obligates NCHE the responsibility to ensure disability inclusion in institutions of higher learning. It also requires that NCHE certify that an institution of higher education has adequate and accessible structures. The Act empowers the NCHE to revoke a provisional licence to an institution if it finds it not meeting the minimum requirements pertaining to infrastructure. Universities and Tertiary Institutions Act (2003) paved the way for the admission of students with disabilities, historically the blind and the deaf and more recently the little people and albinos [42]. At institutional levels, policies have been adjusted to provide for inclusive education [27, p.6]. Institutional policies are also in tandem with national and international policies and guidelines.

14.3.11 The Higher Education Students' Financing Board (HESFB)

The HESFB was established by an Act of Parliament in 2014 as a body corporate - semi-autonomous mandated to provide loans and scholarships to students intending to pursue higher education. The Board was inaugurated on 22 April 2014 and started operations in May 2014. In 2017, it started prioritizing students with disabilities in its loaning system. HESFB offers loans to students pursuing degree and diploma programmes, and such loans can be used to cover functional fees, tuition fees, research fees and assistive devices for students with disabilities.

14.3.12 ICT for People with Disabilities

The Uganda Communications Act 2000 promotes research into the development and use of new communications techniques and technologies. The National IT Authority-Uganda (NITA-U) Act (2009) provided for the establishment of the National IT Authority–Uganda (NITA-U). Section (f) of NITA-U mandated it to promote access to and use of ICT for people with disabilities. It is also responsible for the evaluation of university virtual learning environments for accessibility. A draft ICT for Disability Policy was published in 2017 [43]. It promotes sector-wide interventions to improve the lives of all Ugandans through ICT including inclusive education, which is to be achieved by developing and supplying accessible ICTs and ICT-enabled services in Uganda.

14.3.13　Gaps in the Policy Environment

The Universities and Other Tertiary Institutions Act Section 110 – Revocation of a Charter has no mention of universal design and accessible facilities among the set grounds for such revocation in case a university fails to adhere to its guidelines. Failure by an institution to provide universal design, accessible facilities, reasonable accommodation, appropriate instruction or teaching methods and qualified staff for special needs students is a path to the exclusion of students with LDs in the institution's programmes.

There is partial coverage of access for people with LDs by ICT policies. The relevancy of existing ICT policies to the promotion of ICT-enabled services such as eLearning that are accessible to those with LDs is low. The NITA-U Act 2009, the Uganda Communications Act 2000 and the National ICT Policy 2014–2019 on people with disabilities have identified with youth and women as a special interest group. Due to this generalization, the use of ICT as a means to remove barriers to learning for students with LDs is not highlighted [8]. This limits the relevancy of such policy instruments to the demand for access, equity and quality to educational services. Where specific provisions are made, they are limited to specific types of disability. For example, the Uganda Communications Act 2000 only refers to the promotion of accessibility of communication services to the hearing impaired. But, given the current technological and communications convergence, people with other disabilities particularly LDs are also affected by the telecommunication and broadcasting services regulated by the Communications Act 2000 [43]. The situation is not different from all policies running from international to national policies which are seen to be generic in nature.

There is no policy on eLearning accessibility to people with LDs either as part of other related policies or as a standalone policy. The Special Needs and Inclusive Education policy that was drafted in 2011 has not yet been operationalized although it offers an opportunity for further advocacy and engagements, a situation the Government of Uganda attributes to lack of financial resources for its implementation [19]. But even this draft does not mention eLearning, much less accessible eLearning. In the absence of such policy guidance, consideration of LD-specific needs in the implementation of eLearning in higher education institutions is limited, and in most cases non-existent. This is coupled with the nonexistence of any requirement for educators to be trained, and the education environment is not supported in providing eLearning that is accessible to people with LDs. This greatly disadvantages those living with LDs, especially given the potential benefits of eLearning to them.

Similarly, the ICT and Disability Policy has been in draft form since 2017 for the same reason of inadequate resources and makes no mention of accessible eLearning/education technology. The 2011 draft Policy on Special Needs and Inclusive Education does not specify AT and educational technology as one of the instructional materials, equipment and supportive services. On the other hand, the 2017 second draft Policy on ICT for Disability proposes comprehensive sector-wide interventions to improve the lives of all Ugandans by facilitating and promoting the use of ICTs but does not refer to making eLearning accessible to people with LDs.

While the Persons with Disability Act, 2006 provided for a 10% budget allocation on issues of disability, only less than 1% is being implemented in some sectors and not in higher education [19].

Access to ATs is limited, quality is negatively affected by constraints ranging from a lack of physical capacity, infrastructure, and learning materials to inadequate teacher capacity, negative attitudes and stigma [21], assessment and examination [44] especially right now where some of the universities are still operating in traditional pedagogies. The impact of the policy developments has thus been seen in the increasing enrolment of students with disabilities in higher education with no evident matching infrastructure [19,44]. The situation is even dire for the poor LD students where Zoom video conferencing technology has gained popularity in post-COVID-19, yet is accessed only by students who can afford data costs for the internet and devices such as computers, laptops and smartphones [27]. The disability university policies in Uganda do mention not the LD so there is no legal basis for AT for inclusive education.

14.4 THEORETICAL CONTOURS

A hybrid of a social model of disability and a rhizome framework was used. Drawing from Oliver [45], a social model of disability is "a theoretical understanding of the concept of disablement from a socio-political perspective" (p. 57). The argument is that disability is somewhat imposed on people with disabilities on top of their impairment by an oppressive and discriminating social and institutional structure [46, pp. 3–4]. The social model of disability mostly explains the relationship between people with impairments and their participation in society [45]. The model is hinged on the principle that impairment and disability are distinctively different [45,46]. The implication is that disability is a social oppression, not impairment, and that disability is a social construction, and to a large extent is culturally produced and culturally structured [45]. For equal participation for people with disabilities, the model demands the removal of the society's economic, environmental, cultural and other barriers against people with disabilities. The understanding of the social model of disability, in this chapter brings about the operationalisation of the right to education for persons with disabilities.

In exploring the experiences of students with disability, a metaphor of a rhizome was utilised drawing from philosopher Deleuze and psychoanalyst Guattari [47]. A rhizome is non-linear, non-hierarchical and open-ended and its nodes can be connected to other nodes forming an "assemblage" [47, p. 7]. An assemblage "is the process of making and unmaking…, arranging, organizing and fitting together" [48, p. 263]. Rather than being confined to the single tap roots of traditional Western ways of thinking: '[Rhizomes] affirm what is excluded from western thought and reintroduce reality as dynamic, heterogeneous and non -dichotomous … they propagate, displace, join, circle back, fold' [12, p. 27]. In this perspective, a rhizome was used in analysing various realities by giving voice to those that are hitherto unheard [12,49,50]. We took the understanding of experience as multiplicity. This is because, within rhizomatic thought, experience is viewed as contingent and fluid and a site where various forces interact rather than a fixed set of possibilities for agency within a set structure. In a rhizome, there are no fixed lines, but only lines of flight and lines of articulation that can rupture and at other times reconnect with the rhizome in unpredictable ways [47, p. 89]. This aided in the understanding of the often-times contradictory and ambivalent, ever-shifting realities of the students with LDs as they engaged in AT for university education. Lines of flight can also represent the agency of the students, who sometimes are able to defy our common-sense understanding of their marginalised position, by displaying capacities for agency as they seek to access education. We thus analysed how these students engage with diverse ATs and how their agency interacts with their participation supported by the fieldwork methods.

14.5 FIELD WORK METHODS

A qualitative rapid ethnography design was used to carry out research from January 2023 to August 2023 in university settings in Uganda with ethical clearance from the Uganda National Council for Science and Technology. In Uganda, although there is no data available on students with LDs who transition into university education, our insider positionality in university education as lecturers with the support of snowball sampling enabled us to access students with LDs and their lecturers. In making an assemblage of student experiences, we plugged in as argued by Deleuze and Guattari [47, p. 23] to the 'field of reality' (data, theory and method), 'field of representation' (producing different knowledge, resisting stable meaning) and 'field of subjectivity' (becoming-researcher). In the field of reality, we interviewed thirty female students (connoted as S_{1-30}) and thirty male students (connoted as S_{31-60}) from six universities (public: Makerere, Kyambogo and Busitema; Private: Uganda Technology and Management University, Muteesa I Royal University and Victoria University) because of accessibility to information and existence of ATs from the insider perspectives. Researching in a non-linear way, rapid ethnographic methods of data collection such as participant observations in the classes that we teach, go-along to other classes in other universities as we interviewed using narrative inquiry. The purpose was to understand the lived experiences without interference from the researchers and give a voice to those that were hitherto unheard.

The student study population criteria measure for inclusion was that a student must have been living with an LD. All students were much older (between 25 and 28 years) in relation to the age group of their peers in the same undergraduate classes. This was attributed to especially failure to pass all papers resulting in repeating years and slow learning. The selection was also supported through snowball sampling to allow students with LDs to identify other potential subjects accompanied by the validation of lecturers who teach them. The identified students interviewed and their lecturers were from different programmes and professions of universities including ICT, Computing, Mathematics, Literature, Law, Social Sciences and Education. The intention of the study was not about the college, education and age of a student with an LD, but rather the focus was on the main characteristic of an LD. Parents (2 men: P_{1-2} of 62 and 65 years old; 3 women: P_{3-5} of 52, 57 and 60 years old, respectively) were interviewed. Key informants (2 men: K_{1-2} of 48 and 51 years old; 3 women: K_{1-3} of 44, 47 and 58 years old) and lecturers (3 men: L_{1-3} of 37, 40 and 41 years old; 2 women: L_{4-5} of 35, and 37 years old, respectively) were also interviewed. Consent was sought from all participants with assurance of confidentiality through anonymised numbers. The study was concluded at the point of saturation when no new information was coming to the fore. Secondary data was to get more nuanced voices through studying texts in class notes to understand their experiences and literature. Within the field of representation, we sought voices that, even as partial and incomplete, produced 'multiplicities and excesses of meaning and subjectivities' [48, p. 263]. Rather than seeking stability within and among the data, we drew data that seemed to be about difference and not sameness as presented in the results.

14.6   RESULTS

We lay down the nodes one after the other, with nodes one, two and three revealing AT usage difficulties faced by students with LDs, forms of AT and usage for education, impact of AT on learning, completion of tasks and influence on engagement. Numbering these nodes is not to rank or structure them, but to label them succinctly, so that it is possible to suggest where pathways might connect and chasms separate them on the rhizomatic map. In this context, therefore, we sought to "work [the rhizome] for all its worth" [51, p. 281].

14.6.1   NODE ONE: AT USAGE DIFFICULTIES FACED BY STUDENTS WITH LDs-A BAD START

The choice of node to commence this mapping is random. Rhizomes have no commencement or end, but are "always in the middle, between things, inter being, intermezzo" [47, p. 25]. St. Pierre [51] insists on the vital role of rhizomes in a world in which we are all "becoming": "[W]e…. must learn to live in the middle of things, in the tension of conflict and confusion and possibility; and we must become adept at making do with the messiness of that condition and at finding agency within" (p. 176). In the search for some form of messy agency, we choose to start with the story of a LD student with dyslexia. This story brings home the difficulties faced by many students with LDs; who are not only in universities with traditional pedagogies, policy and institutional failures, but also those who are poor and cannot even meet their AT needs for education. The student explained that

> I failed all my first year semester one examinations during Covid-19 period when the university introduced ODeL and I was made to stay put [repeat the first year]. I mixed up letters within words and so my sentences did not make sense. The cost of the AT that I need is quite high. I also lacked money to pay my assistant to support me, so I had to do it myself. I got difficulties in reading and writing given that the content of the exam was not supported by AT. Moreover, the university ODeL did not have in-built software to support my disability, yet I barely had the skills required for navigating the system in my condition. When the university re-opened after the lockdowns, the university classes had no structures/software to support me. The university physical lecture rooms use black boards and chalk or white boards and marker. When I got a chance to get AT from a friend who graduated, it was not the right one that I needed, and the content was also not supported because the lecturers are actually ignorant about the required ATs for my disability and have no skills to design such content anyway. The lecturers have not been trained, but also there is no policy to support us. Moreover, the disability office has not yet recognised students with LDs which implies that we cannot get any support from them. Lecturers are aware of the plight of students with disabilities, but unfortunately, they have no skills to identify, assess and give us appropriate AT to support required learning (S_1).

This story is reinforced by the first lecturer we interviewed who explained that

> The learning management systems could be having some support ATs, but the challenge is lack of skills to use them for learning. There is a need for specialized training and re-tooling of staff on AT usage for education. There is ineffective and inefficient use of ATs, pedagogical barriers due to lack of diversity and flexibility in teaching and evaluation methods which excludes students with LDs from meaningful learning. Lecturers have no patience, coupled with inappropriate curricular, but also the students have not been introduced to independence through use of ATs. The budget for accessing ATs is also low at the university level and so most the ATs are not accessible for everyone. Even this room accommodates only a few at a time and the ATs especially hardware's are not portable which limits remote learning (L_1).

Given that most universities have not yet started integrating ATs for LDs in Uganda, students with LD difficulties face complex challenges, especially with universities that are still running traditional pedagogies. However, universities that have started integrating ATs (Kyambogo University with leading ATs and expertise), and other universities explained how software, apps and hardware ATs are used for enhancing learning which leads us to node 2.

14.6.2   NODE TWO: FORMS OF AT AND USAGE FOR EDUCATION

Interviews with students, lecturers and parents revealed several software and hardware ATs used to support learning as discussed in the subsequent sections using screenshots taken while collecting primary data from universities starting with JAWS (Job Access with Speech in Figure 14.1).

A student we found at the SNE disability centre from the College of Computing explained how he uses JAWS in Figure 14.1 to learn narrating,

> To start my JAW reading the screen, I press the key combination INSERT+DOWN ARROW [this is the Say All command]. While using the Say All Command, I can press the LEFT or RIGHT ARROW keys to rewind or fast forward through the text or any content. I can also press PAGE UP or PAGE DOWN to increase or decrease the speech rate (S_{31}).

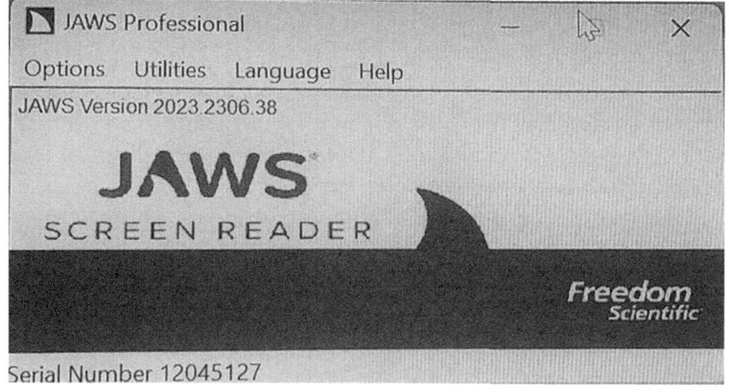

FIGURE 14.1   JAWS.

An expert in charge of SNE at the disability centre from a public university explained the usage of JAWS (Figure 14.1) that

> JAWS is a computer screen reader program for Microsoft Windows that allows not only students with reading difficulties especially for cognitive difficulties (dyslexia) to access content, but also those who have triple disabilities of LD, blind and visually impaired users to read the screen either with a text-to-speech output or by a refreshable Braille display. Jaws supports by translating text to audio. Lecturers give content in soft copies and students can read and attach meaning to learning materials. Students can also search internet and JAWS support them by reading/talking every letter that a student presses on the computer. JAWS has a variety of features, including Braille support, multi-lingual speech synthesis, and multi-screen support which enhances many of the students' performance. However, there are two main constraints; one, users can only access if they can afford to buy a computer/laptop and data given that the centre does not lend out; and two, many of the good books are not in soft copy which again excludes many poor students from learning. The centre has only two staff to support summarizing books to soft copy (L_1).

Voice Notebook (Figure 14.2) was another predominant software that was cited in all research sites of the study.

Voice Notebook usage for learning was explained to us while interviewing a student from computing who had a double difficulty of dyslexia and Attention-Deficit Hyperactivity Disorder. Attention-Deficit Hyperactivity Disorder (ADHD) is not an LD; however, it does make learning difficult. For example, it is pretty hard to learn or acquire a skill when you struggle to focus on what your lecturer is saying or when you seem not to be able to sit down and pay attention to a book. The student illustrated the following.

Voice notebook supports my learning through speech-to-text (TTs) which bypasses the tasks of handwriting and spelling, and allows me to concentrate on developing my ideas and planning my work. TTs allow me to see how words are spelt by reading them out loud (S_2).

JAWS and Voice Notebook software were predominantly used in all universities. The most common ATs were software that could be used as explained to the research

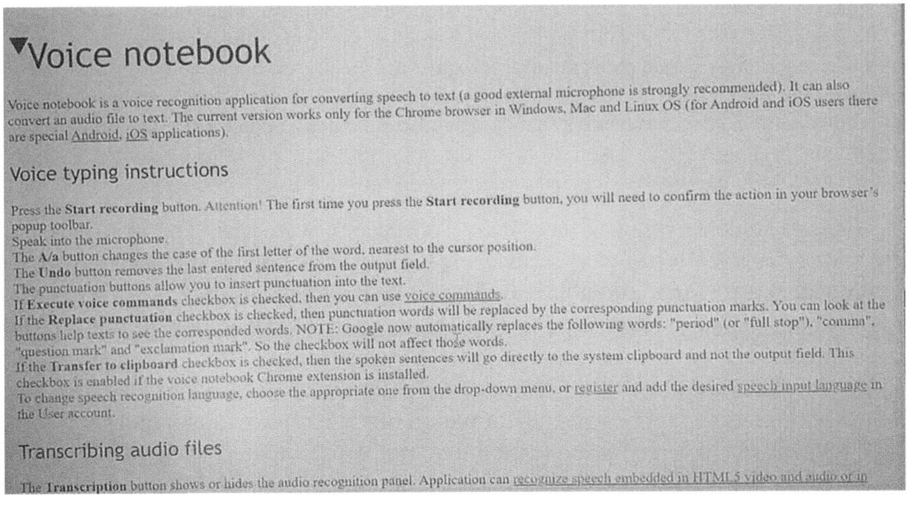

▼Voice notebook

Voice notebook is a voice recognition application for converting speech to text (a good external microphone is strongly recommended). It can also convert an audio file to text. The current version works only for the Chrome browser in Windows, Mac and Linux OS (for Android and iOS users there are special Android, iOS applications).

Voice typing instructions

Press the **Start recording** button. Attention! The first time you press the **Start recording** button, you will need to confirm the action in your browser's popup toolbar.
Speak into the microphone.
The **A/a** button changes the case of the first letter of the word, nearest to the cursor position.
The **Undo** button removes the last entered sentence from the output field.
The punctuation buttons allow you to insert punctuation into the text.
If **Execute voice commands** checkbox is checked, then you can use voice commands.
If the **Replace punctuation** checkbox is checked, then punctuation words will be replaced by the corresponding punctuation marks. You can look at the buttons help texts to see the corresponded words. NOTE: Google now automatically replaces the following words: "period" (or "full stop"), "comma", "question mark" and "exclamation mark". So the checkbox will not affect those words.
If the **Transfer to clipboard** checkbox is checked, then the spoken sentences will go directly to the system clipboard and not the output field. This checkbox is enabled if the voice notebook Chrome extension is installed.
To change speech recognition language, choose the appropriate one from the drop-down menu, or register and add the desired speech input language in the User account.

Transcribing audio files

The **Transcription** button shows or hides the audio recognition panel. Application can recognize speech embedded in HTML5 video and audio or in

FIGURE 14.2   Voice notebook.

team by students, lecturers and parents including smart boards, tablet computers, iPods, laptops, and desktop computers, while projectors, flashcards, videos, and PowerPoint presentations were also used to facilitate learning. Additionally, while students indicated that the ATs were efficient in empowerment and stigma alleviation, lecturers suggested that the learners felt being in a less restrictive environment as they could effectively compete with their peers who were without disability. Moreover, students with LDs indicated that the use of tablets and computers to obtain apps also assisted them in improving their learning processes such as Taptapsee on the lower right side of the screenshot was used by taking a photo and the app describes it in a verbal/talking mode). Other apps (voice-over and Siri for iPhone, and Google TalkBack for Androids) work just like JAWS. For example, a student would ask the app to write a message, time-reminder and read texts for individualized activities and learning.

Inbuilt software functions in ICT College improved students' writing skills as attested by a student:

> Using spell check and grammar features helped me to communicate my ideas and write with confidence knowing that they can easily make changes. In addition, being able to submit my final assignment that is neater and better organized improves my previously lost self-esteem (S_{32}).

In consonance with the student, a lecturer in the ICT department who had students with LDs in his class explicitly explained the forms of AT they use for education.

> Text-to-speech (for example the Kurzweil 3000 convert text into spoken output. The program reads aloud digital text found on computers, smartphones, and tablets. Text to speech is used to read back a student's or lecturers original work, scanned or downloaded worksheets as well as textbooks for comprehension), speech-to-text (e.g., Dragon Naturally Speaking- is voice or speech recognition software that takes a student's or lecturers spoken words and converts them to typed words on a computer or other digital device also called Voice-to-text), Rocket notebooks (with the accompanying App) uploaded to various cloud services such as Google Drive, Dropbox and more, using a smartphone or tablet. This helps students to organize their work and create a digital portfolio especially in Math. Smart pens use audio recording technology and special notebooks to record the lecture's allowing students to take minimal notes. It is helpful to students that struggle to listen and write notes at the same time. Scanning pens use OCR technology which reads aloud text from worksheets, textbooks and novel studies. Scanning pens give students (especially students with dyslexia) greater independence allowing them access to content that cannot always be delivered in a digital format and graphic organizers (e.g., Inspiration) are the main software functions for students who struggle with reading, writing and organising difficulties (L_2).

A literature student explained that 'Text-to-speech software assists me with monitoring and revising my typed work, as hearing the text read aloud assists me in catching grammatical errors that might have otherwise gone unnoticed' (S_3).

A social science student (S_4) boosted about the software support to improve the perception of her work and the ability to write expressively. Other 2 students (S_{33} and S_{34}) from ICT noted that the program decreases the negative emotions they had associated with reading and provides them with a more complete comprehension of the text. Student (S_{35}) from education noted that 'pen top computers such as Live Scribe smart pens provided text-to-speech, and other organizational functions' as also attested by another student (S_6). A student from math (S_6) narrated that 'it is useful in reading, but also provides auditory feedback during composition. It provides reminders such as "don't forget to carry" during multiplication questions'. A lecturer (L_4) explained that 'whenever such students with LDs use speech-to-text, the work is eased because the handwriting, spelling, punctuation, and grammar are catered for in the students' essays as well as high-level composition skills (for example in planning and generating content)'. She elaborated that 'a student who struggles to use printed material might have the choice to use an audiobook as an alternative. Unfortunately, some of the good books are not in soft copy' (L_4).

Graphic organizers helped students with LDs to improve the quality of writing as attested to by a lecturer who has students with LDs. The lecturer narrated that

> Graphic organizers provide an organizational framework that helped students to generate topics and content for research because it helped with the planning and organizational stages of writing, and use concept mapping software which increase the quality and quantity of writing. Using a web-based graphic organizer with procedural prompts also enabled students to produce better organized and higher quality essays, than they could have produced with handwritten organizers (L_5).

Hardware screenshots also supported learning as presented in Figures 14.3–14.11.

Ablecon (Figure 14.3) works more like a learner management system.

A lecturer in charge of the education disability centre explained the usage of Ablecon in Figure 14.3 that

> Ablecon centre has a camera which picks live sessions when the lecturer is teaching. For example, a lecturer can go to you tube and picks videos or records himself /herself, or using sign language, then transfers the recording to a flash drive to add caption for easy access to students. For those students who prefer large spaces to concentrate, they do not have to be sited in front to inconvenience others, but can come and pick a code and have live sessions on their devices [computer or phone] even at the back of the class or a nearby safe area. This AT can be used by all students of different disabilities given that it accommodates them through providing audio tape, content in large print, reduce number of items per page or line, provide a designated reader, and present instructions orally (L_1).

FIGURE 14.3   Ablecon.

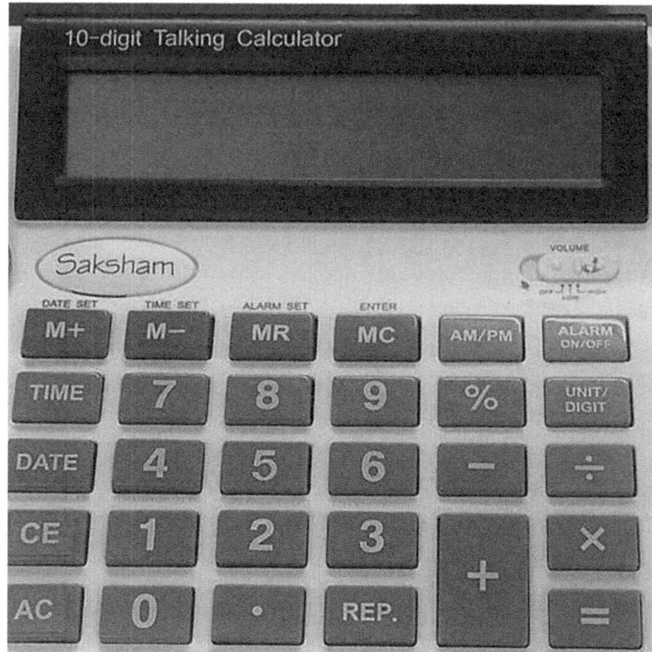

FIGURE 14.4   Talking calculator.

A Talking calculator (Figure 14.4) helps students, especially for math.

Some students (S_{36-41}) in the math department noted that scientific and talking calculators including graphing calculators provided visual conformation of the graph shape which motivated them to learn. A lecturer in the math department at a public university explained the usage of a talking calculator (Figure 14.4) that

> Students who have dyscalculia can benefit greatly from a talking calculator. It makes it easier to check assignments, read numbers and perform calculations. While the talking calculator is a fairly simple tool, it offers an exceptional benefit for students who would otherwise struggle in math classes (L_5).

Digital voice recorder pen (Figure 14.5) can be commanded to write and you move with it.

A student from social science explained the usage of a digital voice recorder pen (Figure 14.5) that

> I have both a LD and a physical disability limiting my fine motor control, so I find the pen useful for learning; especially because note taking is restricted to writing a single word to describe each part of the lecture. I can use the pen to replay just the section of the audio recording associated with each of my keywords (S_{42}).

Omni Reader (Figure 14.6) was designed for LD students, but especially those who also have low vision.

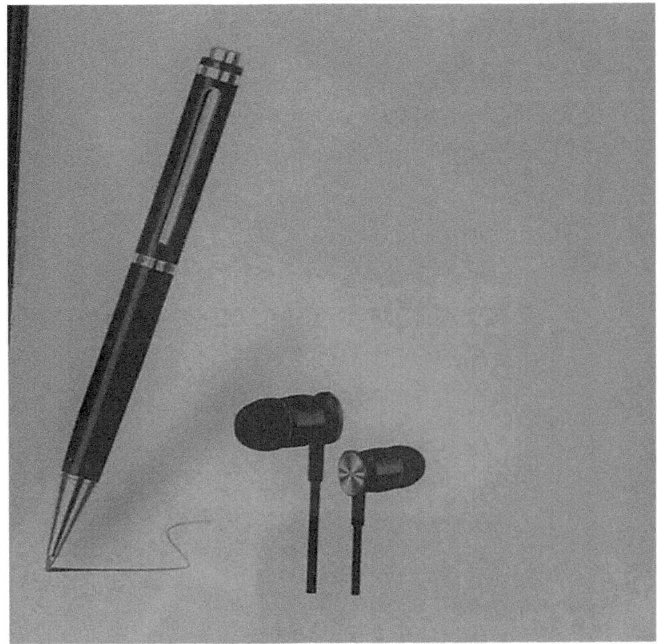

FIGURE 14.5   Digital voice recorder pen.

For Omni reader usage (Figure 14.6), a student from Law illustrated that

I can place a hard copy book below the scanner, and after scanning, I enlarge the words at my convenience. I can also play around with colour and contrast. Its usefulness is with the machine reading out loud for me to hear. The used document could also be stored for future reference given that the device has a USB port which can be saved and transferred to other devices. I can also install an audio book and it begins reading for me. More importantly, I can connect Omni reader to wireless network and access online books (S_7).

Prodigi Merlin ultra CCTV (Figure 14.7) is used by LD students to access information easily.

A student from Law we found using Prodigi Merlin CCTV (Figure 14.7) illustrated how it is used by indicating that

The only disadvantage is that it is not portable, but can serve several students at the same time if learning similar content. It has a camera which transfers information from a book to the screen for me to see and it reads the book loud. It can be tuned to any volume convenient for me at the time, I can also play around with colours and save my materials on USB drive. However, because it has no interaction on the tray, sometimes I dose off (S_{50}).

The Prodigi screenshot in Figure 14.8 was an example captured during data collection showing a constitution being projected on the screen and reading out loud by the Prodigi machine.

FIGURE 14.6 Omni reader.

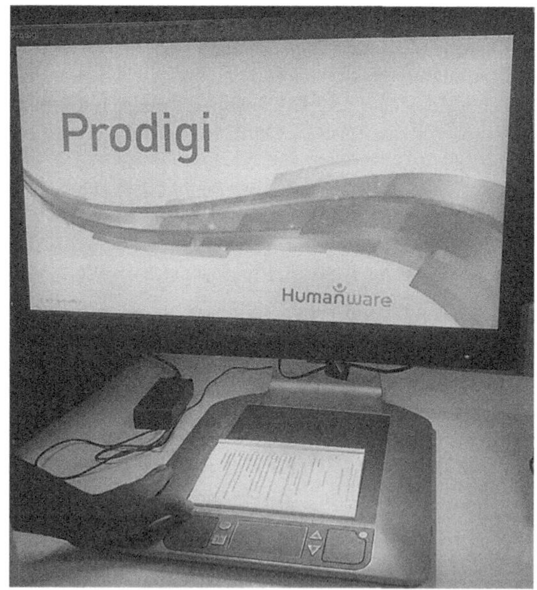

FIGURE 14.7 Prodigi Merlin CCTV.

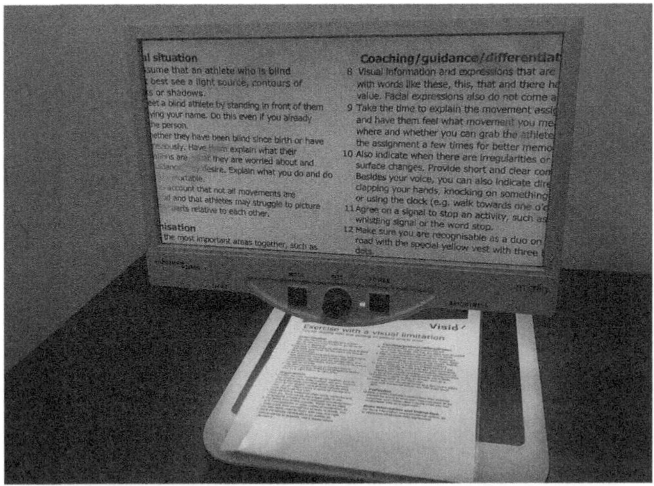

FIGURE 14.8 Prodigi screenshot reading.

The lecturer (L_1) interview brought to the fore an amazing insight by introducing to us the ultra CCTV machine which increases the interaction of students because of the tray which contravenes the students' interview of the Prodigi machine which has no tray interaction. We unfortunately noticed the disadvantage of the ultra CCTV machine of not being able to read out loud although can project information from a book to a screen. Therefore, it is more useful to use Prodigi Merlin CCTV.

A clear reader (Figure 14.9) was used by all disabled students.

S_8 from education explained the usage of the clear reader (Figure 14.9) that

You simply place a book in front of it, take a picture and it will begin reading out loud automatically. No matter how the book is laid out (whether upside down or changed sideways), the device will read it.

A student (S_9) from development studies revealed that 'Perkins braille (Figure 14.10) is mainly for the blind to make braille notes. Albeit the noise made while typing, as a student with a LD, I am also blind and so I use it'.

The lecturer (L_1) narrated a sad story about an incident of Perkins braille (Figure 14.10) where a 'lecturer' stopped a student to enter a lecturer room in 2010 due to fear of bombs that were rampant in Kampala at that time. This story brings home the difficulties faced by not only the LD student who was denied access to class, but many disabled students whose stories are not captured in this chapter, but also exposing the limited knowledge of lecturers about ATs for education.

S_{10} from computing explained that

The Braille embosser (Figure 14.11) helps to change content from soft copy to braille notes for the especially blind with LDs to access learning.

The usage of hardware ATs was in especially only two out of three selected public unities and a few students use software in the third public university. Very few students in private universities use software ATs (results reveal only software and not hardware) which led us to node 3 of AT's impact on education.

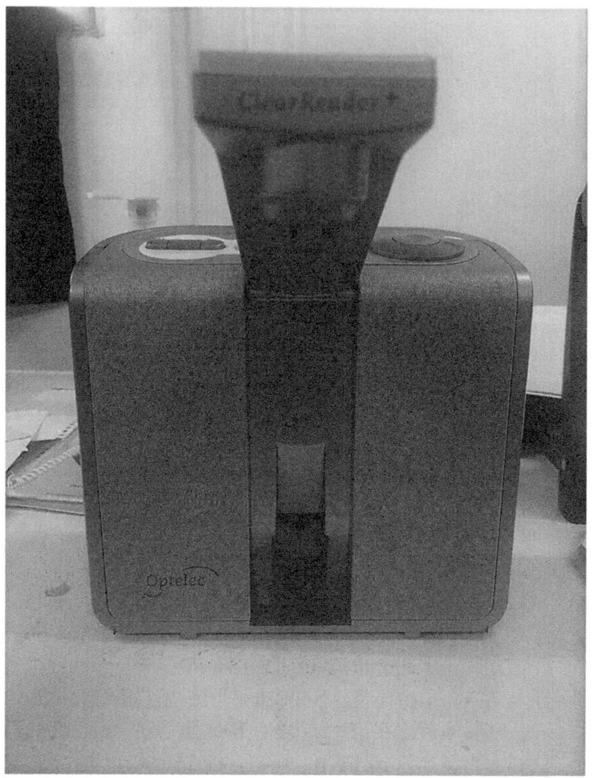

FIGURE 14.9 Clear reader.

FIGURE 14.10 Perkins braille.

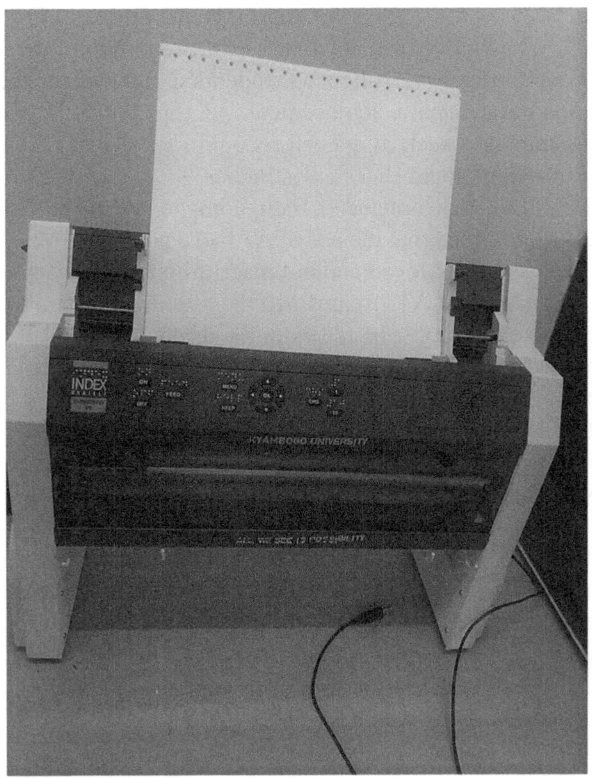

FIGURE 14.11   Braille embosser.

14.6.3   Node Three: Impact of AT

In this node, an assemblage of various factors and relations emerge as students negotiate between different education programmes vis-à-vis their LDs. A focus is on eLearning education and institutional and infrastructural policy discourse in the selected six universities. The experiences faced while in the universities open up a window to demonstrate how some students were able to negotiate the stigma attached to LD and those students who retake amidst many other barriers. The metaphor of love in this chapter is two-sided. For the students with LDs; after stigma with fellow learners at lower levels, acceptance into universities to learn using ATs built their self-image from "academic failures to qualified professionals". The feeling of acceptance is thus referred to as *"love"* in such a setting of an institution referred to as a "shelter". The feeling we labelled our title *'a shelter of love'*.

14.6.4   Impact of AT on Student Learning

Some lecturers and some students supported AT usage for education for all its worth. Each interview discussed specific reasons for supporting AT usage and its impact on LD students' education. The personal insights were based on multiple personal experiences which provided an in-depth understanding of the concept concerned.

Lecturers narrated the AT's ability to attract the attention of LD students, transfer information swiftly, and reduce the amount of effort required for teaching. Moreover, they believed that the use of AT enhanced students' confidence and enabled them to attain a higher level of academic achievement. AT thus built LD students' strengths and worked on challenges such as speech recognition, writing, and numeric calculations. In addition, they noted that their students were able to gain essential skills through the use of AT that enhanced their competence in the learning process. Finally, lecturers highlighted that the use of AT had a positive effect on student learning because at most they feel loved and not discriminated when using their ATs. For example, they explained that AT streamlined the learning process by employing students' multiple sensory abilities. In addition, the use of AT enabled their students to access more comprehensive learning materials than other methods, which broadened their perspectives.

Students attending universities that were embracing ATs in the system and use in classes narrated factors such as improvement in their communication skills by facilitating swift responses, promoting students' attention, and engaging multiple sensory abilities. More importantly, students explained that such ATs prompted note-taking, typing, and listening skills that improved their learning approaches and believed that the use of AT could foster comprehension abilities and learning capacity and facilitate students' academic progress. They however pitied those students with LDs who attend universities without ATs for missing out on the benefits of AT for education. This was attributed to difficulties in accessing for some, economic shocks of the country and lack of funding from the government to support ATs for every student with LDs.

14.6.5 IMPACT OF AT ON THE COMPLETION OF ASSIGNMENTS

Key informants, parents, lecturers and students explained that AT can facilitate their completion of assignments, for instance, by motivating them to complete their tasks efficiently. They also reported that AT could help such students access a vast range of learning materials and improve their focus and concentration. Moreover, the lecturers believed that the study habits necessary for executing academic tasks could be enhanced easily and quickly if students with LDs use AT. However, some lecturers (L_3 & L_5), and key informants (K_{1-5}) noted that AT could also create confusion and act as a source of distraction. Some students and lecturers are not able to use ATs and the selection of appropriate AT poses challenges even if they want to use AT.

14.6.6 INFLUENCE OF AT ON ENGAGEMENT

All those participants who had integrated AT into their education system agreed that AT can foster the engagement of students with LDs. The employment of simplified information and sufficient feedback promoted increased participation in learning. Lecturers explained that ATs nurtured psychological and mental preparation, supported students' attention and enhanced their information retention capacity. Moreover, they explained that AT made the learning process fun, and easy due to the interesting content, and the entertaining and enjoyable educational activities it facilitates. Lecturers elaborated that ATs engaged students' multiple sensory abilities, and they also motivated and increased the students' focus. Nonetheless, some lecturers ($L_{3\,\&\,5}$) entreated that, in as much as the AT made learning engaging and immersive for students with LDs, it had its equal share of distraction.

14.7 DISCUSSIONS

The policy level revealed that Uganda has drafted enabling policies and laws aimed at protecting the interests of people with disabilities and creating equal opportunities for people with disabilities. However, there are still considerable policy gaps and key policies such as the Policy on Special Needs and Inclusive Education and the ICT for Disability Policy that are still in draft form. MoH [19] attributes the delay in passing the policies to a lack of financial resources to implement the policies once they are passed. The delay in passing and implementing inclusion policies shows that the Government of Uganda has yet to give inclusion issues in general, and inclusive education in particular, the priority they deserve.

The lack of substantive policy instruments limits the demand for access, equity and quality to educational services. This can explain why the ODeL guidelines released by the NCHE in July 2020 to guide remote teaching and learning activities during the Covid-19 lockdown are silent on ensuring that all remote learning activities, and in particular the LMS and its content, are accessible to students with disabilities, even though 12.4% of the population have a disability [18]. The global and national antidiscrimination laws and treaties on education and ICT must be accompanied by country-specific statutes that specify the steps that should be taken to facilitate inclusion in education and ODeL [8,15]. It has also been noted that there is a substantial lack of data to inform policymaking [8,52]. The implementation of the existing policy framework is weak attributed to insufficient resources [19]. Thus, despite government efforts, students with LDs continue to face difficulties accessing quality education.

We thus argue that the term "learning disability" and the emphasis on neurodevelopmental disorder do not reflect the complex lived experiences of those who face digital exclusion because of how they perceive and interact with the world around them. In disability studies, especially in this specific one, this tension has been framed as a contrast between a medical and a social model of disability. The legal and policy documents cited in this study are strongly influenced by the medical model of disability, which sees neurodevelopment disorders as defects to be cured. In the medical model, such people are provided with assistance to function in a society designed for able-bodied people. In the social model, on the other hand, society itself has to change and remove barriers to participation for this case universities, parents, parents and students.

Regarding AT usage difficulties faced by students with LDs, the literature reviewed revealed that ATs emerge as suitable instruments for both accessibility and inclusion of students with LDs in universities around the world, as well as for meeting their educational needs during their learning process. However, many studies also report similar difficulties as those faced in Uganda in the usage of ATs to access eLearning. For example, an evaluation of higher education websites for 20 universities across the globe did not comply with WCAG 2.0 guidelines to an acceptable level [53]. Yet students with disabilities face difficulties with contacting and obtaining support from the disability office, which potentially results in a vicious circle of exclusion [54]. From interviews with lecturers from a South African ODeL institution, Zongozzi [55] found that lecturers are well aware of the negative impact that lack of accessibility students face with disabilities, but find it difficult to identify students who have disabilities and to provide appropriate support. The lack of educational resources in alternative formats is a common problem in Higher Education Institutions in Ethiopia [56].

The opening story of this chapter results with a student living with dyslexia reveals that negotiating rights to education starts with the policy of universities, structures, and skills of both staff and students for the appropriate ATs and usage. It is also true that exclusion and challenging participation in education is an entanglement between various non-linear factors. For example, while the constitution [38] policy provides for education for all, many public universities still lack the necessary infrastructure partly because they are not fully supported by the government budget. This is a reflection of a challenging state exclusion of students with LDs from education in the research sites. AT is supposed to improve the functionality of students with LDs for education, but for some students, this future is not guaranteed.

Disguising ourselves as sponsors to provide a right to education through providing devices for her education, one of the lecturers (L_4) that taught her changed the "lack of infrastructure story, limited skills to select the right AT, limited skills to develop content supported by ATs and no policy guidelines to enforce AT implementation". She narrated that, the reason for failure was rather seen as academically weak to pass examinations. In drawing a node in her rhizo-narrative, we connect this to the dissonance between institutional practices, the persons with Disabilities Act, 2019 [43] which promises AT for all, yet with no supportive budget, ICT policy (2014–2019) and Education policy of 2011 [40] which stipulate education for all, but not operationalized to date 12 years down the road.

Online examination is one of the machinery of power used in dominant universities to exclude students with LDs. Such practices to control quality in preparation for final examinations do not consider the flights [47] of students with LDs given that some content is not supported by AT. This injustice of university exclusion within the context of LDs is more accentuated as final examinations are used as bait to make students with LDs fail to sit similar examinations like nondisabled students, thereby failing those who have LDs.

AT Usage Impact in Uganda, through the literature reviewed suggested that there are limited studies on the use and impact of AT for education in university settings for students living with LDs. Research in this field over the last decade is not very relevant in line with this topic of focusing on LD. We can also highlight that the impact and repercussions of these studies are non-existent since there are no articles on this specific topic, yet many are available for general students with disabilities. The more frequently a paper is cited, the more often the scientific community recognizes the influence or impact of the cited topic. The scarce existence of scientific literature and its low impact is one of the main problems that may hinder the implementation of AT in classrooms because this field is underdeveloped in Ugandan universities. Similarly, the limited existence of scientific literature on the use of AT for the education of students with LDs makes it difficult to answer the research question posed. Even so, the scanty findings help us to lay the foundations for working to improve the policies in the education of students with LDs, both by offering AT for education on the market and by working on awareness raising to respond to the call of Molero-Aranda et al. [57]. The study has however been done in other countries and some of this study's findings reinforce previous studies although new information has also come to light.

Results of this study in consonance with Paz and Stringer [58] reveal that AT has the ability to be the great equalizer for many students with LDs to access university education and meaningfully participate in society. Similarly, Hetzroni & Shrieber [59], and Blackhurst [60] perspective of AT is that it increases individual performance and productivity of people with LDs. In other words, the application and employment of ATs enable students with LDs to optimize functional participation in education to independence [61].

The use of AT was found to improve the academic performance of students with disabilities. It helps students to develop autonomy and participation [62,63] which increases the acquisition of social skills [64,65]. AT contributes to the development of skills that provide stimulation, fit instantly to a student's level and provide instant feedback for improved learning [66] while performing tasks or functions that they would otherwise be unable to do [67]. AT improve the performance of students with disabilities and improve educational outcomes [68,69]. Thus, AT provides educational prospects for students with LDs and brings out their cognitive potential [13]. As a result, lecturers and students achieve their educational objectives by being active participants in the learning process.

Computer-assisted instruction helped learn by providing spelling and expressive writing skills [70,71], as such software reduced distractibility [72], and helped students learn to read [73] and achieve other academic outcomes [74]. Practice of math drills was eased [75] due to the ability to memorize math facts more easily and developed a more positive attitude towards math than students who did not use computer-assisted instruction [76]. Inbuilt software functions improved the writing skills of students [77]. Text-to-speech software was beneficial to students for understanding text when unfamiliar words were read to them [78,79], reading fluency and reading comprehension [80,81], as well as writing precisely [74,82]. Therefore, text-to-speech software helped students to retain more information through listening than reading especially in literature classes.

AT thus allows such students to bypass the demands of typing or handwriting; freed from these effortful tasks and their work would contain fewer errors [83]. Thus being essential for students who cannot concentrate in class [84]. AT also provide reminders to accomplish tasks [85,86]. On the contrary, one new user of the software in the literature class (female) was frustrated with the software because of her inability to efficiently edit the program's text output.

New information revealed that ATs nurtured psychological and mental preparation, supported students' attention and enhanced their information retention capacity. Moreover, they explained that AT made the learning process fun and easy due to the interesting content and the entertaining and enjoyable educational activities it facilitates, noting that since such technologies engaged students' multiple sensory abilities, they also motivated and increased the students' focus. Nonetheless, one female lecturer entreated that in as much as the AT made learning engaging and immersive for students with LDs, it had its equal share of distraction.

Finally, it is worth mentioning that, these ATs promote motivation and increase students' attention and learning for education.

14.8 RECOMMENDATIONS: WAYS TO IMPROVE AT FOR EDUCATION

Policy Level: A need to create a policy on AT usage supported eLearning for students with LDs which can either be a standalone or be part of the Special Needs and Inclusive Education Policy, which is before the Ugandan Cabinet for approval. At the minimum, the policy should address requirements by global and national disability laws, and the Web Content Accessibility Guidelines by the World Wide Web Consortium as proposed by Baguma and Walters [8]. This can be achieved by identifying representatives from key stakeholder groups such as educators, administrators, developers, IT support team members and people with LDs. Based on the gaps in the university's infrastructure level of eLearning accessibility, institutional action plans will be needed covering milestones, tasks, and target dates to bridge the gap. Accessibility coordinators can play an important role in the delivery of such action plans. There is thus a need to make people with LD a standalone category of stakeholders in education and ICT policies instead of treating them as one of several special interest groups. In order to achieve this, people with LDs should be closely involved in policymaking. Cover all disabilities in the Communications Act such as in mass media, telecommunication, and the Internet, especially since the borders between those three types of communication channels have been significantly reduced. The convergence of media technologies and the digital forms of access and delivery offer more ways for the audiences to engage with the media, and these forms of access are extensively used in ODeL (video, podcasts and interactive presentations).

Student Level: This study recommends that enhancing the impact of AT in enabling participation requires an individualized and holistic understanding of the value and meaning of AT for the individual in their unique context.

University Level: To ensure accessibility of eLearning content, university administrators, web developers, IT administrators, educators, student leaders and other stakeholders who play an active role in decision-making in universities, developing and maintaining an accessible eLearning platform and developing eLearning content should be sensitized and trained in AT usage supported content systems based on their current skills and attitudes, and what actions need to be taken. In our results, we documented the story of a lecturer with experiential knowledge of AT usage which highlighted several issues that can be addressed with appropriate eLearning systems. For example, videos with captions describing the audio track; audio with a text transcript, and provision of alternative text extend the lecture notes. Lecturers have a primary role in promoting the use of ATs. Therefore they need to acquire the necessary skills and competencies. Such gaps call for interventions that target the social lives of students with disabilities, teaching strategies and instructional materials to address the diverse learning needs of all learners in an equitable manner [87]. Drawing from

UNESCO's (2022) report [88], calls for flexible teaching and learning methods with innovative approaches to teaching learning- aids and equipment as well as the use of ICTs to address learning needs.

A lot of work needs to be done to improve AT research [89]. Little research has been conducted on the use and impact of AT in universities, and only a few researchers are conducting systematic, well-designed research that can lead to confident conclusions on how AT can be used to affect learning [89,70]. Interactive whiteboards as noted by one of the lecturers that support students with LDs are used in particular for students with 'ADHD' which is considered as a less visible disability. So, the utilization of the interactive whiteboard technology that can offer visual stimulation of the lessons delivered in the classroom, can help the students stay focused and undistracted, especially with the use of colourful lesson plans that have the ability to teach them what they really need to learn while entertaining them. Mechling et al. [90] posit that smart boards can improve the sight word recognition and other essential skills of students with LD. Organization and memory can also benefit students via hand-held gadgets that help students with intellectual disabilities, keep track of their notes, schedules, calendars, and tasks. In addition, Mechling et al. [91] go ahead to report that smart boards could help students with LDs match words with specific pictures, thereby improving their grocery lists and shopping experience in the study by enhancing their organization skills.

AT can reduce students' dependence on others to read, write, and organize their work [78,92]. When provided with effective strategy instruction, outlining programmes and concept mapping software can help with planning, and word processing, spell check, word prediction, and speech recognition can offer support for transcription and revision [93–95].

While AT can support struggling learners, MacArthur [93] cautions that technology by itself has little impact on learning. For students to benefit from the technology, educators must have an understanding of AT and how to embed AT within quality instructions [96,97].

Despite the enthusiasm that may surround the application of AT in the classroom, AT is not a panacea. Lack of common vision, limited training, access to support services, insufficient funding, and lack of lecturer time are commonly cited problems regarding the implementation of AT [98–100].

14.9 CONCLUSION

This chapter responds to Sustainable Development Goal Four; other international and national policies on people with disabilities with a more specific call of Baguma and Walters [8] in Uganda to extend the study of ATs to people with neurodiversity. Most importantly, it responds to the CRC Press, Taylor and Francis Group call for book chapters on Interactive Media Technologies for personal and commercial applications and their usability evaluations. This study found that although the Government of Uganda has created an enabling environment through the enacted policies, its response to the education needs especially with innovations of remote learning, and learning platforms does not favour students with LDs. Results inform the education interventionists to benefit students with LDs not only in Uganda but also in other countries where replication can be done especially during these days of uncertainty of global pandemics. We found numerous AT accessibility issues to education, even though we mostly focused on accessibility to people with LD. This is consistent with international results, even though our review of the legal and policy landscape in Uganda identified numerous issues that may be addressed in other countries. There are two areas for future work that we wish to highlight. The first area relates to more research on AT usage for specific difficulties. The definition of disability cited earlier highlights neurodisability impairments of mainly cerebral palsy, autism and epilepsy, yet our definition focused mostly on the general LDs. There is a need to extend the accessibility analysis to cover the needs of people with mobility impairment (e.g., those who have lost the full use of limbs-arms, hands, or legs due to traffic accidents, illness or conflict). The second area relates to a universal learning design in university ODeLs.

DISCLOSURE STATEMENT

No potential conflict of interest was reported by the author(s).
Funding: Personal

REFERENCES

1. World Health Organisation. "Global perspectives on assistive technology: Proceedings of the GReAT consultation." *World Health Organisation* 2 (2019): 1.
2. World Health Organization. "Global report on assistive technology." World Health Organisation, Geneva (2022).
3. World Health Organisation. "Assistive technology fact sheet." World Health Organization, Geneva (2016).
4. Smith, E.M., S. Huff, H. Wescott, R. Daniel, I.D. Ebuenyi, J. O'Donnell, M. Maalim, W. Zhang, C. Khasnabis, and M. MacLachlan. "Assistive technologies are central to the realization of the convention on the rights of persons with disabilities." *Disability and Rehabilitation: Assistive Technology* 19 (2022): 1–6.
5. Layton, N., R.O. Smith, and E.M. Smith. "Global outcomes of assistive technology: What we measure, we can improve." *Assistive Technology* 34, no. 6 (2022): 673–673.
6. Burne, B., V. Knafelc, M. Melonis, and P.C. Heyn. "The use and application of assistive technology to promote literacy in early childhood: A systematic review." *Disability and Rehabilitation: Assistive Technology* 6, no. 3 (2011): 207–213.
7. Ebuenyi, I.D., E.M. Smith, M.Z. Jamali, A. Munthali, and M. MacLachlan. "The ideal process for developing assistive technology policy." *Assistive Technology* 25 (2023): 1–8.
8. Baguma, R., and M.K. Wolters. "Making virtual learning environments accessible to people with disabilities in universities in Uganda." *Frontiers in Computer Science* 3 (2021): 638275.
9. Khasnabis, C., Z. Mirza, and M. MacLachlan. "Opening the GATE to inclusion for people with disabilities." *The Lancet* 386, no. 10010 (2015): 2229–2230.
10. Abdulhameed, H.T. "The implementation of assistive technology (at) for students with learning disabilities vs. universal design for learning." *International Journal of Development Research* 12, no. 2 (2022): 54203–54211.
11. Bond, E. *Childhood, Mobile Technologies and Everyday Experiences: Changing Technologies= Changing Childhoods?* Springer, New York (2014).

12. O'Riley, P.A. *Technology, Culture, and Socioeconomics: A Rhizo Analysis of Educational Discourses.* Peter Lang, New York (2003).

13. Chambers, D., ed. *Assistive Technology to Support Inclusive Education.* Emerald Publishing Limited, Bingley, UK (2020).

14. Go U. "Persons with disabilities Act", UPPC, Kampala (2006).

15. Singal, N., H. Ware, and S.K. Bhutani. *Inclusive Quality Education for Children with Disabilities.* University of Cambridge, Cambridge, UK (2017).

16. Sanjana, N.G. "What Is a Learning Disability? Learning disabilities, types, symptoms, causes, diagnosis and treatment" (2022).

17. Shree, A., and P.C. Shukla. "Intellectual disability: Definition, classification, causes and characteristics." *Learning Community-An International Journal of Educational and Social Development* 7, no. 1 (2016): 9–20.

18. National Population and Housing Census. Main Report, the Republic of Uganda (2014).

19. Ministry of Health. "Assistive Technology Country Capacity Assessment." MoH Division Disability and Rehabilitation, Uganda (2020).

20. Uganda Bureau of Statistics. *The Uganda Functional Difficulties Survey.* Government of Uganda, Uganda (2018).

21. World Bank. "Special needs education in Uganda: Co development Goal (SDG) 4concerns quality and inclusive education." (2020).

22. Fernández-Batanero, J.M., M. Montenegro-Rueda, J. Fernández-Cerero, and I. García-Martínez. "Assistive technology for the inclusion of students with disabilities: A systematic review." *Educational Technology Research and Development, Cultural and Regional Perspectives, AECT,* 70, no. 5 (2022): 1911–1930.

23. Devi, C.R., and R. Sarkar. "Assistive technology for educating persons with intellectual disability." *European Journal of Special Education Research* 4 (2019): 3.

24. Dawson, K., P. Antonenko, H. Lane, and J. Zhu. "Assistive technologies to support students with dyslexia." *Teaching Exceptional Children* 51, no. 3 (2019): 226–239.

25. Erdem, R. "Students with special educational needs and assistive technologies: A literature review." *Turkish Online Journal of Educational Technology-TOJET* 16, no. 1 (2017): 128–146.

26. Leppo, R.H.T., S.W. Cawthon, and M.P. Bond. "Including deaf and hard-of-hearing students with co-occurring disabilities in the accommodations discussion." *Journal of Deaf Studies and Deaf Education* 19, no. 2 (2014): 189–202.

27. Sikoyo, L., B.A. Ezati, D. Nampijja, J.A. Asiimwe, M. Walimbwa, D. Okot, and O. Godfrey. "Preparedness of Uganda's public universities to provide education to students with visual impairment: A situational analysis report." Makerere University, Kyambogo and National Council for Higher Education (2021).

28. Sikoyo, L., B.A. Ezati, D. Nampijja, J.A. Asiimwe, M. Walimbwa, and D. Okot. "Staff capacities for inclusive teaching and learning of students with visual impairment: A case of public universities in Uganda." *East African Journal of Education Studies* 6, no. 3 (2023): 174–191.

29. United Nations. "The Sustainable Development Goals Report." Department of Economic and Social Affairs. United Nations Publications, New York, United States of America (2018).

30. United Nations. "Committee on Economic, Social and Cultural Rights (CESCR), General Comment No. 13: The Right to Education (Art. 13 of the Covenant)", 8 December, E/C.12/1999/10, Para. 6. UNFPA (2013).

31. Forlin, C., and C. Chambers. "Teacher preparation for inclusive education: Increasing knowledge but raising concerns." *Asia-Pacific Journal of Teacher Education* 39, no. 1 (2011): 17–32.

32. UN General Assembly. "Universal declaration of human rights." *UN General Assembly* 302, no. 2 (1948): 14–25.

33. UN General Assembly. "Convention on the rights of persons with disabilities" (CRPD). United Nations Human Rights, Geneva (2006).
34. UN General Assembly. "Convention on the rights of persons with disabilities." *Ga Res* 61 (2006): 106.
35. Disability Rights Fund. *Grants for Disabled Persons Organizations (DPOs)*. New York, USA (2016).
36. Uganda DPOs Present Critical Rights Issues to CRPD Committee. Disability Rights Fund, Boston, Massachusetts (2016).
37. East African Community P. East African Community Policy on Persons with Disabilities, Ministry of East African Community. Nairobi, Kenya (2012).
38. Uganda. *Constitution of the Republic of Uganda*. Uganda Print and Publishing Corporation, Uganda (1995).
39. Go U. Persons with disabilities Act. "Ministry of gender, labour and social development Kampala" (2019).
40. Go U. "Policy on Special Needs and Inclusive Education (Draft)", Ministry of Education and Sports, Kampala (2011).
41. Go U. "Universities and Other Institutions (Amendment) Act", UPPC, Kampala (2003).
42. Sengoba, M. "I hope Makerere implements the disability policy" (2014).
43. Government of Uganda (Go U). "ICT for Disability Policy 2nd draft" (2017).
44. Emong, P., and L. Eron. "Disability inclusion in higher education in Uganda: Status and strategies." *African Journal of Disability* 5, no. 1 (2016): 1–11.
45. Oliver, M. *Understanding Disability: From Theory to Practice*. Bloomsbury Publishing, London (2018).
46. UPIAS. "Fundamental principles of disability, Union of the Physically Impaired against Segregation", London (1976).
47. Deleuze, G., and G. Felix. *A Thousand Plateaus: Capitalism and Schizophrenia*. Trans. By Massumi, B.), University of Minnesota, Minneapolis (1987).
48. Jackson, A.Y., and L.A. Mazzei. "Plugging one text into another: Thinking with theory in qualitative research." *Qualitative Inquiry* 19, no. 4 (2013): 261–271.
49. Grosz, E. "A thousand tiny sexes: Feminism and rhizomatics." In *Gilles Deleuze and the Theater of Philosophy*, Boundas, CV, & Olkowski, D (Eds.) pp. 187–210. Routledge, London (2017).
50. Kamberelis, G. "The rhizome and the pack: Liminal literacy formations with political teeth." *Spatializing Literacy Research and Practice* 41 (2004): 161–197.
51. St. Pierre, E.A. "Guest editorial an introduction to figurations-a poststructural practice of inquiry." *International Journal of Qualitative Studies in Education* 10 (1997): 279–284.
52. Wozniak, S., Moses, O., and Bernard, S. *Uganda's Disability Data Landscape and the Economic Inclusion of Persons with Disabilities Report*. Development Initiatives, Nairobi, Kenya. https://www.devinit.org/ (2020).
53. Acosta-Vargas, P., S. Luján-Mora, and L. Salvador-Ullauri. "Evaluation of the web accessibility of higher-education websites." In *2016 15th International Conference on Information Technology Based Higher Education and Training (ITHET)*, Istanbul, Turkey, pp. 1–6. IEEE, New York (2016).
54. Ro'fahanjarwati, A., and J. Suprihatiningrum. "Is online learning accessible during COVID-19 pandemic voices and experiences of UIN sunan kalijaga students with disabilities." *Nadwa: Jurnal Pendidikan Islam* 14, no. 1 (2020): 1–38.
55. Zongozzi, J.N. "Accessible quality higher education for students with disabilities in a South African open distance and e-learning institution: Challenges." *International Journal of Disability, Development and Education* 69, no. 5 (2022): 1645–1657.
56. Beyene, W.M., A.T. Mekonnen, and G.A. Giannoumis. "Inclusion, access, and accessibility of educational resources in higher education institutions: Exploring the Ethiopian context." *International Journal of Inclusive Education* 27, no. 1 (2023): 18–34.

57. Molero-Aranda, T., J.L.L. Cantabrana, M. Vallverdú-González, and M.G. Cervera. "Tecnologías digitales para la atención de personas con discapacidad intelectual." *RIED-Revista Iberoamericana de Educación a Distancia* 24, no. 1 (2021): 265–283.

58. Paz, Z., and Stringer, S. "Choosing the right assistive technology for students with learning disabilities and Dyslexia" (2021).

59. Hetzroni, O.E., and B. Shrieber. "Word processing as an assistive technology tool for enhancing academic outcomes of students with writing disabilities in the general classroom." *Journal of Learning Disabilities* 37, no. 2 (2004): 143–154.

60. Blackhurst, A.E. "Perspectives on applications of technology in the field of learning disabilities." *Learning Disability Quarterly* 28, no. 2 (2005): 175–178.

61. Parant, A., S. Schiano-Lomoriello, and F. Marchan. "How would I live with a disability? Expectations of bio-psychosocial consequences and assistive technology use." *Disability and Rehabilitation: Assistive Technology* 12, no. 7 (2017): 681–685.

62. Harper, K.A., K. Kurtzworth-Keen, and M.A. Marable. "Assistive technology for students with learning disabilities: A glimpse of the livescribe pen and its impact on homework completion." *Education and Information Technologies* 22 (2017): 2471–2483.

63. McNicholl, A., D. Desmond, and P. Gallagher. "Assistive technologies, educational engagement and psychosocial outcomes among students with disabilities in higher education." *Disability and Rehabilitation: Assistive Technology* 18, no. 1 (2023): 50–58.

64. Ari, I.A., and F.A. Inan. "Assistive technologies for students with disabilities: A survey of access and use in Turkish Universities." *Turkish Online Journal of Educational Technology-TOJET* 9, no. 2 (2010): 40–45.

65. Murry, F. "Using assistive technology to generate social skills use for students with emotional behavior disorders." *Rural Special Education Quarterly* 37, no. 4 (2018): 235–244.

66. Murray, D.W., and D.L. Rabiner. "Teacher use of computer-assisted instruction for young inattentive students: Implications for implementation and teacher preparation." *Journal of Education and Training Studies* 2, no. 2 (2014): 58–66.

67. Sullivan, M., and M. Lewis. "Assistive technology for the very young: Creating responsive environments." *Infants & Young Children* 12, no. 4 (2000): 34–52.

68. Alghayth, K.M.A. *The Use of Assistive Technology with Students with Severe Intellectual and Developmental Disabilities in Saudi Arabia: Parents' Perspectives.* University of South Florida, Tampa, FL (2019).

69 Chukwuemeka, E.J., and D. Samaila. "Parents' perception and factors limiting the use of high-tech assistive technology in special education schools in North-West Nigeria." *Contemporary Educational Technology* 11, no. 1 (2020): 99–109.

70. Wanzek, J., S. Vaughn, J. Wexler, E.A. Swanson, M. Edmonds, and A.-H. Kim. "A synthesis of spelling and reading interventions and their effects on the spelling outcomes of students with LD." *Journal of Learning Disabilities* 39, no. 6 (2006): 528–543.

71. Zhang, Y. "Technology and the writing skills of students with learning disabilities." *Journal of Research on Computing in Education* 32, no. 4 (2000): 467–478.

72. Hecker, L., L. Burns, L. Katz, J. Elkind, and K. Elkind. "Benefits of assistive reading software for students with attention disorders." *Annals of Dyslexia* 52 (2002): 243–272.

73. Lee, Y., and L.A. Vega. "Perceived knowledge, attitudes, and challenges of AT use in special education." *Journal of Special Education Technology* 20, no. 2 (2005): 60.

74. Chiang, H.-Y., and K. Jacobs. "Effect of computer-based instruction on students' self-perception and functional task performance." *Disability and Rehabilitation: Assistive Technology* 4, no. 2 (2009): 106–118.

75. Bouck, E.C., and S. Flanagan. "Assistive technology and mathematics: What is there and where can we go in special education." *Journal of Special Education Technology* 24, no. 2 (2009): 17–30.

76. Parkhurst, J., C.H. Skinner, J. Yaw, B. Poncy, W. Adcock, and E. Luna. "Efficient class-wide remediation: Using technology to identify idiosyncratic math facts for additional automaticity drills." *International Journal of Behavioral Consultation and Therapy* 6, no. 2 (2010): 111.

77. Batorowicz, B., C.A. Missiuna, and N.A. Pollock. "Technology supporting written productivity in children with learning disabilities: A critical review." *Canadian Journal of Occupational Therapy* 79, no. 4 (2012): 211–224.

78. MacArthur, C.A., R.P. Ferretti, C.M. Okolo, and A.R. Cavalier. "Technology applications for students with literacy problems: A critical review." *The Elementary School Journal* 101, no. 3 (2001): 273–301.

79. Raskind, M.H., and E.L. Higgins. "Speaking to read: The effects of speech recognition technology on the reading and spelling performance of children with learning disabilities." *Annals of Dyslexia* 49 (1999): 251–281.

80. Izzo, M.V., A. Yurick, and B. McArrell. "Supported eText: Effects of text-to-speech on access and achievement for high school students with disabilities." *Journal of Special Education Technology* 24, no. 3 (2009): 9–20.

81. Montali, J., and L. Lewandowski. "Bimodal reading: Benefits of a talking computer for average and less skilled readers." *Journal of Learning Disabilities* 29, no. 3 (1996): 271–279.

82. Young, G.D. "Examining assistive technology use, self-concept, and motivation, as students with learning disabilities transition from a demonstration school into inclusive classrooms." The University of Western Ontario (Canada) (2012).

83. Graham, S. "The role of text production skills in writing development: A special issue-I." *Learning Disability Quarterly* 22, no. 2 (1999): 75–77.

84. Ditlhale, T.W., and L.R. Johnson. "Assistive technologies as an ODEL strategy in promoting support for students with disabilities." *Technology and Disability* 34, no. 3 (2022): 153–163.

85. Bouck, E.C., Park, J., and Stenzel, K. "Virtual manipulatives as assistive technology to support students with disabilities with mathematics." *Preventing School Failure* 65, no. 4 (2020): 281–289.

86. Doughty, T.T., E.C. Bouck, L. Bassette, K. Szwed, and S. Flanagan. "Spelling on the fly: Investigating a pentop computer to improve the spelling skills of three elementary students with disabilities." *Assistive Technology* 25, no. 3 (2013): 166–175.

87. Ahmad, F.K. "Use of assistive technology in inclusive education: Making room for diverse learning needs." *Transcience* 6, no. 2 (2015): 62–77.

88. UNESCO IITE, COL & BNU. *Smart Education Strategies for Teaching and Learning: Critical Analytical Framework and Case Studies.* UNESCO IITE, Moscow (2022).

89. Gersten, R., and D. Edyburn. "Defining quality indicators for special education technology research." *Journal of Special Education Technology* 22, no. 3 (2007): 3–18.

90. Mechling, L.C., D.L. Gast, and K.L. Thompson. "Comparison of the effects of SMART board technology and flash card instruction on sight word recognition and observational learning." *Journal of Special Education Technology* 23, no. 1 (2009): 2008–2009.

91. Mechling, L.C., D.L. Gast, and K. Krupa. "Impact of SMART Board technology: An investigation of sight word reading and observational learning." *Journal of Autism and Developmental Disorders* 37 (2007): 1869–1882.

92. Mull, C.A., and P.L. Sitlington. "The role of technology in the transition to postsecondary education of students with learning disabilities: A review of the literature." *The Journal of Special Education* 37, no. 1 (2003): 26–32.

93. MacArthur, C.A. "Reflections on research on writing and technology for struggling writers." *Learning Disabilities Research & Practice* 24, no. 2 (2009): 93–103.

94. MacArthur, C.A. "Word prediction for students with severe spelling problems." *Learning Disability Quarterly* 22, no. 3 (1999): 158–172.

95. Higgins, E.L., and M.H. Raskind. "Speaking to read: A comparison of continuous vs. discrete speech recognition in the remediation of learning disabilities." *Journal of Special Education Technology* 15 (2000): 19–30.

96. Marino, M.T., E.C. Marino, and S.F. Shaw. "Making informed assistive technology decisions for students with high incidence disabilities." *Teaching Exceptional Children* 38, no. 6 (2006): 18–25.

97. Michaels, C.A., and J. McDermott. "Assistive technology integration in special education teacher preparation: Program coordinators' perceptions of current attainment and importance." *Journal of Special Education Technology* 18, no. 3 (2003): 29–44.

98. Flanagan, S., E.C. Bouck, and J. Richardson. "Middle school special education parents' perceptions and use of assistive technology in literacy instruction." *Assistive Technology* 25, no. 1 (2013): 24–30.

99. Morrison, K. "Implementation of assistive computer technology: A model for school systems." *International Journal of Special Education* 22, no. 1 (2007): 83–95.

100. Okolo, C.M., and J. Diedrich. "Twenty-five years later: How is technology used in the education of students with disabilities? Results of a statewide study." *Journal of Special Education Technology* 29, no. 1 (2014): 1–20.

15 ICTs for Information and Communication Portray a Revolution in Education

Abhijit Banubakode and Chhaya Gosavi

15.1 INTRODUCTION

Almost every element of our everyday life is being revolutionized and changed by information technology (IT). Through IT, a lot of calculation and communication take place all around us [1,2]. Our educational system is also being impacted by the IT revolution, and the Indian government has been working to promote information and communication technologies (ICTs) in education for the past 10 years [3,4]. Our educational system is increasingly dependent on ICT. The roadmap for successful teaching, learning, and professional development in an increasingly digital world is found in educational technology standards. We need to reevaluate our curricula, instruction, and training in the digital age. Simply said, no one will benefit from outdated educational approaches [4,5].

The advancement of ever-newer technologies has also affected educational practices. There is an immediate need for new student support systems, teacher-student interactions, assignment management, assessment and evaluation, and feedback. The impact of technology on education, learning, and thinking cannot be denied. Individual-focused to collaborative-focused tasks, curriculum-focused to learner-focused approaches, and passive-to-active learning have all been identified as trends. ICT has revolutionized education, particularly as well as society as a whole [6,7].

ICT is a powerful instrument for providing all societal segments—especially the underprivileged—with more formal and informal educational opportunities. According to the International Institute for Communication and Development (IICD), ICT can be used to improve the development of educational content, support administrative procedures in schools and other educational establishments, and increase access to education for teachers and students via distance learning.ICT is currently being employed in education to a greater extent thanks to two concepts: blended learning and e-learning. This has to do with how ICT is used in the classroom. E-learning, as defined by Tinio, "includes learning at all levels, both formal and non-formal, that uses an information network, such as the internet, an intranet (LAN), or an extra-net (WAN), whether fully or in part, for course delivery, interaction, and or facilitation." Some prefer to use the phrase "online learning," which describes education conducted via a web browser. One subset of e-learning is web-based learning. Another term used in reference to educational technology is blended learning. It explains

DOI: 10.1201/9781032664828-18

teaching methods that blend traditional classroom education with virtual learning resources [8,9]. In a typical classroom, for example, students might be assigned both online and print tasks, take part in chat-based online mentoring sessions with their teachers, and sign up for the class email list. "These technologies have great potential for knowledge dissemination, effective learning, and the development of more efficient education services," said UNESCO. Information and communication technology use has, in fact, resulted in a profound paradigm shift in education (ICT) [10]. It has fundamentally altered the way we educate people, making the conventional educational process more engaging, interactive, and customized. These are a few noteworthy ways that ICT has impacted education [11,12].

- **Information Access:** ICT has tremendously sped up and eased access to a multitude of knowledge. Students and teachers can now access online databases, libraries, and instructional resources from anywhere at any time. As a result, learning has become more equitable and democratic, bridging the knowledge gap and providing equal chances for all pupils.
- **Interactive Learning:** Learning has become more interactive and engaging thanks to ICT tools like multimedia presentations, simulations, and educational apps. Virtual labs, interactive exercises, and gamified learning opportunities allow students to actively participate in their education, improving their comprehension and retention of the subject matter.
- **Personalized Learning:** ICT makes it possible for adaptive learning platforms and intelligent tutoring systems that can adjust the pace and content of a lesson to the needs and preferences of specific students. By focusing on their strengths and weaknesses and learning at their own pace, this individualized approach helps students achieve the best possible learning results.
- **Collaborative Learning:** ICT supports environments for collaborative learning both inside and outside of the classroom. Students can work together on group assignments using cloud-based platforms, communicate and share ideas through discussion forums, and collaborate on projects using online tools. This encourages global cooperation, effective communication, and teamwork.
- **Blended Learning:** ICT has made it easier to combine traditional face-to-face instruction with online learning, giving rise to blended learning models. Teachers can combine the benefits of online learning and interactions with face-to-face instruction to give students a more flexible and personalized learning experience.
- **Assessment and Feedback:** ICT provides a number of tools for assessment that allow for automated grading, immediate feedback, and data-driven insights into students' progress. Teachers receive real-time feedback from online tests, assignments, and digital portfolios, enabling them to spot areas for improvement and adjust their teaching methods accordingly.
- **Professional Development:** ICT has revolutionized educator professional development. Teachers have the chance to improve their knowledge, share ideas, and keep up with the most recent trends in educational theory and practice through online courses, webinars, and virtual communities.

This ongoing professional development raises the standards of instruction and student performance.

- **Global Learning Opportunities:** By bringing together students and teachers from all over the world, ICT has broadened educational horizons. Students can participate in cross-cultural learning activities, gain global perspectives, and hone their intercultural competencies through video conferencing, online collaborations, and virtual exchange programs.

While the paradigm shift brought about by ICT in education has been significant, it is crucial to ensure that all students and educators have equitable access to technology and digital literacy. To maximize its impact and create inclusive learning environments, ICT integration should be accompanied by effective pedagogical practices and ongoing support [13,14].

15.1.1 Standardization of Teaching

The pursuit of truth is what education truly is. It is a never-ending quest for wisdom and enlightenment. Such a journey opens up fresh development opportunities. Education needs to be standardized to enhance the true dignity of the human being. Future professionals must be prepared through professional training for a lifetime of self-directed learning [7–9]. Traditional professional education does not specifically foster a lifelong learning philosophy. With the aid of computer technology, students can learn creatively using the devices available for creative animation. Dropouts can be located, and a positive learning environment and a non-threatening assessment are both possible (Figure 15.1).

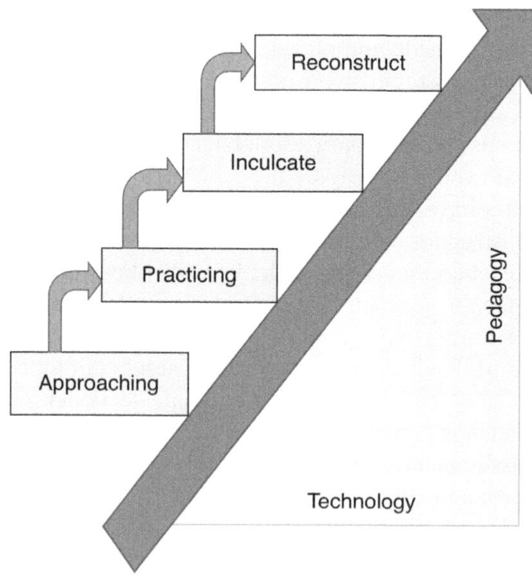

FIGURE 15.1   Different phases of ICT growth.

15.1.2   STAGES OF ICT DEVELOPMENT

The stages of ICT, namely approaching, practicing, inculcating, and ultimately reconstructing, should be followed by educational institutions. New technologies must be developed, put to use, infused, or spread, and then education must be raised to a new level.

15.1.3   LEARNING–TEACHING MANNER AND ICT

Everybody has a preferred method of learning. While some people prefer reading or listening, others prefer heavily visual methods. Offering a variety of learning opportunities is essential because people learn in different ways. Research has shown time and time again that students' motivation, initiative, and academic performance increase when they are able to use their unique styles of learning and information processing. While learner-centered approaches encourage a process of active inquiry, teacher-centered learning approaches frequently favor passive reception of knowledge.

Because learning is an active process, learners are most motivated when they can take ownership of their education. Today, due to the increased use of computers and interactive technologies, students are learning with technology rather than just learning about it. With the help of ICT, teachers can now provide students with effective learning opportunities where they take charge of their education and actively identify their learning needs, locate relevant information, evaluate its quality, expand their own knowledge base, and communicate their findings.

15.2   TEACHERS AND ICT

The use of ICT by teachers should be viewed favorably [1,2]. Teachers' responsibilities have changed as more students use ICT in the classroom, teachers' responsibilities have changed. The teacher's role should be to collaborate with students as they apply their knowledge to real-world situations. Transferring knowledge from teacher to student is no longer the primary focus of instruction. Instead, learning occurs when students conduct independent research, engage in critical thinking, and solve problems using information from various sources, including computers, satellite phones, television, and radio.

For teachers to become self-assured, critical, and creative users of ICT, they must have access to professional development programs that provide them with a variety of skills in both technology use and task design. To support learning, teachers must use models of best practice and knowledge, in addition to their ICT skills. Teachers should be well-trained and possess effective time-management abilities. Teachers must be aware of their students' various learning preferences and accommodate them by providing verbal instructions, in addition to on-screen instructions, and making printed versions of the text available on demand.

Teachers should be able to convert conceptual instructional models into engaging, student-centered teaching strategies. For instance, teachers can include images in their work to improve presentation and facilitate students' comprehension. To teach challenging concepts, teachers can use commercial simulation software. Activities should make learners think critically.

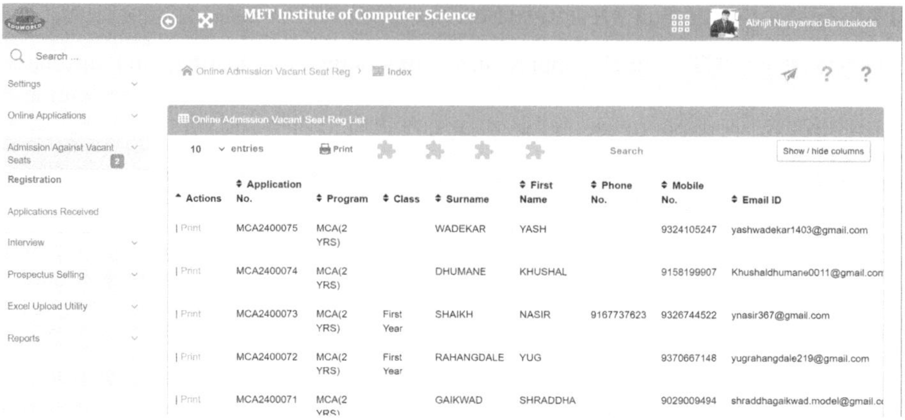

MET Institute of Computer Science — Abhijit Narayanrao Banubakode

Online Admission Vacant Seat Reg › Index

Online Admission Vacant Seat Reg List

10 entries   Print   Search   Show / hide columns

Actions	Application No.	Program	Class	Surname	First Name	Phone No.	Mobile No.	Email ID
| Print	MCA2400075	MCA(2 YRS)		WADEKAR	YASH		9324105247	yashwadekar1403@gmail.com
| Print	MCA2400074	MCA(2 YRS)		DHUMANE	KHUSHAL		9158199907	Khushaldhumane0011@gmail.com
| Print	MCA2400073	MCA(2 YRS)	First Year	SHAIKH	NASIR	9167737623	9326744522	ynasir367@gmail.com
| Print	MCA2400072	MCA(2 YRS)	First Year	RAHANGDALE	YUG		9370667148	yugrahangdale219@gmail.com
| Print	MCA2400071	MCA(2 YRS)		GAIKWAD	SHRADDHA		9029009494	shraddhagaikwad.model@gmail.co

FIGURE 15.2   Admission module-ERP system.

Figure 15.2 shows an ERP Admission Module for educational institutions is a software solution designed specifically to manage the admission process efficiently and effectively. The module provides a centralized platform for managing all aspects of the admission process, including application submission, document management, and communication with applicants. It offers an online application portal where prospective students can easily submit their applications, upload required documents, and track the status of their applications. Educational institutions can customize application forms to collect specific information required for different programs or courses. The module automates various workflows such as application screening, review, and decision-making, reducing manual effort and streamlining the process. Integrated communication tools enable admissions officers to communicate with applicants, send notifications about application status updates, and request additional information when needed. It provides a secure document management system to store, organize, and track all application-related documents such as transcripts, letters of recommendation, and personal statements. Built-in analytics tools provide insights into application trends, demographics of applicants, and conversion rates, helping institutions make data-driven decisions. Seamless integration with the Student Information System (SIS) ensures that admission data are synchronized with student records, facilitating a smooth transition from admission to enrollment. The module may include decision support tools to assist admissions committees in evaluating applications, setting admission criteria, and making informed decisions. After admission, the module helps manage the enrollment process, including accepting admission offers, collecting enrollment deposits, and registering students for classes. It assists institutions in maintaining compliance with regulations and accreditation standards by generating reports on admission metrics, applicant demographics, and other relevant data. The module should be scalable to accommodate fluctuations in application volume and adaptable to meet the unique needs of different types of educational institutions, from K-12 schools to colleges and universities.

Overall, an ERP Admission Module tailored for educational institutions streamlines the admission process, enhances the applicant experience, improves operational efficiency, and supports data-driven decision-making to achieve enrollment goals.

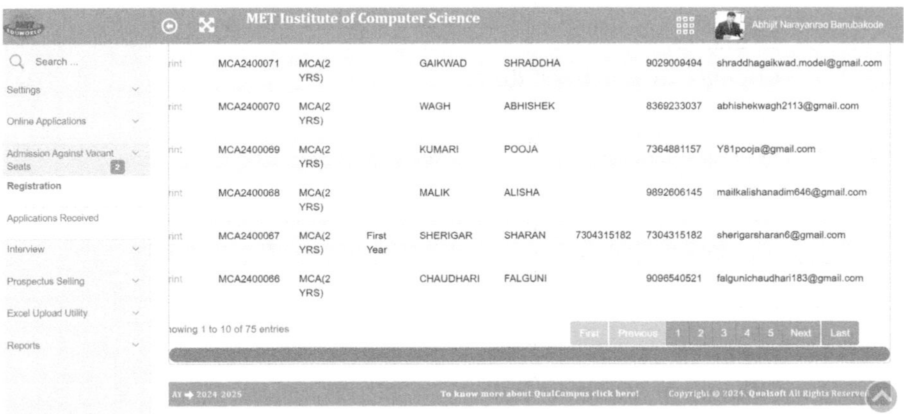

FIGURE 15.3   Students information module-ERP system.

Figure 15.3 shows the Student Information Module within an ERP system for educational institutions is a comprehensive platform designed to manage student data and facilitate various administrative tasks related to student lifecycle management. It serves as a centralized repository for storing and managing student records, including personal information, academic history, attendance records, and disciplinary records.

It tracks students' academic progress, including grades, course credits, and degree requirements, enabling advisors and administrators to monitor student performance and provide support when needed. The module facilitates the tracking of student attendance, helping educators identify patterns of absenteeism and intervene early to support student success. Educational institutions can use the module to manage course offerings, including course scheduling, enrollment limits, prerequisites, and instructor assignments. It automates the generation of official transcripts, allowing administrators to produce accurate and up-to-date transcripts for students upon request. The module simplifies the process of recording and calculating grades, providing tools for entering grades, calculating GPA (Grade Point Average), and generating grade reports. Integrated communication tools enable administrators to communicate with students, send announcements, and share important information about academic deadlines, events, and policies. For institutions offering financial aid, the module helps manage student financial aid applications, awards, disbursements, and compliance reporting. It provides a secure online portal where students can access their academic records, view course schedules, register for classes, and communicate with advisors and instructors. Some systems offer parents or guardians access to student information, allowing them to monitor their child's academic progress, attendance, and communication with the institution.

Overall, an ERP Student Information Module is essential for educational institutions to effectively manage student data, support academic success, and streamline administrative processes throughout the student lifecycle.

15.2.1   ICT HERALD PARADIGM SHIFT FOR EDUCATION

Early on, teaching was done using a difficult traditional method. The method used to transfer content was linear, meaning that it was broken up into manageable modules

TABLE 15.1

Changes in Teachers' Role

Shift From	To
Knowledge transmitter	Learning facilitator, collaborator, coach, knowledge navigator and co-learner
The teacher controls and directs all aspects of learning	Teachers give students more options and responsibilities for their own learning

TABLE 15.2

Changes in Students' Role

Shift From	To
Passive recipient of information	Active participant in the learning process
Reproducing knowledge	Producing knowledge
Learning as a solitary activity	Learning collaboratively with other

and delivered one after the other. The educational landscape has drastically over the last three decades. This idea has changed due to ICT. ICT has given a new definition of learning and quickly closed the gap between learning and teaching. Depending on students' learning capacities, learning can be active, social, neutral, contextualized, and integrative. Teachers must possess ICT skills.

The main focus is no longer on traditional instruction; instead, it is on a virtual learning environment. The new environment has resulted in changes to the roles of both teachers and students, making them more interactive and interesting. This change may make the role of teachers and students more interactive and engaging. Tables 15.1 and 15.2 show how students and teachers have changed over time. The teachers' new role requires a new learning process and way of thinking. To increase access to educational opportunities, teachers should have access to infrastructure and a wide range of tools. There should also be regular interactions with administrators and policymakers.

15.3 ENGINEERING EDUCATION AND ICT IN INDIA

Recent years have seen major developments in engineering education in India. A notable increase in the demand for high-quality education has led to the adoption of ICT to expand the reach of education [14]. A few examples of technology-enhanced efforts already implemented by the Government of India are the National Programme on Technology-Enhanced Learning (NPTEL), the usage of an educational satellite called the EDUSAT, and other initiatives.

- A robust infrastructure is necessary to support the CIT-provided amenities.
- Modern classrooms with computers and projectors should be provided.
- There should be a capacity for financial and strategic planning.
- Opportunities should be created for teachers to regularly collaborate.

- Power should be supplied during working hours.
- Install a stronger security system.
- Develop fresh strategies.
- Teachers should attend frequent workshops or seminars to stay current with technology.

India is making strides in ICT-based education. The ICT for education programs in India have even loftier goals and objectives, including the following:

- Increase the efficacy of learning by utilizing technology-assisted instruction.
- Promote critical thinking and an inquisitive mindset, and make learning enjoyable.
- Enable educators.

Educational establishments in India are equipped with computers, networking devices, printers, and scanners. Moreover, instructional CDs and computer software such as Microsoft Windows, Microsoft Office, CorelDraw, and Microsoft Encarta are available. But internet connectivity is limited to a few cities. There are still a lot of obstacles and challenges to face. Among concerns are infrastructure, software content, a shortage of teachers with the necessary qualifications, and a general lack of desire.

To evaluate every effort made in India, a Management Information System (MIS) is being developed [4]. The following modules are parts of this MIS:

- Information on Students,
- Monitoring Assessments,
- Cooperative,
- Management,
- Content Revision.

15.4   STUDY OBJECTIVES

The main objectives of the study were to look at how students used ICT, how their lecturers used it, how much they knew about different computer applications, and how they felt about using ICT in the classroom.

15.5   RESEARCH METHODOLOGY

This section gives a quick summary of the methods used to achieve the objectives of the study. Surveys (online questionnaires) were used to collect data from primary or secondary sources in accordance with the research needs. A questionnaire was developed after the relevant literature was carefully examined. The education staff at the Mumbai Educational Trust in Bandra, Mumbai, and the Bharati Vidyapeeth (Deemed to be University) College of Engineering in Pune participated in a survey. Out of the 200 staff and faculty personnel we sent the survey to, 175 sent it back. The target population of the study, sampling techniques, sample size, conceptual framework, and data collecting are all examined in the data analysis.

Research Design: Descriptive (quantitative) methods were developed for this explanatory study. There are three stages to conducting a quantitative study that involves questionnaires:

- Performing a survey.
- Disseminating a survey link among the institution's staff and faculty.
- Examining the results of the completed survey after analysis.

15.5.1 PRIMARY DATA

The study's target demographic is the faculty and personnel of the Bharati Vidyapeeth (Deemed to be University) College of Engineering (BVDUCOE) of the Mumbai Educational Trust (MET). MET faculty and staff are deemed to achieve the primary goal and sample if they have utilized or are currently using ICT tools and are aware of the needs for both daily activity and future development.

15.5.2 SAMPLE FRAMEWORK

The model framework for this study is provided by the Bharati Vidyapeeth (Deemed to be University) College of Engineering (BVDUCOE), Pune, and numerous MET institutions. Approximately eight MET and BVDUCOE institutes provided the sample.

15.5.3 SAMPLE SIZE

The sample at MET and BVDUCOE consists of 200 individuals. One hundred and seventy-five people made up the respondents, from whom information was gathered. This example is taken from the survey's questionnaires.

15.6 FINDINGS AND DISCUSSION

We created a survey form and collected data from BVDUCOE and several MET institutes within the MET League of Colleges. We rated the survey on a range of 1–4 depending on how many ICT tools are needed in relation to departmental operations [10]. The following marks were present on the scale:

- Some
- None
- Very Much
- Very Little

No one claims that no one uses the ICT tools we have used for research in academia, and those who do so express great satisfaction. Usage of ICT tools in academia is generally low, as indicated by the small number of users. However, some staff and faculties claim to use these tools. In academic contexts, most faculty members and staff make extensive use of ICT resources. The seven ICT technologies shown in Figure 15.2 include a website, audio, computer, mobile, e-reader, enterprise resource planning, and PowerPoint presentation.

Figure 15.4 shows the results of a survey on the use of ICT tools at MET and BVDUCOE.

Table 15.3 covers all seven ICT survey tools across several departments, categorized as none, very little, some, and very much. After studying and acquiring samples from several departments, we transformed the amount of replies categorized as None, Very Little, Some, and Very Much to percentages. Table 15.3 presents the findings.

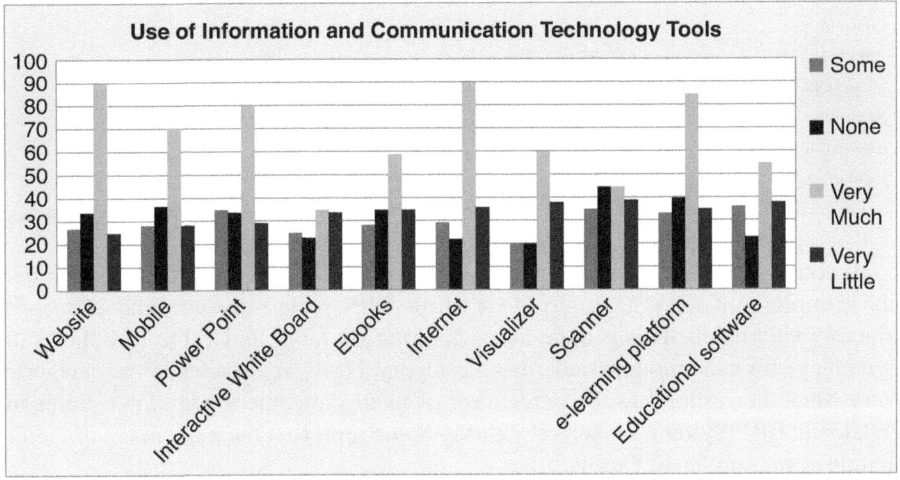

FIGURE 15.4   Survey on the use of ICT tools at MET and BVDUCOE.

TABLE 15.3
Survey Results of ICT Tools Used for Research on Academic Systems

Sr. No.	ICT Devices	Percentage of Faculty Members Who Prefer ICT Tools in the Academic System (%)	Percentage of Faculty Members Who Don't Prefer ICT Tools in Academic System (%)
01	Website	75	25
02	Mobile	72	28
03	Powerpoint	70	30
04	Interactive whiteboard	68	32
05	e-Books	85	15
06	Internet	81	19
07	Visualizer	66	34
08	Scanner	55	20
09	e-Learning platform	61	23
10	Educational software	66	36

TABLE 15.4

Final Survey Results of Seven ICT Tools Used for Research on the Academic System

	Some (%)	None (%)	Very Much (%)	Very Little (%)
Website	27	34	90	25
Mobile	28	37	70	28
Powerpoint	35	34	80	29
Interactive whiteboard	25	23	35	34
e-Books	28	35	59	35
Internet	29	22	90	36
Visualizer	20	20	60	38
Scanner	35	45	45	39
e-Learning platform	33	40	85	35
Educational software	36	23	55	38

According to the first row, on the website ICT Tool, 27% of faculty members chose to prefer none, 34% preferred very little, 90% preferred some, and 25% chose to prefer very much. This is done for each of the six tools, and Table 15.1 shows the percentage for each module under each category. Next, we broadened the categories Very Much correspond to the percentage of faculty members who don't prefer the Academic ERP System. None, Very Little, Some represent the percentage of faculty members who prefer ICT tools.

The final survey findings for two specific categories of the seven ICT tools are displayed in Table 15.4. The Table 15.3 result is derived from the Table 15.4 result. Over 70%–85% of faculty members prefer employing ICT tools in the academic system, while 25%–30% of faculty members do not, according to the results displayed in Table 15.4 for practically all modules. In this modern technology era, instructors and students can greatly benefit from ICT technologies. The fact that more academic staff members enjoy and choose to employ ICT tools in the classroom lends credence to this.

We covered seven ICT instruments for the educational sector in this study. Tablets, smart boards, interactive whiteboards, and flipped classrooms are examples of future ICT instruments that can be used. By doing this, all of the information may be updated in the database, facilitating more comfortable and easy communication.

Principal Research Findings:

- It was found that their instructors occasionally use ICT.
- A significant percentage of respondents said that their department uses ICT tools.

15.7   REASONING BEHIND USING AND NOT USING ICT

The decision to use or not use ICT can depend on various factors. The first factor is Efficiency and Productivity as ICT can streamline processes, automate tasks, and enhance productivity. It enables faster communication, data storage, and retrieval, allowing organizations to work more efficiently. The second factor is Access to Information; in this case, ICT provides access to an abundance of information and knowledge resources. It allows individuals to stay updated, conduct research, and learn new skills. The third factor is Connectivity and Collaboration, and in this case, ICT makes it possible for people to connect and collaborate across geographic boundaries. It improves communication, coordination, and teamwork. The fourth factor is Decision-making and Analysis; in this case, ICT provides tools for data collection, analysis, and visualization. It enables organizations to make data-driven decisions, identify trends, and respond proactively to challenges. The fifth factor is Innovation and Creativity; here, ICT can spur innovation and creativity by providing platforms, tools, and resources for experimentation, prototyping, and problem-solving [15,16].

There are certain reasons for not using ICT. The first reason is Financial Constraints. In this case, implementing and maintaining ICT infrastructure can be expensive, especially for small businesses or resource-constrained organizations. Lack of financial resources may deter their adoption of ICT. The second reason is the Infrastructure and Technology Barrier; in this case, there are some regions or remote areas, where the lack of infrastructure, such as reliable electricity or internet connectivity, may pose challenges for using ICT effectively. The use of ICT may raise concerns about data privacy and security. Organizations need to invest in robust security measures and ensure compliance with regulations to mitigate risks. Effective use of ICT often requires technical skills and knowledge. Lack of training or expertise can hinder the adoption and utilization of ICT. In certain contexts, there may be cultural or social barriers to the adoption of ICT. Resistance to change, skepticism, or reliance on traditional methods can limit the acceptance of ICT [17,18]. In the end, an organization's or a person's unique demands, resources, and constraints will determine whether or not ICT is used. Before making a choice, it's critical to weigh the possible advantages, expenses, and difficulties of implementing ICT.

15.8 CONCLUSION

ICT is causing changes in educational institutions' policies about education. Whiteboards or blackboards have been superseded by this idea. Consequently, there is now greater engagement in the exchanges between students and professors. The quality of education in institutions has been altered by ICT. The use of the internet has increased thanks to ICT. The newest media and technology are available to both teachers and students. Instructors ought to be more vigilant. Both group learning and individualized instruction are made possible by the utilization. Teaching approaches must change for these strategies to be used effectively.

The use of ICT should be in line with the desired learning objectives. ICT helps students and has the power to drastically change how we teach and learn. The study found that because ICT has such a big impact on lectures, educators should use it as much as possible in students' learning.

REFERENCES

1. Chhabra, A. (2014). Information and communication technology (ICT): A paradigm shift in teacher education. *Scholarly Research Journal for Humanity and English Language*, 1, 308–317.

2. Islam, A. Y. M. A., Mok, M. M. C., Gu, X., Spector, J., & Hai-Leng, C. (2019). ICT in higher education: An exploration of practices in Malaysian Universities. *IEEE Access*, 7, 16892–16908.

3. Das, S. R., & Mohapatra, S. (2008). "Social and public impact of ICT enabled education," *2008 International Conference on Information Technology*, Bhubaneswar, pp. 300–303.

4. Vesisenaho, M. (2010). "ICT education and computer science education for development: Impact and contextualization," *2010 IEEE Frontiers in Education Conference (FIE)*, Washington, DC, pp. F4J-1–F4J-6.

5. Belagra, M., Benachaiba, C., & Guemid, B. (2012). "Using ICT in higher education: Teachers of electrical engineering department at the University of Bechar: Case study," *Proceedings of the 2012 IEEE Global Engineering Education Conference (EDUCON)*, Marrakech, Morocco, pp. 1–6.

6. Chanwijit, J., Lomwongpaiboon, W., Dowjam, O., & Tangworakitthaworn, P. (2016). "Decision support system for targeting higher education," *2016 Fifth ICT International Student Project Conference (ICT-ISPC)*, Nakhon Pathom, Thailand, pp. 154–157.

7. Zahariev, P., Bencheva, N., Hristov, G., & Ruseva, Y. (2013). "ICT convergence challenges in education and their impact on both instructors and students," *2013 24th EAEEIE Annual Conference (EAEEIE 2013)*, Chania, Greece, pp. 193–197.

8. Dhandabani, L., & Sukumaran, R. (2014). "Use of ICT in engineering education: A survey report," *2014 IEEE International Conference on Computational Intelligence and Computing Research*, Coimbatore, India, pp. 1–5.

9. Dhotre, S., & Patil, S. (2010). Intelligent e-learning systems using Web 3.0. *Journal of Engineering Research and Studies*, 1(2), 230–232.

10. Faculty of Education at IUB. (2011). Library Philosophy and Practice (e-journal). 677. https://digitalcommons.unl.edu/libphilprac/677

11. Nishant Gunjan, "ICT Based Education: A Paradigm shift in India", Department of Education (B.Ed), M.L.T. College, Saharsa (Bihar), India.

12. Zahariev, P., Bencheva, N., Hristov, G., & Ruseva, Y. (2013). "ICT convergence challenges in education and their impact on both instructors and students," *2013 24th EAEEIE Annual Conference (EAEEIE 2013)*, Chania, Greece, pp. 193–197.

13. Dhandabani, L., & Sukumaran, R. (2014). "Use of ICT in engineering education: A survey report," *2014 IEEE International Conference on Computational Intelligence and Computing Research*, Coimbatore, India, pp. 1–5.

14. Dhotre, S., & Patil, S. (2010). Intelligent e-learning systems using Web 3.0. *Journal of Engineering Research and Studies*, 1(2), 230–232.

15. Toprakci, E. (2006). Obstacles at integration of schools into information and communication technologies by taking into consideration the opinions of the teachers and principals of primary and secondary schoolsin Turkey. *Journal of Instructional Science and Technology (e-JIST)*, 9(1), 1–16.

16. Venkatesh, V., & Davis, F. D. (2000). A theoretical extension of the technology acceptance model: Four longitudinal filed studies. *Management Science*, 46, 186–204. Retrieved from https://pubsonline.informs.org/doi/10.1287/mnsc.46.2.186.11926.

17. Ali, Z. W., Nor, H. M., Hamzah, A., & Alwi, H. (2009). The conditions and level of ICT integration in Malaysian Smart Schools. Retrieved from https://www.learntechlib.org/p/42320/.

18. Watson, D. (Ed.) (1993). *ImpacT – An Evaluation of the Impact of the Information Technology on Children's Achievements in Primary and Secondary School*. London: King's College London.

16 e-Learning Adoption Model
A Literature Review

*Chhaya Gosavi, Madhuri Tasgaonkar,
and Abhijit Banubakode*

16.1 INTRODUCTION

E-learning is also known as remote learning, online education, and e-learning. E-learning is a tool that can help the teaching and learning process become more flexible, inventive, and student-centered [1]. Ref. [2] asserts that making educational materials accessible at all times and allowing for flexible access speeds up learning. The eLearning Adoption model aids educational institutions, corporate, and private tutors in comprehending and evaluating the adoption of e-learning techniques and technologies. It offers a methodical method for organizing, carrying out, and reviewing e-learning projects in an educational environment. In the current digital era, where technology-enhanced learning is becoming more and more vital for both formal education and corporate training, this concept is especially pertinent [3].

Depending on the particular model being used, the e-Learning Adoption Model often comprises numerous stages or phases. However, a typical framework includes the components as depicted in Figure 16.1.

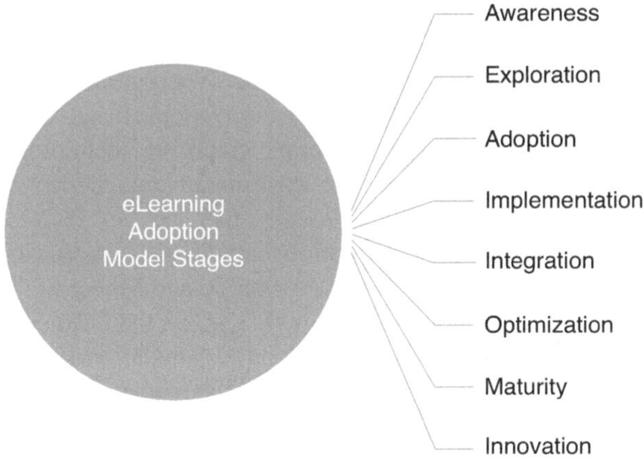

FIGURE 16.1 E-Learning Adoption Model stages.

DOI: 10.1201/9781032664828-19

e-Learning Adoption Model Stages are briefly described as follows:

1. **Awareness:** Organizations become aware of the potential advantages of e-learning. Recognizing the need for more adaptable and scalable training solutions or the need to reach a larger audience at an affordable cost is necessary for this.
2. **Exploration:** During this stage, businesses or organizations begin to research the different e-learning solutions and technologies on the market. To learn more about the options, they might study research, go to workshops, or try out small-scale e-learning projects.
3. **Adoption:** At this point, firms finalize the decision to make e-learning a crucial part of their learning and development plan. This entails deciding the technology and spending money on the tools, resources, and infrastructure that are required. Training courses may be created from scratch or revised to include e-learning components.
4. **Implementation:** The real deployment of e-learning programs and solutions takes place during the implementation phase. Key tasks at this level include content generation, platform selection, and user training. Additionally, organizations set up procedures for managing and monitoring their online learning resources.
5. **Integration:** The seamless integration of e-learning into a larger organizational or educational setting is referred to as integration. This may involve porting the available data to the e-learning platform, integrating e-learning data with other systems, connecting e-learning with current curriculum or training programs, and making sure e-learning supports business objectives.
6. **Optimization:** The e-Learning Adoption Model must be optimized for ongoing progress. To improve the value and impact of their e-learning offerings, organizations routinely evaluate the success of their e-learning programmers, solicit student input, and make the required improvements.
7. **Maturity:** Over time, organizations reach a level of maturity in their e-learning adoption. This stage signifies that e-learning is an established and integral part of their learning and development strategy. It is well-integrated, continuously improved, and aligned with organizational objectives.
8. **Innovation:** Some models include an innovation phase, where organizations actively seek out and incorporate emerging technologies and best practices to stay at the forefront of e-learning advancements.

In recent years, numerous e-learning adoption technologies, frameworks, models, and theories have been created. The Technology Acceptance Model (TAM), the Unified Theory of Acceptance and Use of Technology (UTAUT), the Diffusion of Innovations (DoI), Five Rogers' Factors, the Technology-Organization-Environment (TOE) Framework, and the Social Cognitive Theory (SCT) are some of the frameworks.

Depending on the particular requirements, objectives, and expected outcomes of the company or institution, the complexity of the e-Learning Adoption Model can change. It offers a defined roadmap for businesses to follow as they switch from conventional learning strategies to more technologically advanced ones. The ultimate goal is to use e-learning to improve educational projects' overall efficacy, boost accessibility, and improve learning outcomes.

16.2 LITERATURE REVIEW

Examining and summarizing current studies, theories, and frameworks about the adoption and implementation of e-learning in various educational and organizational settings are key components of this chapter. In the area of e-learning adoption, this study can be used to highlight common themes, difficulties, and best practices.

According to Muneer Abbad's research, external factors including subjective norms, Internet usage, system interaction, self-efficacy, and technical support have an impact on how well people accept new technology [1]. Based on e-learning technologies adoption factors such as perceived ease of use, perceived usefulness, compatibility, complexity, relative advantage, observability, trialability and intention to use, actual usage, and patronage factor, Kimwise Alone analyzed the Theory of Reasoned Action (TRA), the origin TAM, the TOE Framework, and the revised TAM models. He presented a new model, an extended TAM model, based on this discussion [2]. Md. Mahmudul Hasan Khan et al.'s framework [3] can serve as a guide for higher education institutions as well as elementary educational institutions in developing their e-learning environments. This framework ensures the quality of the institutions by increasing knowledge, altering attitudes, and extending expectations that can serve as a guide for institutional practitioners in various fields by identifying the scope and degree of important features. To explore e-learning adoption intention in China from an innovation adoption perspective and better understand the e-learning preference in China, Lingxian Zhang et al. identified the important influence elements of e-learning adoption on Chinese undergraduates [4]. The computation's output revealed the relative importance of the 33 evaluation index layer influence factors to the overall goal of e-learning adoption, with pricing having the biggest impact, education quality coming in second, and perceived general benefits having the smallest impact [5]. Amer Mutrik Sayaf [6] showed that user satisfaction and collaborative participation have a positive effect on the use of e-learning systems. He thus advocated for colleges to support e-learning as a long-term educational plan. Mei-Hui Peng et al. [7] presented a research approach to look at how motivated students are to use social media platforms for e-learning. They found that students' motivation to use a resource was highly influenced by perceived utility, perceived ease of use (PEOU), perceived cost, perceived effectiveness, and self-efficacy. In higher education, evaluation, perceived susceptibility, and perceived harshness are not important predictors of students' motivation to use e-learning.

After reviewing the literature on e-learning adoption and opposition in developing nations, Eduardo Bizzo chose, examined, and categorized 44 research into the following categories: (i) level of education, (ii) user, (iii) method (TAM, UTAUT, crucial success factor, and others), (iv) nation (region and income group), and (v) whether or not studies point out particularities for developing countries [8]. These models are summarized in Table 16.1.

TABLE 16.1

Overview of e-Learning Adoption Models in Literature

Authors	Model/Framework	Field	Advantages/Findings	Limitations/Future Scope
Muneer Abbad [1]	Conceptual Model/ Technology Acceptance Model (TAM)	IT	Determines the important factors that may affect students' adoption of e-learning systems	Proposed Model II were not present in the TAM, it might be able to provide a more complete understanding of usage in e-learning contexts
Md. Mahmudul Hasan Khan et al. [3]	Concerns-Based Adoption Model (CBAM)	Primary education institute	The framework can be taken as basis for developing a national framework for assuring and enhancing quality in primary educational institutes	More development research studies are needed to build up the foundation for a robust and at the same time contextualized framework
Carol Russell [4]	Cognitive mapping	University education	National, institutional and disciplinary influences on e-learning adoption	Further research in other contexts is needed to validate the broader applicability of this approach
Lingxian Zhang et al. [5]	The hierarchical structural model	Chinese undergraduates	The research results show the importance sequence of influence factors from smallest to largest	Needs to improve on many aspects
Amer Mutrik Sayaf [6]	AMOS 23 of the AMOS paradigm		Peer interaction, interactive lectures, information quality, system quality, and service quality are the characteristics that influence users' satisfaction and collaborative behavior of distance learners	Not mentioned
Mei-Hui Peng [7]	Technology Acceptance Model (TAM) and Technology Threat Avoidance Theory (TTAT)	262 students in Taiwan University	Perceived usefulness, Perceived Ease of Use, perceived cost, perceived effectiveness, and self-efficacy significantly influenced students' motivation to use	Future studies could use more than one institute to explore cross-culture findings
Chugh et al. [8]	Examination of factors in accepting e-learning by learners	Survey, data collection and quantitative analysis of 384 e-learners from HEIs in India	The quality of instructor and contents influences the most along with course structure, learner's attitude, technology used, ease of use of the platform selected	Only learner's perspective analyzed while parting with teacher's context, availability of infrastructure and other resources not discussed

16.3 TECHNOLOGY ACCEPTANCE MODEL

The literature available on the TAM provides an overview of the key concepts, developments, and research findings related to TAM and its variations. TAM is a widely recognized theoretical framework that helps understand how users adopt and accept new technology. The review summarizes the essential aspects of TAM and highlights some of the notable research in this field [9].

16.3.1 INTRODUCTION TO TAM

The TAM was first introduced by Fred Davis in 1989 as an information systems theory designed to predict and explain user acceptance and adoption of technology. Since its initial proposal, TAM has undergone a number of modifications and extensions, each of which aims to increase our understanding of technology adoption behavior. The history of TAM's creation, its essential elements, important additions, and its uses in diverse situations will all be covered in this examination of the literature [10].

16.3.2 DEVELOPMENT OF TAM

The original TAM proposed by Davis (1989) postulates that PEOU and perceived usefulness (PU) are key determinants of the behavioral intentions (BI) of users to use technology. It implies that if a system is simple to use and users feel it will improve their performance, they are more likely to adopt and use it.

16.3.3 CORE COMPONENTS OF TAM

PEOU measures how simple a user thinks a given technology is to use. It is frequently linked to elements like system usability, user-friendliness, and the amount of work needed to master and use the technology.

PU is the term used to describe how the user feels the technology can help them perform better or accomplish certain objectives. If users believe a piece of technology is useful, they are more inclined to accept it.

The user's intent to adopt or reject a technology is reflected in their BI. To connect PEOU and PU to actual usage behavior, it is regarded as a critical intermediary component.

Actual System Use (AU) is a measurement of how the user actually uses the technology. It's frequently used as a stand-in for technological adoption. The association between the components and the outcome of the actual use of technology is depicted in Figure 16.2. However, the usage of technology cannot be overlooked by the choice of the user. It can be well supported by developing technical efficacy, subjective norms, BI, and last but not least attitude toward use [11].

16.3.4 EXTENSIONS AND VARIATIONS OF TAM

TAM 2: By incorporating social impact and cognitive instrumental processes, Davis (1993) expanded the original TAM. Subjective norm (SN) and image (IM) were included in the extended model as new variables affecting user approval. Venkatesh et al. (2003) introduced the UTAUT, which unifies several acceptance models,

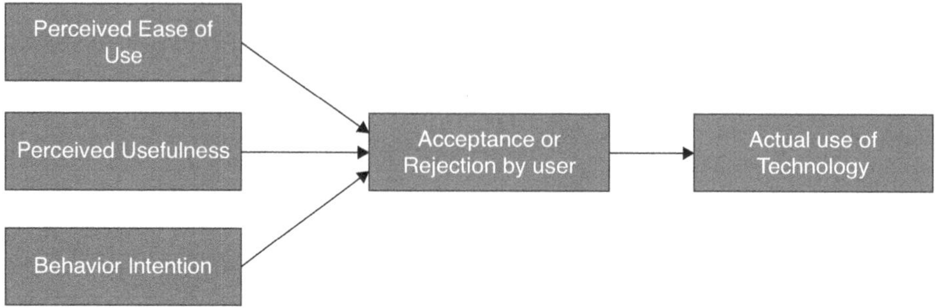

FIGURE 16.2 Core components and working of TAM.

including TAM, and adds new dimensions including performance expectancy, effort expectancy, social influence, and facilitating factors. The UTAUT sought to offer a thorough foundation for comprehending technology adoption.

TAM 3: TAM 3 was developed by Venkatesh and Davis (2000) and combined TAM with the TRA and Theory of Planned Behavior (TPB). Variables like SN, attitude, and perceived behavioral control were included in this extension. Extended TAM for Mobile Commerce (TAM-2M): In 2005, Lai and Li suggested an extended TAM model that was designed exclusively for the adoption of mobile commerce. It included trust and perceived danger as important determinants of acceptance of mobile technologies [12].

Table 16.2 summarizes the TAM and its variations. TAM has offered a useful framework for researchers and practitioners attempting to understand and foresee user acceptance of technology from its basic version and up to the more current UTAUT and mobile commerce modifications. TAM and its variations are still useful tools for researching and encouraging technology adoption even as technology advances. These models may be improved and expanded in future studies to take user behavior changes and technical advancements into account.

16.4 UNIFIED THEORY OF ACCEPTABILITY AND USE OF TECHNOLOGY

A well-known and influential model in the area of technology adoption and user acceptability is the UTAUT. The UTAUT was created by Venkatesh, Morris, Davis, and Davis in 2003 with the goal of integrating and expanding several technology acceptance models. It offered a thorough framework for comprehending the variables that affect technology adoption and use. This review of the literature gives a general overview of UTAUT, including its essential elements, extensions, and significant empirical discoveries as of the September 2021 knowledge cutoff date [15].

16.4.1 INTRODUCTION TO UTAUT

The UTAUT is a theoretical framework that seeks to explain and predict user behavior regarding technology adoption and usage. It builds on prior models like the TAM,

TABLE 16.2

Overview of Technology Acceptance Model

Authors	Model/Framework	Field	Advantages/Findings	Limitations/Future Scope
Qingxiong Ma et al. [10]	Technology Acceptance Model (TAM)	Conducted a meta-analysis based on 26 selected empirical studies	The results suggest that both the correlation between usefulness and acceptance, and that between usefulness and ease of use are somewhat strong	Relationship between PEOU and TA is uncertain, more future studies are needed to resolve this uncertainty
Yawen Su et al. [12]	Technology Acceptance Model (TAM)	Online entrepreneurship education	Ease of use and usefulness are two important factors that affect the acceptance and use of online entrepreneurship education platforms by new entrepreneurs	Limitation of the QS samples. Mutual relationship among various factors has not been fully verified
Opoku [13]	Relevance of the Technology Acceptance Model (TAM) in information management research	Survey of TAM model	TAM is still recognized as the right model for quantitative-based information management research, and to a lesser extent qualitative information management research and desk studies	TAM was too limited in the areas of theoretical assumptions and practical effectiveness
Mulugeta Hayelom Kalayou et al. [14]	Modified Technology Acceptance Model (TAM)	384 healthcare professionals	Perceived usefulness and Technical infrastructure has a significant influence on attitude and intention to use eHealth	Effect of IT experience on the intention to use eHealth was not significant. So, implementers should give priority in enhancing the organizations technical infrastructure, staff's IT skill, and their attitude toward eHealth by giving continuous support

the TRA, and the TPB while introducing new constructs to account for the complexity of technology adoption in different contexts.

16.4.2 Core Components of UTAUT

The UTAUT identifies several key constructs that influence user acceptance and use of technology:

- **Performance Expectancy (PE):** It is the user's perception of how using a technology will improve their job performance or task outcomes.
- **Effort Expectancy (EE):** It is the ease with which users can use the technology.
- **Social Influence (SI):** It is the influence of peers and other social media with which the user adopts a technology.
- **Facilitating Conditions (FC):** It is the degree to which users believe that organizational and technical infrastructure is in place to support technology use.
- **Behavioral Intention (BI):** Similar to TAM, it indicates the user's motive to adopt and use the technology voluntarily.
- **Actual Use (AU):** Once adopting the technology it indicates the actual usage by the user. Figure 16.3 shows the association among these components.

16.4.3 Extensions and Variations of UTAUT

Since its introduction, the UTAUT has undergone some variations and adaptations to fit specific contexts and technologies. Some notable extensions include:

- **UTAUT2:** Venkatesh et al. (2012) proposed an extension of UTAUT called UTAUT2, which includes constructs like hedonic motivation and price value. It was designed to address technology adoption in consumer contexts.

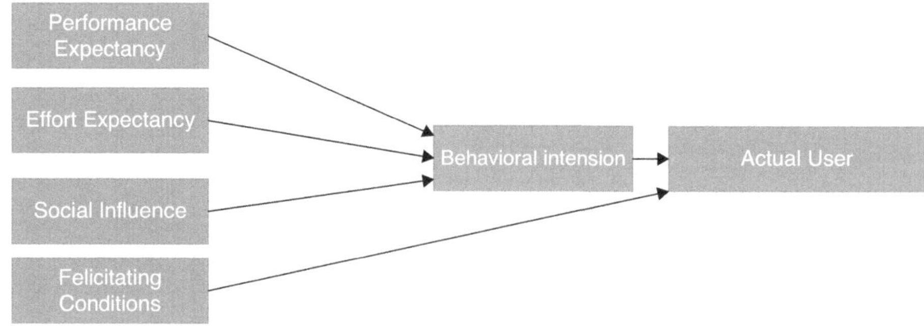

FIGURE 16.3   Core components of UTAUT.

- **Mobile UTAUT:** Various studies have adapted UTAUT for mobile technology adoption, recognizing the unique characteristics and user behaviors associated with mobile devices and apps.
- **Healthcare UTAUT:** Researchers have applied UTAUT to understand the adoption of health information technology and electronic health records (EHR) in the healthcare sector.

16.4.4 Key Findings

Some of the key findings from empirical studies applying UTAUT include:

- PE and EE consistently emerge as significant predictors of BI and actual use.
- SI plays a crucial role, particularly in settings where peer recommendations and network effects are strong.
- FC are essential, as users must feel that the necessary infrastructure and support are available to use the technology effectively.
- Insightful information about the adoption of technology in particular contexts, such as mobile technology, healthcare, and e-commerce, has been gleaned via UTAUT variants [15].

The UTAUT has established itself as a reliable and adaptable framework for comprehending the adoption of technology in a variety of disciplines is summarized in Table 16.3.

The UTAUT has become a useful tool for researchers and practitioners looking to understand and forecast user behavior with reference to technology uptake and usage due to its capacity to combine and enhance existing models. The UTAUT and its modifications are still useful for researching and encouraging technology acceptance in a variety of settings as technology continues to advance. This model may be improved upon and expanded upon in subsequent studies in order to handle new user behaviors and technical advancements.

16.5 DIFFUSION OF INNOVATIONS

In the fields of communication, sociology, and innovation studies, Everett M. Rogers created the DoI hypothesis in 1962. It talks about why innovations propagate through social systems and how they do it. A survey of the Diffusion of Innovations (DoI) theory's literature is provided here, highlighting its main ideas, important models, and noteworthy applications in various industries [18].

16.5.1 Introduction to DoI

The DoI hypothesis is concerned with how novel concepts, methods, goods, or practices are incorporated into and propagated throughout a certain social system. According to the theory, the process of adopting an innovation is not arbitrary or accidental but rather follows a predictable pattern that is impacted by specific traits of both the innovation and the people or groups adopting it.

TABLE 16.3

Overview of Unified Theory of Acceptance and Use of Technology (UTAUT)

Authors	Model/Framework	Field	Advantages/Findings	Limitations/Future Scope
Alaa M. Momani [14]	Unified Theory of Acceptance and Use of Technology (UTAUT)	Review of its development and structure	This detailed approach was created to look at how people really use the technology and gauge their happiness with it	Model can be extended based on its tested technology and the proposed research sample
Che-Hung Liu [15]	Unified Theory of Acceptance and Use of Technology (UTAUT)	PX Pay mobile payment app for PX Mart. online payment industry and academia	Consumers' behavioral intentions are enhanced by performance expectations, ease of use expectations, and social effect, and behavioral intentions strongly affect usage behavior	Based on an online survey, the sample survey was carried out. The honesty and intent of the questionnaire are still up for debate
Ahmet Ayaz [16]	Unified Theory of Acceptance and Use of Technology (UTAUT)	Electronic Document Management System (EDMS) in Bartın University	Factors of performance expectancy and social influence has a positive effect on the intention of use but of EE factor does not have a positive effect	It was seen that the effort expectation factor did not have a significant effect on the intention to use EDMS. This subject can be examined in more detail and the reasons can be determined with concrete data
Thathsarani Hewavitharana [17]	Modified Unified Theory of Acceptance and Use of Technology (UTAUT)	Construction industry	The conceptual model consisted of four primary constraints are Perceived Easy to Use (PE), Perceived Usefulness (PU), Subjective Norms (SN), and Facilitating Conditions (FC)	The research would like to propose longitudinal studies than cross-sectional research to obtain precise results for future studies

16.5.2 Core Concepts of DoI Theory

The DoI components are:

- **Innovation:** A unique idea, practice, technology, or product is considered innovative by a group of potential users. It could be a material object or an abstract idea.
- **Adoption:** The choice to use an innovation, usually as a trial, by a person or organization is referred to as adoption.
- Diffusion is the process by which an innovation spreads over time among the individuals within a social system.
- **Relative Advantage:** The extent to which an innovation is thought to be superior to the theory or method it replaces. The degree to which a new innovation is viewed as being congruent with the values, customs, and requirements of potential users.
- **Complexity:** The perception of a barrier to using and understanding innovation.
- **Trialability:** The degree to which the idea can be tested out in a small-scale setting.
- Observability refers to how easily other people can see the benefits of employing an innovation.

16.5.3 Key Models and Concepts within DoI Theory

Rogers' Adoption Curve: The adoption process is represented by Rogers' bell-shaped adoption curve, which comprises innovators, early adopters, early majority, late majority, and laggards. This model aids in the explanation of the population-level diffusion process.

The Invention-Decision Process: According to Rogers, adopting an invention normally involves five steps: knowledge, persuasion, decision, execution, and confirmation. This procedure aids in illuminating the cognitive and behavioral phases of adoption.

Critical Mass: The theory contends that before an innovation can become widely accepted, it must first reach a critical mass of adopters. knowledge tipping points in the diffusion process requires a knowledge of this idea.

16.5.4 Applications and Empirical Studies

Marketing, public health, agriculture, education, and technology adoption are just a few of the areas where the DoI idea has been extensively used. A few noteworthy applications are:

- **Public Health:** Interventions in public health Public health initiatives including immunization campaigns, smoking cessation programs, and the adoption of healthy behaviors have all been planned and carried out using the DoI principle.

- **Agricultural Innovation:** The theory of DoI has been applied in agricultural settings to encourage farmers to accept new farming techniques, crop types, and technologies.
- **Technology Adoption:** A lot of research has been done using this idea to understand how new technologies like cellphones, social media, and e-commerce are adopted.
- **Educational Innovations:** The adoption of educational technologies, instructional strategies, and curriculum modifications in educational institutions have all been examined by researchers using the DoI hypothesis [19].

16.6 TECHNOLOGY-ORGANIZATION-ENVIRONMENT FRAMEWORK

A well-known theoretical paradigm for innovation and technology management is the TOE paradigm. It aids in the understanding of the variables affecting technological innovation and adoption inside organizations by researchers and practitioners. The TOE Framework's essential elements, applications, and significant research findings are highlighted in the literature review that follows [20].

16.6.1 INTRODUCTION TO TOE

In order to investigate the factors influencing technology adoption inside organizations, Tornatzky and Fleischer created the TOE Framework in the late 1990s. It offers a methodical way to examine how three important dimensions—technology-related factors, organizational factors, and environmental factors—have an impact on technological advancements.

16.6.2 CORE COMPONENTS OF THE TOE FRAMEWORK

Figure 16.4 shows core components of the TOE and their relationship with each other.

Technology Dimension: This aspect is concerned with the traits of innovation or technology itself. Key elements consist of:

- *Technology Readiness:* Technology maturity and stability are indicators of technology readiness.
- *Technology Complexity:* It is the degree to which a technology is difficult to comprehend and use.
- *Technological Compatibility:* The degree to which the technology aligns with the organization's existing systems and practices.

Organization Dimension: This dimension considers the internal factors within an organization that affect technology adoption. Core factors include:

- *Organizational Readiness:* The organization's preparedness and willingness to adopt and assimilate new technology.

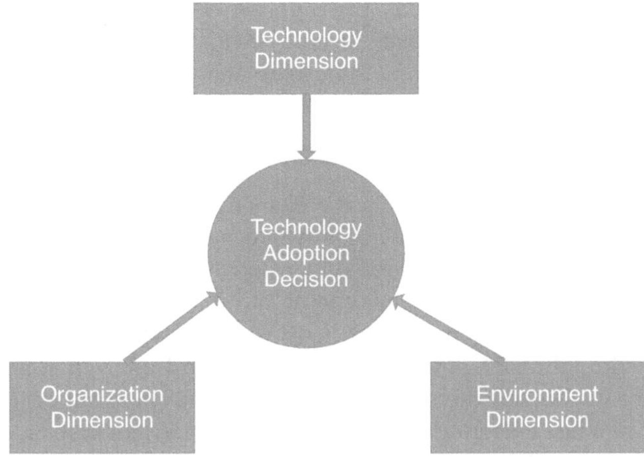

FIGURE 16.4   Core components of TOE.

- *Organizational Capabilities:* The resources, skills, and capabilities available within the organization.
- *Top Management Support:* The extent to which senior management supports and champions technology adoption efforts.

Environment Dimension: The environment dimension explores external factors that can influence technology adoption. These factors include:

- *Competitive Pressure:* The pressure from competitors adopting similar technologies.
- *Regulatory Environment:* Laws and regulations that may encourage or hinder technology adoption.
- *Market Dynamism:* The rate of change and volatility in the industry or market.

16.6.3 APPLICATIONS AND EMPIRICAL STUDIES

Some notable applications and research findings include:

- **Manufacturing Sector:** The adoption of advanced manufacturing technologies, such as Industry 4.0 technologies, within manufacturing enterprises has been the subject of numerous studies using the TOE Framework. Research has emphasized the significance of organizational competencies, market dynamism, and technology preparedness in this environment.
- **Healthcare Sector:** The use of telemedicine and EHR technology in healthcare organizations has been studied using the TOE Framework. The results highlight the importance of organizational preparation, regulatory issues, and top management support in the implementation of EHRs.

- **Small and Medium-sized Enterprises (SMEs):** Using the TOE Framework, researchers looked at how SMEs embrace and adapt to new technology. For SMEs looking to develop and maintain their competitiveness, factors including organizational capacity and technical compatibility are crucial.

A thorough framework for examining the elements influencing technology uptake and innovation within organizations is provided by the TOE Framework which is summarized in Table 16.4. TOEs' three-dimensional methodology takes into account both internal and external aspects, making it an important tool for researchers and technology management practitioners. The TOE Framework continues to be useful for comprehending the factors that influence successful innovation and technology implementation as organizations continue to negotiate the challenges of technology adoption in a fast changing environment. This paradigm may be expanded and improved in future studies to handle new problems and technology trends.

16.7 SOCIAL COGNITIVE THEORY

Albert Bandura created the Social Cognitive Theory (SCT), a thorough psychological framework that emphasizes the significance of social interactions, cognitive functions, and observational learning in influencing human behavior. Numerous sectors, including education, psychology, health, and communication have found extensive use for this notion. We shall examine the foundational ideas, essential elements, and noteworthy applications of SCT in this overview of the literature [23].

16.7.1 INTRODUCTION TO SCT

Albert Bandura first proposed the Social Learning Theory (SLT) in the 1960s, which is often referred to as the SCT. It asserts that people pick up skills from their social environment, especially through other people and through imitation and engagement. Cognitive functions, self-regulation, and self-efficacy beliefs are heavily emphasized in SCT.

16.7.2 CORE CONCEPTS OF SCT

- **Observational Learning:** The premise that people learn through seeing the behaviors, activities, and results of others is at the heart of SCT. Modeling and imitation are common terms used to describe this process.
- **Reciprocal Determinism:** SCT suggests a dynamic relationship between environmental influences, behavior, and human elements (cognitive processes, personal qualities). These three factors have a two-way impact on one another.
- Self-efficacy is a key concept in SCT and refers to a person's confidence in their ability to carry out a certain job or realize a specific objective. Higher levels of self-efficacy are linked to more drive and perseverance.
- **Cognitive Processes:** SCT places a strong emphasis on cognitive abilities that are crucial to learning, including attention, memory, and thinking [24].

TABLE 16.4

Overview of Technology-Organization-Environment (TOE)

Authors	Model/Framework	Field	Advantages/Findings	Limitations/Future Scope
Peggy et al. [19]	Evaluation of technological competence, Organizational support, work flexibility, Perceived behavioral control for e-working	Data collected from 238 remotely working people from Hong Kong	TOE and TPB framework analysis for e-working, Analyses the government and organizational support for e-working, analyses the technical support needed for remote operations	E-learning study is not covered, limited sample size which also lacks cross regional samples
Duggal [20]	Statistical and structural equation modeling using SPSS and AMOS for factors affecting the e-learning in India, study based on UTAUT	331 e-learning adults	Infrastructure dependability, effectiveness of course design and content, and students' competency are the main factors influencing acceptance of e-learning	Limited sample size considering the diversity of availability of resources in India, data from urban users collected
Anke et al. [21]	Study of factors affecting adoption of e-learning for sustainable management	By using 374 articles available on Google with search keywords as "e-learning" and "Technology-Organization-Environment" and interview of German experts from automotive and healthcare domain	Systematic discussion of factors involving Technology, Organization and Environment, Analysis of challenges faced in adoption of e-learning	Lacks cross regional and cross domain experts, most data is qualitative
Abdulsaeed et al. [22]	Evaluation of critical factors in adopting cloud computing-based e-learning based on TOE and TAM frameworks	Study based on survey of literature study in the context of Iraqi higher education organizations	14 critical factors to use cloud computing for e-learning in higher educational institutes, infrastructural development, government and organizational support are also involved with influential factors like compatibility, security, cost, etc.	The study is void of real data collection and analysis

16.7.3 Key Components of SCT

- **Behavioral Capability:** People must possess the information and abilities required to carry out a behavior successfully. SCT contends that learning occurs both directly from one's own experiences and indirectly through watching others.
- **Expectations and Self-Efficacy:** People's motivation and behavior are greatly influenced by their expectations for the results of their behavior as well as their confidence in their capacity to carry it out.
- **Environmental Factors:** Reinforcement and Punishment, SCT recognizes the importance of both reinforcement (reward) and punishment (consequences) in influencing behavior. Bandura emphasized that observational learning can still take place even in the absence of explicit reward. Figure 16.5 shows the relationship between these components.

16.7.4 Notable Applications and Empirical Studies

- **Education**: SCT has been used to better understand how children pick up knowledge by seeing their teachers and peers in the classroom. It has also been utilized to create instructional tactics that improve students' motivation and sense of self-efficacy.
- **Health Behavior Change:** SCT has been used in public health initiatives to encourage healthy habits like quitting smoking, eating well, and exercising. In order to modify behavior, it emphasizes the value of modeling and giving role models.
- **Media and Communication:** SCT has been used in the study of media to examine how media content affects society, particularly how characters and public figures serve as role models for behavior.

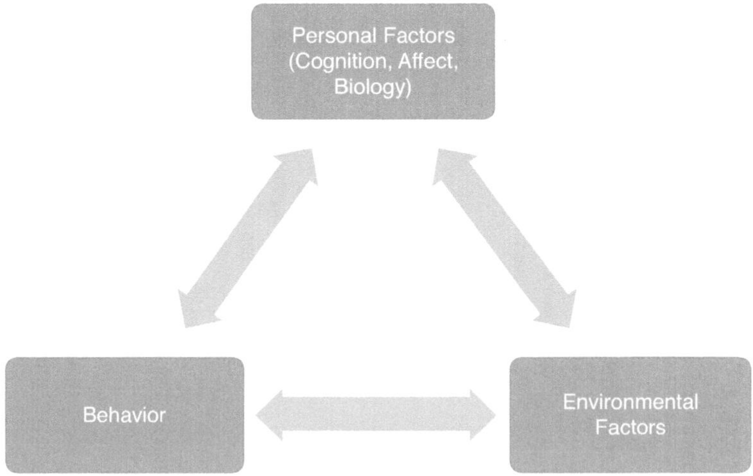

FIGURE 16.5 Key components of SCT.

- **Therapy and Counseling:** By focusing on self-efficacy and self-regulation through cognitive-behavioral therapies, SCT has been utilized to address a variety of psychological illnesses in clinical psychology and counseling [25].

Table 16.5 shows an overview of SCT. The SCT emphasizes the functions of social interactions, observational learning, cognitive processes, and self-regulation to provide a thorough framework for understanding human behavior. Its applications in psychology, communication, health, education, and other fields have helped us better understand how people form and modify behaviors. SCT is a relevant and flexible theory for examining the dynamics of learning and behavior in various contexts as technology and communication advance. The SCT may be further developed and expanded in the future to address new issues and opportunities that the digital age presents.

16.8 DISCUSSION

Technology adoption frames discussed in this chapter are summarized in Figure 16.6. TAM, DoI, TOE and SOT are mostly e-learning adoption technologies in various organizations. Variations of very popular framework TAM like UTAUT, TAM2, TAM3 and TAM-2M used in different applications as per the need. More specific variations of TAM like UTAUT2, Mobile UTAUT and Healthcare UTAUT have shown significant results in the application-specific technology adoptions.

TAM and its variations have been used by researchers to examine the adoption of technology across a range of industries, including e-commerce, healthcare, education, and information systems. The usefulness of TAM in forecasting user behavior and comprehending the variables that affect technology acceptance has repeatedly been demonstrated by empirical investigations.

The UTAUT has been extensively applied in empirical research across various domains, including e-commerce, education, healthcare, and information systems. Researchers have used the UTAUT to identify the critical factors affecting user acceptance and to predict adoption rates.

Everett M. Rogers' DoI theory offers a useful framework for comprehending how innovations spread throughout social systems. Designing successful strategies for fostering the acceptance and adoption of innovations has made extensive use of its fundamental concepts, models, and stages of adoption. The DoI hypothesis is still significant for scholars and practitioners who want to understand and have some control over how new ideas and technologies spread throughout our society as they continue to develop and change. The idea may be improved upon and modified in the future study to address new trends and difficulties in the acceptance of innovations.

The TOE Framework has been applied in various contexts and industries to understand technology adoption and innovation.

SCT focuses on cognitive functions, self-regulation, and self-efficacy.

To conclude, the six e-learning models discussed are summarized in Table 16.6.

TABLE 16.5
Overview of Social Cognitive Theory

Authors	Model/Framework	Field	Advantages/Findings	Limitations/ Future Scope
Mamolo [24]	Studies the effect on students' behavior due to online learning of math students, survey conducted by teacher	31 students of grade 11 from STEM under k-12 curriculum in Philippines	The paper reports the students' excitement in the beginning but later reduced self motivation. The paper also surveyed the difficulties faced by students for online working	Very limited number of students, third party survey can help, cross regional survey would result in broader outcomes
Santos [25]	Motivational factors of students to study martial arts in online mode	Interview with 12 teachers who selected distant after-school learning for their students	The study produced recommendations for how to improve and modify the policies and rules that are now in place for after-school programs for government officials, school administrators, policymakers, parents, students, and researchers	Interviews with limited numbers, other methodology for data and experience collection can be recommended
Moor and Miller [26]	Articles with empirical analysis of cognitive processes selected to analyze motivation, satisfaction, and learning in online education	Systematic review of 24 articles published during 2008–2020	Community of Inquiry (CoI) and Practical Inquiry Model (PIM) were selected as the models to examine the cognitive presence in online learning environments and the awareness of instructors about it	Instructions only for instructors are given for cognitive presence in online teaching-learning
Chang and Tsai [27]	Study examines the effects of students' academic accomplishment for English online lessons in Shanghai on their emotional intelligence, learning motivation, and self-efficacy	Survey of 450 students using questionnaires from 10 universities in Shanghai	AVE, correlation and SEM-based statistical analysis of 432 student's survey data. Relationships among Self-efficacy, Emotional Intelligence (EI), Learning motivations, and academic achievement developed empirically	EI measurement should be done using ability tests and academic achievement in co-curricular or extracurricular activities is missing
Dahri et al. [28]	Evaluation of challenges faced in providing online development programs to teachers	Feedback from 35 participants from Pakistan and Saudi Arabia	A framework of certification development of mobile-based, usefulness of mobile-based training approach in limited resources	

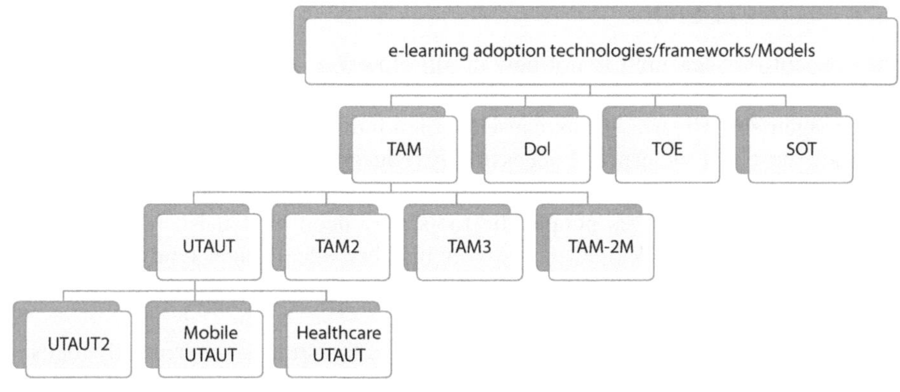

FIGURE 16.6	E-learning adoption frameworks.

TABLE 16.6
Summary of e-Learning Models

Model/ Parameters	Focus	Elements	Applications
Technology Acceptance Model (TAM)	Individual users' perceptions and attitudes toward technology adoption	Perceived usefulness and Perceived Ease of Use	To understand the usefulness and ease of use of e-learning technologies
Unified Theory of Acceptance and Use of Technology (UTAUT)	Incorporating social influences, facilitating conditions, and performance expectancy	Performance expectancy, effort expectancy, social influence, and facilitating conditions	Influencing e-learning adoption, considering social and environmental factors
Diffusion of Innovations (DoI)	Emphasizes the spread of innovations over time and across different user groups	Innovativeness, communication channels, social system, time, and the rate of adoption	e-learning innovations are adopted and spread across educational settings
Five Rogers' Factors	Based on the work of Everett Rogers, these factors influence the rate of adoption of innovations	Relative advantage, compatibility, complexity, trialability, and observability	Can assess how e-learning technologies compare to existing methods and their ease of trial and observation
Technology- Organization- Environment (TOE) Framework	Considers the interplay of technological, organizational, and environmental factors in the adoption process	Technological context, organizational context, and environmental context	Offers a thorough perspective on factors affecting e-learning adoption, encompassing technology, organizational policies, and the external environment
Social Cognitive Theory	Emphasizes the role of observational learning and social influence in shaping behavior	Observational learning, self-efficacy, outcome expectations, and reinforcements	Can be applied to understand how learners observe and learn from others in the e-learning context, influencing their adoption behavior

16.9　CONCLUSION

The concept of e-learning is not new at all. However, the Covid-19 pandemic has made us adopt this "New Normal." Offering flexibility, accessibility, customization, and engagement, e-learning has caused a paradigm change in education. Because of its low cost and widespread accessibility, education is more inclusive and gives students from all backgrounds more influence. E-learning encourages a culture of lifelong learning and gives people the tools they need to flourish in a constantly changing environment. With its data-driven insights, e-learning is ready to transform the future of education by offering students throughout the world a dynamic and engaging learning environment.

Like said earlier in the chapter, TAM is suitable for studies that primarily focus on individual attitudes and perceptions toward e-learning technologies. The UTAUT is suitable for a broader understanding of technology adoption, considering both individual and social factors. DoI is beneficial for studies focusing on the diffusion of e-learning innovations across a larger population or organizational context. Rogers' factors are suitable for studies that want to emphasize the specific characteristics of an e-learning technology and how these factors influence its adoption. The TOE Framework is suitable for studies that aim to understand the interplay between e-learning technologies, organizational structures, and external influences. The Social Cognitive Theory is useful for studies that want to explore the social aspects of e-learning adoption and the influence of role models and social norms.

Choosing the right framework is completely dependent on the objective and scope of the utilization. Some frameworks are more individual-focused, while others consider organizational or societal factors. Ultimately, the "best" framework depends on the research objectives and the context of the study. It may also be beneficial to integrate elements from multiple frameworks to provide a comprehensive understanding of e-learning adoption. The blended learning combines traditional face-to-face instruction with online learning components. This approach seeks to leverage the strengths of both in-person and online learning to create a more flexible and effective educational experience. Flipped classroom, rotational models, flex models are some examples of hybrid e-learnings. Moreover, they suggest technology infrastructure, training and support for both instructors and students, proper content design, clear communication channels for students and alignment of assessments with the learning objectives.

REFERENCES

1. M. Abbad, "A conceptual model of e-learning adoption", *IJET*, 6(2): EDUCON2011, May 2011.
2. K. Alone, "Adoption of e-learning technologies in education institutes/organizations: A literature review", *Asian Journal of Educational Research*, 5: 4, 2017.
3. M. M. Hasan Khan, M. A. Rahman, M. Ahmed, "Introducing a tentative framework for adopting e-learning systems in developing Countries", Master's (one year) thesis in Informatics (Autumn 2010:MI15), University of Borås.
4. C. Russell, "A systemic framework for managing e-learning adoption in campus universities: Individual strategies in context", *ALT-J, Research in Learning Technology*, 17(1): 3–19, March 2009.

5. L. Zhang, H. Wen, D. Li, Z. Fu, S. Cui, "E-learning adoption intention and its key influence factors based on innovation adoption theory", *Mathematical and Computer Modelling*, 51: 1428–1432, 2010.

6. A. M. Sayaf, "Adoption of e-learning systems: An integration of ISSM and constructivism theories in higher education", *Heliyon*, 9: e13014, 2023.

7. M.-H. Peng, H.-G. Hwang, "An empirical study to explore the adoption of e-learning social media platform in Taiwan: An integrated conceptual adoption framework based on technology acceptance model and technology threat avoidance theory", *Sustainability*, 13: 9946, 2021.

8. E. Bizzo, "Acceptance and resistance to e-learning adoption in developing countries: A literature review", *Essay: Evaluation and Public Policies in Education*, 30: 115, 2022.

9. L. A. Mamolo, "Online learning and students' mathematics motivation, self-efficacy, and anxiety in the 'new normal'", *Education Research International, Hindawi Publications*, 2022: 9439634, 2022.

10. Q. Ma, L. Liu, "The technology acceptance model: A meta-analysis of empirical findings", *Journal of Organizational and End User Computing*, 16(1): 59, 2004.

11. M. Almulla, "Technology acceptance model (TAM) and E-learning system use for education sustainability", *Academy of Strategic Management Journal*, 20: 4, 2021.

12. Y. Su, M. Li, "Applying technology acceptance model in online entrepreneurship education for new entrepreneurs", *Frontiers in Psychology*, 2021:12, 2021.

13. M.O. Opoku, F. Enu-Kwesi, "Relevance of the technology acceptance model (TAM) in information management research: A review of selected empirical evidence", *Research Journal of Business and Management (RJBM)*, 7(1): 34–44, 2019.

14. M. H. Kalayou, B. F. Endehabtu, B. Tilahun, "The applicability of the modified technology acceptance model (TAM) on the sustainable adoption of e-health systems in resource-limited settings", *Journal of Multidisciplinary Healthcare*, 13: 1827–1837, 2020.

15. A. M. Momani, "The unified theory of acceptance and use of technology: A new approach in technology acceptance", *International Journal of Sociotechnology and Knowledge Development*, 12: 3, 2020.

16. Y.-C. Chang, Y.-T. Tsai, "The effect of university students' emotional intelligence, learning motivation and self-efficacy on their academic achievement: Online English courses", *Frontiers in Psychology*, 13: 818929, 2022.

17. A. Ayaz, M. Yanartas. "An analysis on the unified theory of acceptance and use of technology theory (UTAUT): Acceptance of electronic document management system (EDMS)", *Computers in Human Behavior Reports*, 2: 100032, 2020.

18. T. Hewavitharana, S. Nanayakkara, A. Perera, P. Perera, "Modifying the unified theory of acceptance and use of technology (UTAUT) model for the digital transformation of the construction industry from the user perspective", *Informatics*, 8: 81, 2021.

19. S. M. Alshahrani, H. Mohamed, M. Mukhtar, U. A. Mokhtar, "The adoption of the e-portfolio management system in the technical and vocational training corporation (TVTC) in Saudi Arabia", *International Journal of Information Management Data Insights*, 3(1): 100148, 2023.

20. P. M. L. Ng, K. K. Lit, C. T. Y. Cheung, "Remote work as a new normal. The technology-organization-environment (TOE) context", *Technology in Society*, 70: 102022, 2022.

21. S. Duggal, "Factors impacting acceptance of e-learning in India: Learners' perspective", *Asian Association of Open Universities Journal*, 17(2): 101–119, 2022.

22. J. M. Anke, S. Hobert, M. Schumann. "What drives a successful adoption of e-learning modules for sustainable management? An empirical investigation of influencing factors and challenges", *Proceedings of the 52nd Hawaii International Conference on System Sciences,* Maui, Hawaii, 2019.

23. A. A. Abdulsaeed, M. A. Burhanuddin, M. K. A. Ghani, M. M. Jaber, "Critical factors that affecting adoption of e-learning based on cloud computing for Iraqi higher education organizations", *International Journal of Health Sciences*, 6(S1): 4357–4373, 2022.
24. M. Chugh, R. Upadhyay, N. Chugh, "An empirical investigation of critical factors affecting acceptance of e-learning platforms: A learner's perspective", *SN Computer Science*, 4: 240, 2023.
25. M. D. Díaz-Noguera, C. Hervás-Gómez, A. M. De la Calle-Cabrera, E. López-Meneses, "Autonomy, motivation, and digital pedagogy are key factors in the perceptions of Spanish higher-education students toward online learning during the COVID-19 pandemic", *International Journal of Environmental Research and Public Health*, 19(2): 654, 2022.
26. D. Santos, L. Miguel, "Learning taekwondo martial arts lessons online: The perspectives of social cognitive career and motivation theory", *International Journal of Instruction*, 15(1): 1065–1080, 2022.
27. R. L. Moore, C. N. Miller, "Fostering cognitive presence in online courses: A systematic review (2008–2020)", *Online Learning*, 26(1): 130–149, 2022.
28. C.-H. Liu, Y. T. Chen, S. Kittikowit, T. Hongsuchon, Y. J. Chen, "Using unified theory of acceptance and use of technology to evaluate the impact of a mobile payment app on the shopping intention and usage behavior of middle-aged customers", *Frontiers in Psychology*, 3(13): 830–842, 2022.

Index